T0320394

Witten
Index

Other World Scientific Titles by the Author

Lectures on Quantum Chromodynamics
ISBN: 978-981-02-4331-9

Differential Geometry through Supersymmetric Glasses
ISBN: 978-981-12-0677-1

Witten
Index

Andrei Smilga

University of Nantes, France

World Scientific

NEW JERSEY · LONDON · SINGAPORE · BEIJING · SHANGHAI · HONG KONG · TAIPEI · CHENNAI · TOKYO

Published by

World Scientific Publishing Co. Pte. Ltd.
5 Toh Tuck Link, Singapore 596224
USA office: 27 Warren Street, Suite 401-402, Hackensack, NJ 07601
UK office: 57 Shelton Street, Covent Garden, London WC2H 9HE

Library of Congress Control Number: 2024938785

British Library Cataloguing-in-Publication Data
A catalogue record for this book is available from the British Library.

WITTEN INDEX

ISBN 978-981-12-9317-7 (hardcover)
ISBN 978-981-12-9318-4 (ebook for institutions)
ISBN 978-981-12-9319-1 (ebook for individuals)

For any available supplementary material, please visit
https://www.worldscientific.com/worldscibooks/10.1142/13839#t=suppl

Desk Editor: Carmen Teo Bin Jie

Typeset by Stallion Press
Email: enquiries@stallionpress.com

Printed in Singapore

Preface

The reader has definitely already heard the word *supersymmetry*. And s/he has heard that supersymmetry is a very beautiful mathematical concept involving symmetry between bosons and fermions in quantum field theory. S/he also knows that the world that we live in does not look supersymmetric — the spectra of the elementary bosons and elementary fermions are quite different. For example, everybody saw massless photons, but nobody observed their fermionic partners — the photinos, which should also be massless in supersymmetric theory.

Still the beauty of supersymmetric field theories (a lot of them have been constructed and studied by theorists) and certain important physical indications — the so-called *hierarchy problem*, the observed baryon asymmetry of the Universe, which can be explained only assuming nonconservation of baryon number, which in turn indicates together with the curves of the evolution of the coupling constants of the Standard Model that the observed $SU(3) \times SU(2) \times U(1)$ interactions are all unified at the scale $E \sim 10^{16}$ GeV and that the unifying theory is supersymmetric[1] — all that led many physicists to believe that the fundamental Theory of Everything is supersymmetric.

The fact that we do not see supersymmetry (an essential point of building the Large Hadron Collider at CERN was the hope of finding supersymmetric partners of the particles in the Standard Model at the energy range $E \lesssim 1$ TeV, the hope that unfortunately has not come true) means that, even if supersymmetry is there at the energy scale of grand unification, it is *broken* at low energies where experiments can be done.

As was already mentioned, the massless photon would have a massless fermion partner called photino in a supersymmetric theory. Its absence

---

[1] Wait for Chapter 3 for a detailed discussion of all these points.

means that photino (if it exists) carries a mass ($\sim$50 GeV or more, as experimentalists tell us), i.e. supersymmetry is broken.

There are two known mechanisms for breaking any symmetry including supersymmetry: *explicit* breaking and *spontaneous* breaking. Explicit breaking means that the theory is in fact not supersymmetric — its Lagrangian involves the terms that are not invariant under supersymmetry transformations. Spontaneous breaking of supersymmetry is more aesthetically appealing. In this case, the Lagrangian stays supersymmetric, but the physical vacuum states do *not*.

Now, the Witten index, to which this book is devoted, is a powerful theoretical tool. It allows one to find out whether supersymmetry breaks spontaneously in a given theory or not.

Speaking briefly (we will explain this later in all the details), supersymmetry is not broken IFF the vacuum states in the spectrum of a supersymmetric quantum system are annihilated by the action of the generators of supersymmetry transformations — the supercharges. The energy of such vacuum states (call them supersymmetric) is zero.

The Witten index is defined as the difference between the number of bosonic and fermionic supersymmetric vacuum states:

$$I_W = n_B^{\text{vac}} - n_F^{\text{vac}}. \tag{0.1}$$

This index has a topological nature meaning that it is *invariant* under smooth modifications of the parameters of the theory. And in many cases this allows one to calculate it by deforming the theory in interest in such a way that the analysis of its vacuum dynamics becomes feasible, and hence determine whether supersymmetry is spontaneously broken or not. This is the main physical interest of the Witten index.

The Witten index also has many beautiful purely mathematical connotations. It represents a variant of the so-called *equivariant index* introduced by Cartan back in 1950 [1]. It is also closely related to the Atiyah-Singer index [2]. Supersymmetric language allows one to describe in a rather transparent way some known facts of differential geometry and derive also new results in this field [3,4].

This book is mostly addressed to the experts in quantum field theory. However, I have tried to write the first part of the book (its first five chapters and to some extent also Chapter 6) in a way that might be accessible also to mathematicians interested in the mathematical aspects of the Witten index. In particular, Chapter 1, where the basic notions of QFT are briefly reviewed, is written for *them*.

In Chapter 2, we introduce the notion of supersymmetry and start with presenting simple supersymmetric quantum-mechanical (SQM) systems.

In Chapter 3, we present supersymmetric field theories whose dynamics will be discussed later in the book.

Chapter 4 is devoted to general aspects of the Witten index concept. In particular, we show how to evaluate I_W by calculating a proper path integral.

Chapter 5 is mathematical. We show how many well-known notions of differential geometry (like de Rham complex or Dolbeault complex) may be described in a rather simple way using supersymmetry.

In Chapter 6, we start our discussion of low-energy dynamics of $4D$ supersymmetric field theories. We calculate the indices for supersymmetric electrodynamics and pure supersymmetric Yang-Mills (SYM) theories.

Chapter 7 is devoted to nonchiral non-Abelian supersymmetric gauge theories involving matter multiplets.

In Chapter 8, we discuss *chiral* supersymmetric gauge theories with left-right asymmetric matter content.

Chapter 9 is devoted to the SQM system obtained by dimensional reduction of a maximally supersymmetric $\mathcal{N} = 1$ $10D$ SYM theory.

In Chapter 10, we calculate the indices for $3D$ supersymmetric Yang-Mills-Chern-Simons theories with and without matter.

Finally, in Chapter 11, we discuss some relatives of the Witten index such as Römelsberger's index for weak supersymmetric systems and the CFIV index.

Notation.

We will use almost exclusively the natural units, $\hbar = c = 1$. Talking about field theories, we will use the Minkowski metric with the signature $g_{\mu\nu} = \text{diag}(1, -1, -1, -1)$ and similarly in other dimensions. We will write the vector indices at two levels to distinguish between covariant and contravariant vectors and tensors, even though in flat Minkowski space it is often not so relevant.

Whenever we care to distinguish time and space coordinates, our convention is $x^\mu = (t, \boldsymbol{x})$ and $\partial_\mu = \left(\frac{\partial}{\partial t}, \frac{\partial}{\partial \mathbf{x}}\right)$. For the other 4-vectors we adopt the convention $V^\mu = (V_0, \boldsymbol{V})$ and hence $V_\mu = (V_0, -\boldsymbol{V})$. The spinor indices[2], both undotted and dotted [they correspond to two $SU(2)$ subgroups of $SO(3,1)$], will also be written at two levels. They will be raised

---

[2]We will reserve for them the beginning of the Greek alphabet, while its middle will be reserved for the vector indices.

and lowered according to

$$X_\alpha = \varepsilon_{\alpha\beta} X^\beta, \quad X^\alpha = \varepsilon^{\alpha\beta} X_\beta \tag{0.2}$$

with $\varepsilon_{\alpha\beta} = -\varepsilon^{\alpha\beta}$ and $\varepsilon_{12} = 1$, and the same for the dotted indices of complex conjugated spinors.

We will also meet two matrix-valued 4-vectors:

$$\sigma^\mu_{\alpha\dot\alpha} = (\mathbb{1}, \boldsymbol{\sigma})_{\alpha\dot\alpha}, \quad (\bar\sigma^\mu)^{\dot\alpha\alpha} = (\mathbb{1}, -\boldsymbol{\sigma})^{\dot\alpha\alpha}, \tag{0.3}$$

where $\boldsymbol{\sigma}$ are the Pauli matrices,[3] and the Dirac 4×4 gamma matrices

$$\gamma^\mu = \begin{pmatrix} 0 & \bar\sigma^\mu \\ \sigma^\mu & 0 \end{pmatrix}, \tag{0.5}$$

which satisfy the Clifford algebra $\gamma^\mu\gamma^\nu + \gamma^\nu\gamma^\mu = 2\eta^{\mu\nu}\mathbb{1}$.

Our conjugation symbols policy is the following. The star * will denote complex conjugation of ordinary numbers and bosonic fields. The bar $^-$ will mark antiholomorphic indices, will be used for complex conjugation of Grassmann numbers and fermion fields and mostly used for Hermitian conjugation of supercharges and other fermion operators. To avoid confusion, the Dirac conjugation of bispinors will be denoted by a long bar: as $\overline{\psi}$ rather than $\bar\psi$. We will also use the long bar to denote complex conjugation of *superfields*, both bosonic and fermionic. The dagger † will be used for Hermitian conjugation of bosonic operators and *occasionally* also of fermionic ones.

And the hats are mostly used for the operators (like $\hat{H}$), but also for the matrix-valued gauge fields ($\hat{A}_\mu = A^a_\mu t^a$).

Acknowledgements.   I am indebted to D. Bykov, N. Dorey, K. Konishi, Z. Komargodski and M. Vysotsky, to whom I showed the draft of the book, for many illuminating discussions. Special thanks are due to Carmen Teo Bin Jie for her meticulous editing work.

---

[3]One can observe that $\text{Tr}\{\sigma^\mu\bar\sigma^\nu\} = 2\eta^{\mu\nu}$ and

$$\varepsilon^{\alpha\beta}\varepsilon^{\dot\alpha\dot\beta}(\sigma^\mu)_{\beta\dot\beta} = (\bar\sigma^\mu)^{\dot\alpha\alpha}. \tag{0.4}$$

Contents

Chapter 1

Bosons and fermions

There are many good books on quantum field theory. I believe that most readers of this book are experts on QFT and do not really need extra explanations. But, as was noted in the *Preface*, the notion of Witten index has also pure mathematical aspects, and it is not absolutely excluded that a mathematician, who has a working knowledge of the ordinary quantum mechanics, but might be not so savvy with the standard QFT wisdom, would open our book. This chapter is written for them. But it is written in a very concise manner and to better learn this subject, the reader can look into my book [5] written for non-experts, into the book of A. Zee [6], which is also addressed to non-experts, but involves more mathematical details, and finally into the textbooks like Peskin's and Schröder's book [7].

1.1 Scalar fields

Consider the action describing free complex $(3+1)$-dimensional scalar field $\phi(\boldsymbol{x}, t)$,

$$S_{\text{free}} = \int d^3\boldsymbol{x}\,dt\,\left[(\partial_\mu \phi^*)(\partial^\mu \phi) - m^2 \phi^* \phi\right].\qquad(1.1)$$

The action is dimensionless, hence ϕ has a dimension of mass.

This system has a continuous number of degrees of freedom, and such systems are not tractable by conventional functional analysis methods. We can, however, regularize it in the infrared by putting it in a large spatial box of length L and imposing periodic boundary conditions,

$$\phi(x+L, y, z) = \phi(x, y+L, z) = \phi(x, y, z+L) = \phi(x, y, z).\quad(1.2)$$

We can then expand $\phi(x)$ in the Fourier series

$$\phi(x) = \frac{1}{L} \sum_n c_n \exp\left[\frac{2\pi i n x}{L}\right] \qquad (1.3)$$

with the sum running over all integer-valued vectors n. Then the continuous dynamical variable $\phi(x)$ is traded for the discrete set of Fourier coefficients $\{c_n\}$, carrying no dimension.

The canonical quantum Hamiltonian corresponding to the action (1.1) represents an infinite sum of harmonic oscillators,

$$\hat{H} = \sum_n \hat{H}_n \qquad (1.4)$$

with

$$\hat{H}_n = -\frac{1}{L}\frac{\partial^2}{\partial c_n \partial c_n^*} + L\left[m^2 + \left(\frac{2\pi n}{L}\right)^2\right] c_n c_n^*. \qquad (1.5)$$

The ground (vacuum) state of this Hamiltonian has the wave function[1]

$$\Psi_{\text{vac}} \propto \prod_n \exp\{-L\omega_n c_n c_n^*\} \qquad (1.6)$$

with

$$\omega_n = \sqrt{m^2 + \left(\frac{2\pi n}{L}\right)^2}. \qquad (1.7)$$

The excitations of $\hat{H}_n$ have the meaning of the physical particles carrying the energy ω_n.

The Hamiltonian (1.4) commutes with the operator

$$\hat{P} = \sum_n \hat{p}_n = \frac{2\pi}{L} \sum_n n \left(c_n^* \frac{\partial}{\partial c_n^*} - c_n \frac{\partial}{\partial c_n}\right). \qquad (1.8)$$

Its obvious meaning is the momentum of the system.

The coefficients c_n are complex and the system has two degrees of freedom for each n. Correspondingly, there are two one-particle states with a given momentum $p_n = 2\pi n/L$: a particle and an antiparticle.[2] Their wave

---

[1] The energy of this state $E = \frac{1}{2}\sum_n \omega_n$ is infinite. If one ignores gravity, an infinite constant can be subtracted from the Hamiltonian without changing the dynamics, and if gravity is present, we are facing a difficult and unresolved problem of the infinite cosmological constant.

[2] In the theory of real scalar field ϕ with the action

$$S = \frac{1}{2}\int d^4x \left[(\partial_\mu \phi)^2 - m^2\phi^2\right]$$

the particles and antiparticles coincide, and for each value of momentum p_n there is a single state.

functions read

$$\Psi_{\text{particle}} \propto c_n^* \prod_m \exp\{-L\omega_m c_m c_m^*\},$$

$$\Psi_{\text{antiparticle}} \propto c_{-n} \prod_m \exp\{-L\omega_m c_m c_m^*\}. \tag{1.9}$$

The minus in c_{-n} is associated with the minus in the second term in (1.8).

The convention according to which the wave function of a particle involves the factor c^* rather than c has historical reasons. The variables c_n^* and c_n correspond to the quantum operators $\hat{c}_n^\dagger$ and $\hat{c}_n$. The operator $\hat{c}_n^\dagger$ is the creation operator — it creates the particle with momentum p_n or annihilates the antiparticle with momentum $-p_n$. And the annihilation operator $\hat{c}_n$ annihilates the particle with momentum p_n or creates the antiparticle with momentum $-p_n$.

These particles are *bosons*, meaning that there can be an arbitrary number of particles (or antiparticles) in the same quantum state.

A nontrivial theory is obtained if one adds to the free action (1.1) an interaction term, for example,

$$S_{\text{int}} = -\frac{\lambda}{4} \int d^4x \, (\phi^*\phi)^2. \tag{1.10}$$

The theory with the action

$$S_\phi = S_{\text{free}} + S_{\text{int}}$$

is not exactly solvable anymore, but it can be treated perturbatively when the coupling constant λ is small. It is renormalizable, which means that the perturbative series for the scattering amplitudes (the first few terms of this series can be determined by calculating appropriate Feynman graphs) makes sense.

1.2   Gauge fields

Consider the electromagnetic vector field A_μ. In the absence of interactions, its Lagrangian reads

$$L_{\text{phot}} = -\frac{1}{4} \int d^3x \, F_{\mu\nu} F^{\mu\nu}, \tag{1.11}$$

where $F_{\mu\nu} = \partial_\mu A_\nu - \partial_\nu A_\mu$. This Lagrangian has important distinguishing features. One can make two observations.

(1) The field $A_0(t, \boldsymbol{x})$ enters the Lagrangian without time derivatives. It is not a dynamical variable, but has the meaning of a *Lagrange multiplier*. The variation of the Lagrangian with respect to A_0 gives the *constraint*

$$\partial_j E_j \;=\; 0\,, \tag{1.12}$$

where the components of the electric field,

$$E_j \;=\; \frac{\delta L}{\delta \dot{A}_j} \;=\; \dot{A}_j - \partial_j A_0\,, \tag{1.13}$$

are the canonical momenta corresponding to the dynamical variables A_j. The canonical Hamiltonian corresponding to the Lagrangian (1.11) is

$$H \;=\; \frac{1}{2} \int d^3x \,(E_j^2 + B_j^2)\,, \tag{1.14}$$

where $B_j = \frac{1}{2}\varepsilon_{jkl}F_{kl}$ is the magnetic field.

In quantum theory, we only have to take into consideration the eigenstates of the Hamiltonian (1.14) whose wave functions satisfy the quantum constraints:

$$\hat{G}\,\Psi(\boldsymbol{A}) \;=\; -i\partial_j \frac{\partial}{\partial A_j}\,\Psi(\boldsymbol{A}) \;=\; 0\,. \tag{1.15}$$

We are then left with the *transverse* photons whose polarizations lie in the plane orthogonal to the direction of their flight — two states with energy $\epsilon_{\boldsymbol{n}} = |\boldsymbol{p_n}|$ for a given momentum $\boldsymbol{p_n}$.

(2) The Lagrangian (1.11) is invariant under *gauge transformations*

$$A_\mu \;\rightarrow\; A_\mu + \partial_\mu \chi(x) \tag{1.16}$$

with an arbitrary function $\chi(x)$.

Actually, the two mentioned facts are closely related. A gauge theory is a theory that is formulated with including extra unphysical variables — in our case, the variables $A_0(x)$ describing "scalar photons" and the longitudinal components of $A_j(x)$. In simple cases (like free photon theory), one can get rid of these extra variables and formulate the theory in terms of physical fields only. But in more complicated cases, for theories involving nontrivial interactions, it is very difficult (as it is in the case of quantum electrodynamics including besides photons also charged fields with which photons interact) or just impossible (as it is for non-Abelian theories).

For practical perturbative calculations in interacting theories, one should accurately *fix the gauge* using the Gupta-Bleuler or, in a non-Abelian case, Faddeev-Popov quantization procedure, however we will not pursue further this subject here. Many our readers, experts in QFT, know it as well as the author. Those who do not know it so well and wish to learn more may have a look, for example, at Lecture 7 in Ref. [8].

The Lagrangian[3] for the *scalar* electrodynamics where the electromagnetic field couples to the charged scalar field reads

$$\mathcal{L} = -\frac{1}{4} F_{\mu\nu} F^{\mu\nu} + (\mathcal{D}_\mu \phi)^* (\mathcal{D}^\mu \phi), \qquad (1.17)$$

where[4] $\mathcal{D}_\mu = \partial_\mu + ieA_\mu$. One can also add the mass term $-m^2 \phi^* \phi$ and the quartic interaction term $\propto (\phi^* \phi)^2$. This Lagrangian is invariant under the gauge transformations, which now also act on the scalar fields:

$$A_\mu \rightarrow A_\mu + \partial_\mu \chi, \qquad \phi \rightarrow e^{-ie\chi} \phi. \qquad (1.18)$$

The group of gauge transformations leaving the Lagrangians (1.11) and (1.17) invariant is infinite-dimensional, but locally (in a particular spacetime point) it is $U(1)$. The Lagrangian of the Standard Model, which describes our world, has also gauge symmetry, but this symmetry is larger. Locally, it is $SU(3) \times SU(2) \times U(1)$.

This Lagrangian includes non-Abelian gauge fields $\hat{A}_\mu = A_\mu t^a$ and $\hat{W}_\mu = W_\mu^a t^a$, where t^a are the generators of $SU(3)$ and $SU(2)$, correspondingly. A_μ^a are the gluon strongly interacting fields, the components $W_\mu^1 \pm iW_\mu^2$ are weakly interacting W-bosons and the field W_μ^3 represents a certain linear combination of the fields describing the Z-boson and the photon. The Lagrangian of a pure non-Abelian gauge field was first written by Yang and Mills back in[5] 1954. It reads[6]

$$\boxed{\mathcal{L} = -\frac{1}{2} \text{Tr} \left\{ \hat{F}_{\mu\nu} \hat{F}^{\mu\nu} \right\},} \qquad (1.19)$$

---

[3] Actually, the Lagrangian *spatial density*, but, using sloppy physical terminology, we will use the same name for $\mathcal{L}$ and for $L = \int d\boldsymbol{x} \mathcal{L}$.

[4] Bearing in mind our conventions spelled out on p. vii, this means that $\mathcal{D}_0 = \partial_0 + ieA_0$ and $\mathcal{D}_i = \partial_i - ieA_i$.

[5] They had, of course, no idea about gluons or W-bosons. They wrote the Lagrangian from purely mathematical premises and tried then to use it for the description of ρ mesons, which did not work.

[6] Another popular convention consists in redefining $gA_\mu \rightarrow A_\mu$ so that the coupling constant shows up only in the common factor in front of the Lagrangian,

$$\mathcal{L} = -\frac{1}{2g^2} \text{Tr} \left\{ \hat{F}_{\mu\nu} \hat{F}^{\mu\nu} \right\},$$

but not in the definition of $\hat{F}_{\mu\nu}$.

where

$$\hat{F}_{\mu\nu} = \partial_\mu \hat{A}_\nu - \partial_\nu \hat{A}_\mu - ig[\hat{A}_\mu, \hat{A}_\nu]. \qquad (1.20)$$

$\hat{F}_{\mu\nu}$ belongs, as $\hat{A}_\mu$ does, to the adjoint representation of the group and can be written as $\hat{F}_{\mu\nu} = F^a_{\mu\nu} t^a$ with

$$F^a_{\mu\nu} = \partial_\mu A^a_\nu - \partial_\nu A^a_\mu + g f^{abc} A^b_\mu A^c_\nu, \qquad (1.21)$$

where f^{abc} are the structure constants of the gauge group G in interest and g is a real number having the physical meaning of the coupling constant.

The Lagrangian (1.19) is invariant with respect to the following gauge transformations:

$$\hat{A}_\mu \to \Omega \hat{A}_\mu \Omega^{-1} - \frac{i}{g}(\partial_\mu \Omega)\Omega^{-1}, \qquad (1.22)$$

with $\Omega \in G$. Then $\hat{F}_{\mu\nu} \to \Omega \hat{F}_{\mu\nu} \Omega^{-1}$ and the trace (1.19) is invariant indeed.

It is noteworthy that the notion of gauge fields is also known to mathematitians, but their terminology is different. The gauge potential $\hat{A}_\mu$ is called the *connection of a principal fiber bundle* or *principal G-connection* in the mathematical language. And the field density $\hat{F}_{\mu\nu}$ is the *curvature form of a principal G-connection.*[7] But Yang and Mills were unaware of these mathematical findings. Neither were they aware of the Standard Model and tried to use the Lagrangian (1.19) to describe the interactions of ρ mesons...

One can couple the Lagrangian (1.19) to other fields belonging to fundamental and other representations of the group G in a way that the gauge invariance of the Lagrangian is preserved. For scalar fields in the fundamental representation, one may write in analogy with (1.17)

$$\mathcal{L}_{\text{YM+scalars}} = -\frac{1}{2}\text{Tr}\left\{\hat{F}_{\mu\nu}\hat{F}^{\mu\nu}\right\} + (\mathcal{D}_\mu \phi)^*(\mathcal{D}^\mu \phi) - V(\phi^*, \phi) \quad (1.23)$$

with[8] $\mathcal{D}_\mu = \partial_\mu - ig\hat{A}_\mu$. A variant of this Lagrangian describes the weak and electromagnetic sector in the Standard Model (its bosonic part).

The Lagrangian of quantum chromodynamics involves besides the gauge fields in the adjoint representation of $SU(3)$ (*gluon* fields) also fermionic quark fields in the fundamental representation. We will present its Lagrangian in the next section.

---

[7] As far as I know, the first mathematician who discovered this idea was Elie Cartan back in 1926 [9]. He did it for tangent bundles [the tangent bundle on a Riemann manifold of dimension d can be treated as a principal bundle with the group $SO(d)$]. And C. Ehresmann was probably the first to generalize this concept for all other groups [10].

[8] See the footnote on p. 16.

The Lagrangian (1.19) includes the quadratic term and also cubic and quartic interaction terms. In the quadratic approximation, the spectrum of the corresponding Hamiltonian includes d vector massless particles with the same properties as photons (d being the dimension of the gauge group).

However, the presence of nonlinear interactions modifies the physical picture completely. It is an experimental fact that the physical spectrum in QCD does not involve gluons and quarks, but only hadrons which are color singlets. The same *confinement* scenario is expected to be realized in pure Yang-Mills theory — its spectrum must involve only *glueballs* — massive color-singlet states. There are heuristic theoretic arguments allowing one to understand why it happens, but a rigourous proof of confinement is absent. One of the "Millennium prizes" [11] awaits a scholar who would construct it.

1.3 Fermion fields

The first subatomic particle discovered in experiment was the electron. The electrons carry spin 1/2 and they are fermions, which means that they obey the Fermi-Dirac statistics so that two electrons cannot occupy one and the same quantum state. Fierz and Pauli noticed [12] that this property (*Pauli exclusion principle*) can be achieved if one postulates the basic *anticommutation* relations between the electronic creation and annihilation operators, like

$$\boxed{\{c_{\boldsymbol{p}}^\dagger, c_{\boldsymbol{q}}\} \ \propto \ \delta(\boldsymbol{p} - \boldsymbol{q}),} \tag{1.24}$$

rather than the commutation relations that are operative for scalar particles and for the photons. They also noticed that, if not doing so and keeping the commutation relations also for the electrons or any other particles with half-integer spin, the spectrum of their Hamiltonian would not have a bottom, but would involve the states with arbitrarily low energies. But the question of where these nontrivial anticommutators come from (the standard quantum mechanics involves only commutators, like $[\hat{p}, x] = -i\hbar$) was not answered at that time.

Full understanding came only in the nineteen sixties when it was understood [13] that, in the *classical* mechanics of fermion fields, the dynamical variables are not ordinary real or complex numbers, but Grassmann anticommuting numbers. They were introduced in physics by Felix Berezin [14].

1.3.1 *Grassmann algebra*

The main definitions are the following:

- Let $\{a_i\}$ be a set of n basic anticommuting variables: $a_i a_j + a_j a_i = 0$. The elements of the Grassmann algebra can be written as the functions $f(a_i) = c_0 + c_i a_i + c_{ij} a_i a_j + \ldots$ where the coefficents $c_0, c_i, \ldots$ are ordinary (real or complex) numbers. The series terminates at the n-th term: the $(n+1)$ - th term of this series would involve a product of two identical anticommuting variables (say, a_1^2), which is zero. Note that even though a Grassmann number like $a_1 a_2$ commutes with any other number, it still cannot be treated as an ordinary number but represents instead an *even element* of the Grassmann algebra. There are, of course, also odd anticommuting elements. The variables $\{a_i\}$ are called the *generators* of the algebra.

- One can add Grassmann numbers, $f(a_i) + g(a_i) = c_0 + d_0 + (c_i + d_i)a_i + \ldots$, as well as multiply them. For example,

$$(1 + a_1 + a_2)(1 + a_1 - a_2) = 1 + 2a_1 - 2a_1 a_2$$

  (the anticommutation property of a_i was used).

- One can also differentiate the functions $f(a_i)$ with respect to Grassmann variables:

$$\partial/\partial a_i (1) \stackrel{\text{def}}{=} 0, \quad \partial/\partial a_i (a_j) \stackrel{\text{def}}{=} \delta_{ij},$$

  the derivative of a sum is the sum of derivatives, and the derivative of a product of two functions satisfies the Leibniz rule, except that the operator $\partial/\partial a_i$ should be thought of as a Grassmann variable, which sometimes leads to a sign change when $\partial/\partial a_i$ is pulled through to annihilate a_i in the product. For example, $\partial/\partial a_1(a_2 a_3 a_1) = a_2 a_3$, but $\partial/\partial a_1(a_2 a_1 a_3) = -a_2 a_3$.

- One can also integrate over Grassmann variables. In contrast to the ordinary case, the integral cannot be obtained here as a limit of integral sums, cannot be calculated numerically through "finite limits" (this makes no sense for Grassmann numbers) with Simpson's method, and so on. What one can do, however, is to integrate over a Grassmann variable in the "whole range" ("from $-\infty$ to ∞" if you will, though this is, again, meaningless). The definition (due to Berezin) is

$$\int da_j\, f(a) \stackrel{\text{def}}{=} \frac{\partial}{\partial a_j} f(a). \tag{1.25}$$

We have in particular

$$\int da_i\, a_j \;=\; \delta_{ij}\,. \tag{1.26}$$

An obvious (and important for the following) corollary of (1.25) is

$$\int da_j\, \frac{\partial f}{\partial a_j} \;=\; 0 \tag{1.27}$$

(no summation over j) for any $f(a)$.

- If the Grassmann algebra involves an even number of generators $2n$, one can divide them into two equal parts, $\{a_{j=1,\dots,2n}\} \to \{a_{j=1,\dots,n}, \bar{a}_{j=1,\dots,n}\}$ and introduce an involution, $a_j \leftrightarrow \bar{a}_j$, which we will associate with complex conjugation. We will assume in particular that, simultaneously with the involution of the generators, the ordinary numbers c_0, etc. are replaced by their complex conjugates:

$$f(a) = c_0 + c_i a_i + d_i \bar{a}_i + \dots \;\longrightarrow\; \overline{f(a)} = c_0^* + c_i^* \bar{a}_i + d_i^* a_i + \dots \tag{1.28}$$

It is also convenient to assume that, for any two elements f, g of the Grassmann algebra, $\overline{fg} = \bar{g}\bar{f}$, as for the Hermitian conjugation (the order of the factors in (1.28) is, of course, not relevant).

The following important relation holds:

$$\boxed{\int \prod_{i=1}^{n} da_i d\bar{a}_i\, \exp\{M_{jk}\bar{a}_j a_k\} \;=\; \det(M)\,.} \tag{1.29}$$

1.3.2   *Neutrino*

We start by writing down the simplest expression for the relativistic Lagrangian describing massless fields of spin $\frac{1}{2}$ that is invariant under Lorentz transformations.

As is well known, the Lorentz group $SO(3,1)$ represents, in some sense, a product of two $SU(2)$ groups. Its representations are thus labelled by two numbers (j_L, j_R), each of which can be integer or half-integer. There are two different spinor representations, the left-handed spinors ξ_α carrying the undotted indices $\alpha = 1, 2$ and belonging to the representation $(\frac{1}{2}, 0)$ and the right-handed spinors $\zeta_{\dot\alpha}$ carrying the dotted indices and belonging to the representation $(0, \frac{1}{2})$.[9]

---

[9]Using physical conventions, we call "representation" what mathematicians usually call "representation space". The names "left-handed" and "right-handed" are also physically motivated: the particles that the field $\xi_\alpha(x)$ describes have negative helicity (the direction of the spin is opposite to the direction of the momentum) and resemble thus left-handed screws. And the particles that the field $\zeta_{\dot\alpha}$ describes have positive helicity and resemble right-handed screws.

Take one of them, say, the left-handed spinor field $\xi_\alpha(x)$ and assume it to be an odd element of the Grassmann algebra. The spinor that is complex conjugate to ξ_α belongs to the representation $(0, \frac{1}{2})$ and can be denoted $\bar{\xi}_{\dot\alpha}$. The simplest Lorentz-invariant Lagrangian for the field $\xi_\alpha(x)$ involving the derivative operator is

$$\mathcal{L}_{\text{kin}}^{\text{ferm}} = i\xi^\alpha \partial_{\alpha\dot\beta} \bar{\xi}^{\dot\beta} \equiv i\xi^\alpha (\sigma^\mu)_{\alpha\dot\beta} \, \partial_\mu \bar{\xi}^{\dot\beta}, \tag{1.30}$$

with the notation σ_μ introduced in (0.3). The spinor indices are raised and lowered according to (0.2), and the fields ξ^α and $\bar{\xi}^{\dot\alpha}$ are mutually conjugate. The canonical dimension of $\xi^\alpha(x)$ is $[\xi] = m^{3/2}$.

Let us treat (1.30) in the same way as we did for the scalar Lagrangian in (1.1). We introduce a finite spatial box of size L and impose periodic boundary conditions on $\xi_\alpha(\boldsymbol{x}, t)$:

$$\xi_\alpha(x + L, y, z) = \xi_\alpha(x, y + L, z) = \xi_\alpha(x, y, z + L) = \xi_\alpha(x, y, z). \tag{1.31}$$

We expand $\xi_\alpha(\boldsymbol{x})$ in a Fourier series

$$\xi_\alpha(\boldsymbol{x}) = \frac{1}{L^{3/2}} \sum_{\boldsymbol{n}} \xi_\alpha^{\boldsymbol{n}} \exp\left\{ \frac{2\pi i \boldsymbol{n} \cdot \boldsymbol{x}}{L} \right\}, \tag{1.32}$$

substitute it in (1.30) and integrate over $\int d\boldsymbol{x}$. We obtain[10]

$$L = \frac{1}{L} \sum_{\boldsymbol{n}} \left[i\xi_{\boldsymbol{n}}^\alpha \delta_{\alpha\dot\beta} \frac{d}{dt} \left(\bar{\xi}_{\boldsymbol{n}}^{\dot\beta} \right) + 2\pi \boldsymbol{n} \, \xi_{\boldsymbol{n}}^\alpha \boldsymbol{\sigma}_{\alpha\dot\beta} \, \bar{\xi}_{\boldsymbol{n}}^{\dot\beta} \right]. \tag{1.33}$$

The corresponding canonical Hamiltonian is

$$H = -\sum_{\boldsymbol{n}} \frac{2\pi \boldsymbol{n}}{L} \, \xi_{\boldsymbol{n}}^\alpha \boldsymbol{\sigma}_{\alpha\dot\beta} \, \bar{\xi}_{\boldsymbol{n}}^{\dot\beta}. \tag{1.34}$$

As it was the case for the free scalars, the Hamiltonian represents an infinite sum of independent simple Hamiltonians in the sectors with a given $\boldsymbol{n}$. To simplify things, set $L = 1$ and choose e.g. $\boldsymbol{n} = (0, 0, 1)$. The corresponding contribution in (1.34) is

$$H_{(001)} = 2\pi(\xi^2 \bar{\xi}^2 - \xi^1 \bar{\xi}^1). \tag{1.35}$$

It represents a difference of two elementary *Grassmann oscillators*.

The classical Hamiltonian of the Grassmann oscillator is

$$H = \omega \xi \bar{\xi}. \tag{1.36}$$

---

[10] Hopefully, the same notation L for the Lagrangian and for the length of the box will not result in any confusion.

This expression reminds the expression for the ordinary oscillator Hamiltonian $H = \omega a a^*$, where

$$a = \frac{\omega q + ip}{\sqrt{2\omega}}, \qquad a^* = \frac{\omega q - ip}{\sqrt{2\omega}}. \tag{1.37}$$

are the holomorphic variables, but ξ and $\bar{\xi}$ are now anticommuting Grassmann numbers canonically conjugated to one another.

After quantization, ξ and $\bar{\xi}$ become nilpotent operators, which do not anticommute anymore, but instead satisfy the relation

$$\{\hat{\xi}, \hat{\bar{\xi}}\} = 1. \tag{1.38}$$

In other words, the simplest Grassmann algebra becomes the simplest Clifford algebra after quantization. This general rule works also for more complicated systems involving Grassmann variables.

The algebra (1.38) has two natural representations. There is an obvious matrix representation:

$$\hat{\bar{\xi}} = \sigma^+ = \begin{pmatrix} 0 & 1 \\ 0 & 0 \end{pmatrix}, \qquad \hat{\xi} = \sigma^- = \begin{pmatrix} 0 & 0 \\ 1 & 0 \end{pmatrix}. \tag{1.39}$$

But the representation

$$\hat{\xi} = \xi, \quad \hat{\bar{\xi}} = \frac{\partial}{\partial \xi} \tag{1.40}$$

is more universal (it can be easily generalized for systems with several Grassmann variables, especially for field theory systems). Substituting (1.40) in (1.36), we derive the quantum Hamiltonian:

$$\hat{H} = \omega \xi \frac{\partial}{\partial \xi}. \tag{1.41}$$

The Hilbert space for this system is very simple. It involves the wave functions

$$\Psi(\xi) = a + b\xi. \tag{1.42}$$

All the higher-order terms in the Taylor expansion of $\Psi(\xi)$ vanish due to the property $\xi^2 = 0$!

The Hamiltonian (1.41) has two eigenfunctions: $\Psi(\xi) = 1$ with the eigenvalue $E = 0$ and $\Psi(\xi) = \xi$ with the eigenvalue ω. Thus, in contrast to the usual harmonic oscillator, our Grassmann oscillator does not have an infinite tower of equidistant states, it has only two states. This is where the Pauli principle comes from!

The quantum counterpart of the Hamiltonian (1.35) is

$$\hat{H} = \frac{2\pi}{L}\left(\xi^2\frac{\partial}{\partial\xi^2} - \xi^1\frac{\partial}{\partial\xi^1}\right) \tag{1.43}$$

(we restored the dimensional factor $1/L$). It acts on the wave functions

$$\Psi(\xi^1,\xi^2) = a + b\xi^1 + c\xi^2 + d\xi^1\xi^2. \tag{1.44}$$

It has four eigenstates:

$$A)\ \Psi = \xi^1 \qquad \text{with} \qquad E = -\frac{2\pi}{L};$$

$$B)\ \Psi = 1 \quad \text{and} \quad C)\ \Psi = \xi^1\xi^2 \qquad \text{with} \qquad E = 0;$$

$$D)\ \Psi = \xi^2 \qquad \text{with} \qquad E = \frac{2\pi}{L}. \tag{1.45}$$

The state A is the ground state.

Note that the same Hamiltonian (1.43) with the variables $\xi^{1,2}$ treated as ordinary rather than Grassmann numbers would not have a bottom. That is why the Grassmann nature of the fermion fields is not an option, but a necessity.[11]

The variables ξ^α in (1.43) describe the Fourier modes with $\boldsymbol{n} = (0,0,1)$, but we can do the same analysis for any other choice of $\boldsymbol{n}$. Again, we obtain four states in the spectrum:

$$A)\ \Psi = \xi_n^{(+)} \qquad \text{with} \qquad E = -\frac{2\pi|\boldsymbol{n}|}{L};$$

$$B)\ \Psi = 1 \quad \text{and} \quad C)\ \Psi = \xi_n^{(+)}\xi_n^{(-)} \qquad \text{with} \qquad E = 0;$$

$$D)\ \Psi = \xi_n^{(-)} \qquad \text{with} \qquad E = \frac{2\pi|\boldsymbol{n}|}{L}, \tag{1.46}$$

where $\xi_n^{(+)}$ and $\xi_n^{(-)}$ are the positive and negative helicity spinors satisfying

$$(\boldsymbol{\sigma}\cdot\boldsymbol{n})\,\xi_n^{(+)} = |\boldsymbol{n}|\xi_n^{(+)},$$
$$(\boldsymbol{\sigma}\cdot\boldsymbol{n})\,\xi_n^{(-)} = -|\boldsymbol{n}|\xi_n^{(-)}. \tag{1.47}$$

---

[11]The free Hamiltonian (1.43) with ordinary $\xi^{1,2}$ acting in the Hilbert space with polynomial $\Psi(\xi^1,\xi^2)$ is actually quite benign. Though the ground state is absent, the Hamiltonian is Hermitian and the evolution operator is unitary. But including interactions usually brings about troubles — the quantum collapse occurs so that probability leaks into singularity, breaking unitarity. There are some nontrivial interacting systems involving ghosts (meaning that their Hamiltonians have no ground states) where unitarity is still preserved [15], but they are rather special. In Chapters 2,11 we will briefly mention without going into much detail a supersymmetric system having this feature. But this subject is marginal for our book.

This pattern of the spectrum can be interpreted in the *Dirac sea* spirit. The left-handed modes $\xi_n^{(-)}$ have positive energies. In the vacuum, these levels are empty. When a level $\xi_n^{(-)}$ is occupied (the wave function involves the factor $\xi_n^{(-)}$), it means the presence of the physical left-handed particle with momentum[12] $p_n = 2\pi n/L$ and energy $\epsilon_n = |p_n|$. On the other hand, the right-handed modes $\xi_n^{(+)}$ all have negative energies and belong to the sea. In the vacuum, all these levels are occupied. An excited state of the Hamiltonian with an empty sea level (a "hole") describes the physical right-handed antiparticle with momentum $p_n = -2\pi n/L$ and $\epsilon_n = |p_n|$.

In other words, the Lagrangian (1.30) describes free massless left-handed neutrinos and right-handed antineutrinos (though one has to have in mind, of course, that the attributions "left" and "right", "particle" and "antiparticle" to the physical states depend on conventions).

1.3.3 *Electron □ Quarks*

Let us try to generalize (1.30) by including the interaction with A_μ in the same way as we did so for scalar particles and write

$$\mathcal{L} = -\frac{1}{4}F_{\mu\nu}F^{\mu\nu} + i\xi^\alpha(\sigma^\mu)_{\alpha\dot\beta}(\partial_\mu - ieA_\mu)\bar\xi^{\dot\beta}. \tag{1.48}$$

This, however, does not work. It turns out that the Lagrangian (1.48) is gauge-invariant only at the classical level. One cannot keep this invariance in quantum theory due to the so-called *chiral anomaly*.[13] And the breaking of gauge invariance makes the theory non-renormalizable and inconsistent. Thus, neutrinos do not carry electric charge.

We wish, however, also to be able to describe electrons, quarks and other charged fermions. How can we do that?

The answer is the following. We have to double the number of fermion degrees of freedom[14] and to introduce, along with $\xi_\alpha(x)$, *another* similar fermion field $\eta_\alpha(x)$. We further assume that the electric charge of η_α is *opposite* to the electric charge of ξ_α. Then the fermion part of the Lagrangian

---

[12]The corresponding quantum operator is

$$\hat P = \sum_n \frac{2\pi n}{L}\left(\xi_n^\alpha \frac{\partial}{\partial\xi_n^\alpha} - 1\right).$$

[13]We will not explain here what it means addressing the reader to the textbooks on QFT. S/he may consult, e.g. Lectures 12 and 14 in Ref. [8].

[14]And we need to do so irrespectively of the anomaly troubles. Indeed, the quantization of the Lagrangian (1.30) provided us with only two physical states for each value of momentum: left-handed particle and right-handed antiparticle. But the electrons might be both left- and right-handed. And the same concerns positrons.

reads

$$\mathcal{L}_{\xi\eta}^{\text{massless}} = i\xi^{\alpha}(\sigma^{\mu})_{\alpha\dot{\beta}}(\partial_{\mu} - ieA_{\mu})\bar{\xi}^{\dot{\beta}} + i\eta^{\alpha}(\sigma^{\mu})_{\alpha\dot{\beta}}(\partial_{\mu} + ieA_{\mu})\bar{\eta}^{\dot{\beta}}. \quad (1.49)$$

It is invariant under gauge transformations

$$A_{\mu} \to A_{\mu} + \partial_{\mu}\chi,$$
$$\xi \to e^{-ie\chi}\xi,$$
$$\eta \to e^{ie\chi}\eta, \quad (1.50)$$

and this invariance is not broken by quantum effects (the anomaly cancels). The quantization of the field ξ_{α} gives the particles e_L^- and e_R^+ (left-handed electron and right-handed *positron*). The quantization of the field η_{α} gives the left-handed positron e_L^+ and right-handed electron e_R^-.

With two different fermion fields of opposite charge, we can also add to the Lagrangian a gauge-invariant and Lorentz-invariant potential term which gives a mass to the electron:

$$\mathcal{L}_m = -m\left(\xi^{\alpha}\eta_{\alpha} + \bar{\eta}_{\dot{\alpha}}\bar{\xi}^{\dot{\alpha}}\right). \quad (1.51)$$

Due to the presence of mass, the spectrum of the free fermion Hamiltonian involves a gap (see Fig. 1.1).

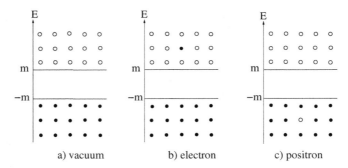

a) vacuum   b) electron   c) positron

Fig. 1.1: Dirac sea. Black and white circles correspond to the filled and empty levels, respectively.

Note that one cannot write a mass term for a single charged fermion field ξ_{α}, the structure $m\xi_{\alpha}\xi^{\alpha}$ (it is called a *Majorana mass term*) is Lorentz-invariant, but not gauge-invariant! On the other hand, one can well endow with a Majorana mass a single *neutral* fermion not participating in gauge interactions. In particular, it is not experimentally excluded that the masses of the *right-handed* neutrinos have partially a Majorana nature.

Traditionally, the fermion part of the Lagrangian is written in a somewhat more compact way. Let us describe it.

The two left-handed two-component fields ξ_α, η_α can be united to form a 4-component *Dirac bispinor*

$$\psi = \begin{pmatrix} \bar{\eta}^{\dot{\alpha}} \\ \xi_\alpha \end{pmatrix}. \tag{1.52}$$

Note that the upper and the lower components of this bispinor carry the same charge e, so that ψ is gauge-transformed as $\psi \to e^{-ie\chi}\psi$.

We also introduce four *Dirac matrices*

$$\gamma^\mu = \begin{pmatrix} 0 & \bar{\sigma}^\mu \\ \sigma^\mu & 0 \end{pmatrix}. \tag{1.53}$$

They satisfy the Clifford algebra,

$$\gamma^\mu\gamma^\nu + \gamma^\nu\gamma^\mu = 2\eta^{\mu\nu}\mathbb{1}_4. \tag{1.54}$$

Consider now the matrix

$$\gamma^5 = i\gamma^0\gamma^1\gamma^2\gamma^3 = \begin{pmatrix} \mathbb{1}_2 & 0 \\ 0 & -\mathbb{1}_2 \end{pmatrix} \tag{1.55}$$

which anticommutes with all γ^μ. It plays a special role, allowing one to build the projectors onto the fields with definite *chirality*. When acting on ψ, they cook up left-handed and right-handed electron bispinor fields: [15]

$$\psi_L = \frac{1 - \gamma^5}{2}\psi = \begin{pmatrix} 0 \\ \xi \end{pmatrix}, \quad \psi_R = \frac{1 + \gamma^5}{2}\psi = \begin{pmatrix} \bar{\eta} \\ 0 \end{pmatrix}. \tag{1.56}$$

Finally, we define *Dirac conjugation* according to $\bar{\psi} = \psi^\dagger\gamma^0$ (with the long bar!), $\psi^\dagger$ being a Hermitially conjugated bispinor. Explicitly,

$$\bar{\psi} = (\bar{\xi}_{\dot{\alpha}}, \eta^\alpha). \tag{1.57}$$

It is easy to see that

$$\bar{\psi}\psi = \bar{\xi}_{\dot{\alpha}}\bar{\eta}^{\dot{\alpha}} + \eta^\alpha\xi_\alpha \tag{1.58}$$

---

[15] Very often people do not distinguish between the notions of chirality and helicity. It looks natural, bearing in mind that, when translated from Greek, chirality means "handedness", and we call the particles carrying negative helicity left-handed and positive helicity right-handed. But there is actually some distinction. Helicity is a property of a *particle state*, as defined in (1.47), while chirality is the property of a *field*. We call the fields with undotted indices *left chiral* and the fields with dotted indices *right chiral*. Then, as was discussed above, the left chiral field ξ_α entering (1.52) describes both left-handed electrons and right-handed positrons, while the right chiral field $\bar{\eta}^{\dot{\alpha}}$ describes the right-handed electrons and left-handed positrons.

is a Lorentz scalar. One can further observe that $\overline{\psi}\gamma^5\psi$ is a pseudoscalar (changes sign under spatial reflection), $\overline{\psi}\gamma^\mu\psi$ is a vector and $\overline{\psi}\gamma^\mu\gamma^5\psi$ is an axial vector.

In these terms, the Lagrangian of quantum electrodynamics reads

$$\mathcal{L} = -\frac{1}{4}F_{\mu\nu}F^{\mu\nu} + i\overline{\psi}\gamma^\mu(\partial_\mu + ieA_\mu)\psi - m\overline{\psi}\psi. \tag{1.59}$$

It is invariant under

$$A_\mu \to A_\mu + \partial_\mu\chi,$$
$$\psi \to e^{-ie\chi}\psi. \tag{1.60}$$

The Lagrangian of quantum chromodynamics — the theory of strong interaction — involves the Yang-Mills fields with the gauge group $SU(3)$ coupled to massive fermions lying in the fundamental representation of the group. There are six types (flavours) of quarks. The QCD Lagrangian reads[16]

$$\mathcal{L}_{\text{QCD}} = -\frac{1}{2}\text{Tr}\left\{\hat{F}_{\mu\nu}\hat{F}^{\mu\nu}\right\} +$$
$$\sum_{f=1}^{6}\left[i\overline{\psi}_f\gamma^\mu(\partial_\mu - ig\hat{A}_\mu)\psi_f - m_f\overline{\psi}_f\psi_f\right]. \tag{1.61}$$

As was mentioned above, the physical spectrum of this theory involves colorless hadrons, while quarks and gluons are confined.

---

[16]We follow the usual convention when the signs of the coupling constant in the fermion kinetic term in (1.61) and (1.59) are opposite. This mess can be traced back to the fact that the electron charge is negative, $e = -|e|$. Unfortunately, when Benjamin Franklin defined two and a half centuries ago which charge is positive and which is negative, he knew nothing about electrons.

Chapter 2

Supersymmetric quantum mechanics

Supersymmetric field theory[1] is the theory where the spectra of bosonic and fermionic excitations coincide. This is a physical definition. But one can give a more general mathematical definition which applies to *all* supersymmetric quantum systems including supersymmetric quantum-mechanical (SQM) systems with a finite number of degrees of freedom.

Definition 2.1. A quantum system is called supersymmetric if its Hamiltonian can be represented as the anticommutator of two nilpotent mutually conjugated operators called *supercharges*:

$$\boxed{\hat{Q}^2 = (\hat{Q}^\dagger)^2 = 0, \qquad \{\hat{Q}, \hat{Q}^\dagger\} = 2\hat{H}.} \tag{2.1}$$

Eq. (2.1) defines the simplest possible supersymmetry *algebra*. Its immediate corollary is the fact that the supercharges $\hat{Q}$ and $\hat{Q}^\dagger \equiv \hat{\bar{Q}}$ commute with the Hamiltonian, and are the integrals of motion.

Representing the complex supercharge $\hat{Q}$ as the sum

$$\hat{Q} = \frac{\hat{\mathcal{Q}}_1 + i\hat{\mathcal{Q}}_2}{\sqrt{2}} \tag{2.2}$$

with Hermitian $\hat{\mathcal{Q}}_{1,2}$, one can rewrite (2.1) as

$$\hat{\mathcal{Q}}_1^2 = \hat{\mathcal{Q}}_2^2 = \hat{H}, \qquad \{\hat{\mathcal{Q}}_1, \hat{\mathcal{Q}}_2\} = 0. \tag{2.3}$$

In other words, the Hamiltonian admits in this case *two different* anticommuting Hermitian square roots. Note that the existence of a single square root is a trivial property of any Hamiltonian whose spectrum is bounded from below. To be more precise, one has first to bring the ground state energy to a positive or zero value by adding, if necessary, a constant to the

---

[1] The first example of such theory was constructed more than 50 years ago by Golfand and Lichtman [16].

17

Hamiltonian and to extract the square root afterwards. But the presence of *two* different anticommuting square roots imposes nontrivial constraints on the spectrum.

Theorem 2.1. *The eigenstates of a supersymmetric Hamiltonian have non-negative energies. If a state Ψ_0 with zero energy exists, it is annihilated by the action of the supercharges:*

$$\hat{Q}\,\Psi_0 \;=\; \hat{\bar{Q}}\,\Psi_0 \;=\; 0\,. \tag{2.4}$$

Proof. Suppose that (2.4) does not hold. Then $\hat{Q}_1\Psi_0 \neq 0$ or $\hat{Q}_2\Psi_0 \neq 0$. Let $\hat{Q}_1\Psi_0 \neq 0$. The Hermitian operator $\hat{Q}_1$ commutes with the Hamiltonian and hence these two operators can be simultaneously diagonalized. Hence Ψ_0 can be chosen to be an eigenstate of $\hat{Q}_1$. Then $\hat{Q}_1\Psi_0 = \lambda\Psi_0$ with nonzero λ. And then it follows from the first relation in (2.3) that $\hat{H}\Psi_0 = \lambda^2\Psi_0$ and the energy is strictly positive. $\qquad\square$

Remark. I cannot hold back at this juncture a side remark concerning *ghosts*.

It was essential for the proof that the operators $\hat{Q}$ and $\hat{Q}^\dagger$ entering (2.1) are Hermitially conjugate to one another so that $\hat{Q}_1$ and $\hat{Q}_2$ are Hermitian. If this condition is not fulfilled, the spectrum may include negative energies. It seems at the first sight that in that case one would obtain a non-Hermitian Hamiltonian that can have *any* eigenvalues, not necessarily real. Unexpectedly, there exist a certain *special* generalized supersymmetric system whose algebra includes the supercharges $\hat{Q}$ and $\hat{\bar{Q}} \neq \hat{Q}^\dagger$. Seemingly, its Hamiltonian is not Hermitian, but actually it belongs to the class of so-called *pseudo-Hermitian* or, better to say, *crypto-Hermitian* Hamiltonians [17]. Its spectrum is real, but it includes both positive and negative energies. This system represents an example of the system with *benign ghosts*: a ground state is absent, but still the evolution operator is unitary [18].

The contact between the physical definition of supersymmetry (boson-fermion degeneracy) and its mathematical definition above is established by the following theorem.

Theorem 2.2. *All eigenstates of a supersymmetric Hamiltonian with $E \neq 0$ are double degenerate.*

Proof. Choose the basis in our Hilbert space including the states that are eigenstates of $\hat{H}$ and simultaneously the eigenstates of $\hat{Q}_1$. Pick one of such states with positive energy E. The eigenvalue of $\hat{\mathcal{Q}}_1$ may in this case be equal to $\lambda = \pm\sqrt{E}$. Let for definiteness $\lambda = \sqrt{E}$. Then the state $\Psi' = \hat{\mathcal{Q}}_2\Psi/\sqrt{E}$ is also an eigenstate of $\hat{\mathcal{Q}}_1$ with the eigenvalue $\lambda' = -\sqrt{E}$. Indeed, bearing in mind that $\hat{\mathcal{Q}}_1$ and $\hat{\mathcal{Q}}_2$ anticommute, we derive

$$\hat{\mathcal{Q}}_1\Psi' = -\frac{\hat{\mathcal{Q}}_2\hat{\mathcal{Q}}_1\Psi}{\sqrt{E}} = -\sqrt{E}\,\Psi'. \tag{2.5}$$

The eigenstates Ψ and Ψ' of the operator $\hat{\mathcal{Q}}_1$ have different eigenvalues (and hence they are orthogonal), but the same energies and the same norm. They are obtained from one another by the action of $\hat{\mathcal{Q}}_2$ (note that $\hat{\mathcal{Q}}_2\Psi' = \sqrt{E}\,\Psi$) and represent a double degenerate pair in the spectrum of the Hamiltonian. Had we picked up in the beginning another eigenstate of $\hat{\mathcal{Q}}_1$, we would obtain another such degenerate pair (or maybe exactly the same if the state Ψ' had been chosen). $\qquad\square$

Consider the states $\Psi_{\pm} = \Psi \pm i\Psi'$. It is easy to see that

$$\begin{aligned} \hat{Q}\,\Psi_- &= \sqrt{2E}\,\Psi_+\,, & \hat{\bar{Q}}\,\Psi_- &= 0\,, \\ \hat{Q}\,\Psi_+ &= 0\,, & \hat{\bar{Q}}\,\Psi_+ &= \sqrt{2E}\,\Psi_-\,. \end{aligned} \tag{2.6}$$

In other words, all the excited states can be divided into two sectors. The states Ψ_- are annihilated by the operator $\hat{\bar{Q}}$ and go over to Ψ_+ under the action of $\hat{Q}$. The states Ψ_+ are annihilated by the operator $\hat{Q}$ and go over to Ψ_- under the action of $\hat{\bar{Q}}$ (Fig. 2.1).

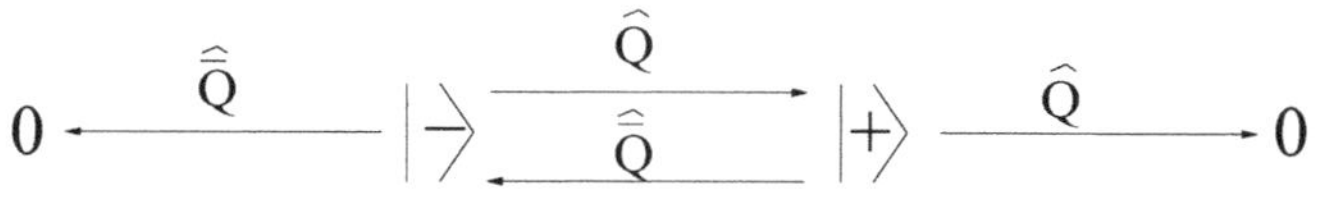

Fig. 2.1: A supersymmetric doublet

And the zero-energy states are annihilated by both $\hat{Q}$ and $\hat{\bar{Q}}$ and are not paired. As we will see soon, it often makes sense to attribute some of these states to the sector $|-\rangle$ and some others to the sector $|+\rangle$, but to do so, one has to concretize our discussion, to write down supersymmetric Hamiltonians for some particular systems and mark out their peculiarities and common features.

2.1   Witten's model

The simplest nontrivial SQM model was constructed by Witten [19]. The wave functions in this model depend on a single ordinary variable x and a holomorphic Grassmann variable ψ. The supercharges and Hamiltonian read

$$\hat{Q} = \psi \left[\hat{p} - i\mathcal{W}'(x)\right],$$
$$\hat{\bar{Q}} = \hat{\bar{\psi}} \left[\hat{p} + i\mathcal{W}'(x)\right]; \tag{2.7}$$

$$\hat{H} = \frac{\hat{p}^2 + [\mathcal{W}'(x)]^2}{2} + \frac{1}{2}\mathcal{W}''(x)(\psi\hat{\bar{\psi}} - \hat{\bar{\psi}}\psi), \tag{2.8}$$

where $\hat{p}$ and $\hat{\bar{\psi}}$ are the ordinary and Grassmann momentum operators[2] — $\hat{p} = -i\partial/\partial x$ and $\hat{\bar{\psi}} = i\hat{\Pi}_\psi = \partial/\partial\psi$ — and $\mathcal{W}(x)$ is an arbitrary real function called the *superpotential*. In the matrix formulation, the Hamiltonian (2.8) describes the one-dimensional motion of a spin $\frac{1}{2}$ particle in the external potential $V = \frac{1}{2}[\mathcal{W}'(x)]^2$ and the magnetic field $\mathcal{W}''(x)$.

In the particular case $\mathcal{W}(x) = \omega x^2/2$, the Hamiltonian acquires the form

$$\hat{H} = \frac{\hat{p}^2 + \omega^2 x^2}{2} + \frac{\omega}{2}(\psi\hat{\bar{\psi}} - \hat{\bar{\psi}}\psi). \tag{2.9}$$

This Hamiltonian describes the *supersymmetric oscillator*. The spectrum of the Hamiltonian (2.9) as well as of the Hamiltonian (2.8) includes two sectors: the wave functions $\Psi^B(x, \psi) = a_0(x)$ and the functions $\Psi^F(x, \psi) = \psi a_1(x)$ including an extra factor ψ. It is natural to call the first sector *bosonic* and the second one *fermionic*. Of course, we are now discussing quantum mechanics and talking about the levels of the Hamiltonian, not about physical particles. However, the presence of the Grassmann factor in the wave function gives to the state a "fermionic flavor". One can introduce the operator

$$\hat{F} = \psi\hat{\bar{\psi}}, \tag{2.10}$$

which commutes with the Hamiltonian and can be called the operator of *fermion charge*. Then the states in the bosonic sector have fermion charge 0 and the states in the fermionic sector have fermion charge 1.

---

[2] Alternatively, one could go into the momentum representation where the wave functions depend on p and the coordinate represents a differential operator $\hat{x} = i\partial/\partial p$ or to the Grassmann antiholomorphic representation where the wave functions depend on $\bar{\psi}$ and $\hat{\psi} = \partial/\partial\bar{\psi}$.

The spectrum of the Hamiltonian (2.9) can easily be found. There are two towers of states, bosonic and fermionic. Their energies are

$$E_n^B = n\omega, \qquad E_n^F = (n+1)\omega \tag{2.11}$$

with $n = 0, 1, \ldots$ The ground state is bosonic and has zero energy. The excited states are double degenerate. At each energy level $E = n\omega$ with $n > 0$, there are two states, bosonic and fermionic. This agrees with the general statement of Theorem 2.2.

For a generic $\mathcal{W}(x)$, one cannot solve the complete spectral problem analytically, but one can always find explicit expressions for the wave function of the zero-energy state if such a state exists.

Theorem 2.3. *A normalized bosonic zero-energy solution to the Schrödinger equation with the Hamiltonian (2.8) exists iff*

$$\lim_{x \to \pm\infty} \frac{\mathcal{W}(x)}{\sqrt{\ln|x|}} = \infty. \tag{2.12}$$

A normalized fermionic zero-energy solution to the Schrödinger equation with the Hamiltonian (2.8) exists iff

$$\lim_{x \to \pm\infty} \frac{\mathcal{W}(x)}{\sqrt{\ln|x|}} = -\infty. \tag{2.13}$$

Proof. All the bosonic states are clearly annihilated by $\hat{\bar{Q}}$ and all the fermionic states are annihilated by $\hat{Q}$. A zero-energy state in the bosonic sector should in addition be annihilated by $\hat{Q}$, as Theorem 2.1 dictates. This gives the equation

$$\left[\frac{\partial}{\partial x} + \mathcal{W}'(x) \right] a_0(x) = 0. \tag{2.14}$$

Its formal solution reads

$$a_0(x) = \exp\{-\mathcal{W}(x)\}. \tag{2.15}$$

The condition (2.12) is the condition for its normalizability.

Likewise, the fermion zero modes $\Psi_0^F(\psi, x) = \psi\Psi_0^F(x)$ should satisfy the equation

$$\left[\frac{\partial}{\partial x} - \mathcal{W}'(x) \right] a_1(x) = 0 \tag{2.16}$$

with the formal solution

$$a_1(x) = \exp\{\mathcal{W}(x)\}. \tag{2.17}$$

It is normalizable iff the condition (2.13) is fulfilled. $\qquad\square$

If $\mathcal{W}(\infty) = \pm\infty$, but $\mathcal{W}(-\infty) = \mp\infty$, the zero-energy states are absent.[3]
The simplest example is given by the cubic superpotential $\mathcal{W}(x) = \lambda x^3$.

The ground states of the Hamiltonian have in this case nonzero energy and are *not* annihilated by the action of all the supercharges. They are paired in supersymmetric doublets as the excited states do. One can say in this case that supersymmetry is *broken spontaneously*. Indeed, the algebra (2.1) holds and the Hamiltonian is supersymmetric. On the other hand, supersymmetric vacuum state(s) which would be invariant under supersymmetry transformations are absent in the spectrum.

2.2   Electrons in a magnetic field

Historically, it was the first supersymmetric system constructed and studied. This was done by Landau[4] back in 1930 [20].

Consider the planar motion of a charged particle in the (x, y) plane in a magnetic field orthogonal to the plane. Landau assumed that the field is homogeneous, but we will allow for an arbitrary dependence $B_z(x, y)$. The corresponding vector potential $\boldsymbol{A}$ whose curl $\boldsymbol{\nabla} \times \boldsymbol{A}$ is equal to the magnetic field has only x and y components. The Hamiltonian reads

$$\hat{H} = \frac{\left[\boldsymbol{\sigma} \cdot \left(\hat{\boldsymbol{P}} - \frac{e}{c}\boldsymbol{A}\right)\right]^2}{2m}, \tag{2.18}$$

where c is the speed of light, m is the electron mass and e is the electron charge. For simplicity, we will set in the following $\hbar = c = m = 1, e = -1$ (the electron charge is negative — see the footnote on p. 16 !) and disregard the free motion along the magnetic field, so that both $\boldsymbol{A}$ and $\hat{\boldsymbol{P}} = -i\partial/\partial\boldsymbol{x}$ are 2-dimensional. Now, $\boldsymbol{\sigma}$ are the Pauli matrices,

$$\sigma_x \equiv \sigma^1 = \begin{pmatrix} 0 & 1 \\ 1 & 0 \end{pmatrix}, \qquad \sigma_y \equiv \sigma^2 = \begin{pmatrix} 0 & -i \\ i & 0 \end{pmatrix}. \tag{2.19}$$

The properties

$$\{\sigma^1, \sigma^2\} = 0 \quad \text{and} \quad [\sigma^1, \sigma^2] = 2i\sigma^3 = 2i\begin{pmatrix} 1 & 0 \\ 0 & -1 \end{pmatrix}$$

hold.

---

[3] A special case is when $\mathcal{W}(\infty)$ or $\mathcal{W}(-\infty)$ is a constant. In this case, the spectrum is continuous, the spectral problem should be regularized and the zero modes may or may not appear as a result of this regularization.

[4] Of course, Landau did not call it supersymmetry—the word did not exist yet. Landau can be compared in this respect to Mr. Jourdain, the *bourgeois gentilhomme*, who spoke prose without knowing what prose is.

We may rewrite (2.18) as

$$\hat{H} = \frac{\left(\hat{\boldsymbol{P}} + \boldsymbol{A}\right)^2}{2} + \frac{B(x,y)}{2}\sigma^3. \tag{2.20}$$

The Hamiltonians (2.18) and (2.20) act on spinor wave functions

$$\Psi(\boldsymbol{x}) = \begin{pmatrix} b_+(\boldsymbol{x}) \\ b_-(\boldsymbol{x}) \end{pmatrix}. \tag{2.21}$$

The second term in (2.20) describes the interaction of the electron magnetic moment with an external magnetic field.

The first term in (2.20) is the quantum Hamiltonian of a spinless particle in a magnetic field. Its classical counterpart [the classical counterpart of the full Hamiltonian (2.20) also exists, we will present it somewhat later] is $(\boldsymbol{P} + \boldsymbol{A})^2/2$, where $\boldsymbol{P}$ is the *canonical* momentum, to distinguish from the *kinetic* momentum $\boldsymbol{p} = \boldsymbol{P} + \boldsymbol{A} = \dot{\boldsymbol{x}}$.

Consider the operators

$$\hat{Q} = \boldsymbol{\sigma} \cdot \left(\hat{\boldsymbol{P}} + \boldsymbol{A}\right) \frac{1 - \sigma^3}{2} \tag{2.22}$$

and

$$\hat{\bar{Q}} = \boldsymbol{\sigma} \cdot \left(\hat{\boldsymbol{P}} + \boldsymbol{A}\right) \frac{1 + \sigma^3}{2}. \tag{2.23}$$

Capitalizing on the fact that $\sigma^{1,2}$ anticommute with σ^3, it is easy to check that these supercharges satisfy the algebra (2.1) and the problem is supersymmetric!

It is so for any $B(x,y)$. But for a homogeneous magnetic field, one can solve the Schrödinger problem exactly. The vector potential can be chosen in this case as $\boldsymbol{A} = (-By/2, Bx/2)$. The Hamiltonian and the supercharges acquire the form

$$\hat{H} = \frac{1}{2}\left[\left(-i\frac{\partial}{\partial x} - \frac{By}{2}\right)^2 + \left(-i\frac{\partial}{\partial y} + \frac{Bx}{2}\right)^2\right] + \frac{B}{2}\sigma^3, \tag{2.24}$$

$$\hat{Q} = -i\sqrt{2}\left(\frac{\partial}{\partial z} + \frac{Bz^*}{2}\right)\sigma^+,$$

$$\hat{\bar{Q}} = -i\sqrt{2}\left(\frac{\partial}{\partial z^*} - \frac{Bz}{2}\right)\sigma^-, \tag{2.25}$$

where[5] $z = (x + iy)/\sqrt{2}$ and $\sigma^\pm$ were defined in (1.39).

---

[5]Do not confuse it with the third spatial coordinate, which we left unnamed!

In agreement with general theorems, the whole Hilbert space (2.21) splits into two subspaces: subspace $|+\rangle$ with positive electron spin projection along the third axis and subspace $|-\rangle$ with negative spin projection. The spectrum of positive-energy states splits into degenerate doublets and the action of the supercharges follow the pattern displayed in Fig. 2.1.

In both sectors the Hamiltonian (2.24) reduces to the Hamiltonian of the harmonic oscillator. The spectrum is thus equally spaced, the spacing between the levels being equal to B (we assume, for definiteness, that $B > 0$). All the levels are infinitely degenerate due to the symmetry under translations in the (x, y) plane. The difference between the two sectors is that the spin-down states are shifted down with respect to the spin-up states by B, which exactly coincides with the spacing between the oscillator levels (Fig. 2.2).

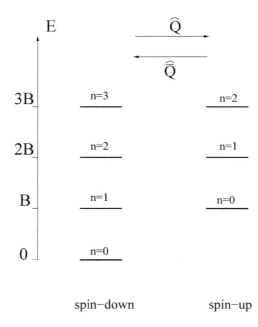

Fig. 2.2: Spectrum of the Landau Hamiltonian $(B > 0)$. $n = 0, 1, \ldots$ mark the oscillator levels

This spectrum is very similar to the spectrum (2.11) of the supersymmetric oscillator with the only difference that all the levels are infinitely degenerate.

The same concerns the zero-energy states. They belong to the spin-down sector and have the wave functions

$$\Psi(\boldsymbol{x}) \propto \begin{pmatrix} 0 \\ \exp\left[-\frac{Bz^*(z+c)}{2}\right] \end{pmatrix}. \tag{2.26}$$

Different complex c correspond in the classical limit to different positions of the center of the circular electron orbits in the (x, y) plane.

But the Landau problem can also be generalized to the case when the magnetic field is a nontrivial function of coordinates [21]. Locally, the algebra (2.1) with the supercharges (2.22) still holds. In Chapter 4 we will show that, if the magnetic flux

$$\boxed{\Phi = \int B(x, y)\, dx dy} \tag{2.27}$$

is finite and represents an integer multiple of 2π, the supersymmetric quantum problem is well defined also globally and the number of the normalized zero energy states (alias zero modes) is equal to $|\Phi|/(2\pi)$.

As was also the case for Witten's model, one can reformulate the Landau problem and the generalized Landau problem using Grassmann variables. The dictionary of translations from one language into another was given in Eq. (1.39). Replacing the matrices σ^+ and σ^- in (2.25) by ψ and $\hat{\bar{\psi}} = \partial/\partial\psi$, respectively, we obtain the expressions

$$\hat{Q} = -i\sqrt{2}\,\psi\left[\frac{\partial}{\partial z} + iA^*(z, z^*)\right],$$

$$\hat{\bar{Q}} = -i\sqrt{2}\,\frac{\partial}{\partial\psi}\left[\frac{\partial}{\partial z^*} + iA(z, z^*)\right] \tag{2.28}$$

(with $A = (A_x + iA_y)/\sqrt{2}$), and the Hamiltonian is given by the expression (2.24), where the matrix σ^3 is traded for

$$\sigma^3 = [\sigma^+, \sigma^-] \longrightarrow \psi\frac{\partial}{\partial\psi} - \frac{\partial}{\partial\psi}\psi.$$

and $B(z, z^*) = i(\partial A^*/\partial - \partial A/\partial z)$. The wave functions depend on z, z^* and the holomorphic variable ψ. There are the states of the form $\Psi(z, z^*; \psi) = b_-(z, z^*)$ (they are what we called before spin-down states) and the states of the form $\Psi(z, z^*; \psi) = \psi b_+(z, z^*)$ (the spin-up states).

2.3 Classical supersymmetry

In the first half of this chapter, we defined the quantum SQM algebra (2.1) and discussed two simplest examples — the Witten and Landau models.

But we have not posed a question yet: What are their classical counter-parts?

We will do so now. We will write the classical Hamiltonians correspond-ing to (2.8) and (2.24). Then we will perform the Legendre transformation to determine the Lagrangians and we will find the supersymmetry trans-formations leaving the classical action invariant. Obviously, this is only possible to carry out if one uses the *Grassmann* description.

Consider Witten's model first. The classical expressions for the super-charges and the Hamiltonian can be written immediately. They simply coincide with (2.7) and (2.8), where now p and $\bar{\psi}$ should be understood as the phase space variables rather than differential operators. The quantum algebra (2.1) corresponds to the classical algebra

$$\{Q, Q\}_P = \{\bar{Q}, \bar{Q}\}_P \;=\; 0, \qquad \{Q, \bar{Q}\}_P \;=\; 2iH \,, \tag{2.29}$$

where the Poisson bracket in a system involving both ordinary phase space variables x_i, p_i and Grassmann phase space variables $\psi_\alpha, \bar{\psi}_\alpha$ is defined ac-cording to

$$\{A, B\}_P = \left[\frac{\partial^2}{\partial p_i \partial Q_i} - \frac{\partial^2}{\partial P_i \partial q_i} - i \left(\frac{\partial^2}{\partial \psi_\alpha \partial \bar{\Psi}_\alpha} + \frac{\partial^2}{\partial \bar{\psi}_\alpha \partial \Psi_\alpha} \right) \right]$$
$$A(p, q;\, \psi, \bar{\psi}) B(P, Q;\, \Psi, \bar{\Psi}) \big|_{p=P, q=Q;\, \psi=\Psi, \bar{\psi}=\bar{\Psi}} \,. \tag{2.30}$$

The corresponding classical Lagrangian is

$$L \;=\; \frac{\dot{x}^2 - [\mathcal{W}'(x)]^2}{2} - i\bar{\psi}\dot{\psi} - \mathcal{W}''(x)\psi\bar{\psi} \,. \tag{2.31}$$

It is invariant up to a total derivative under the following supersymme-try transformations:

$$\delta x = \epsilon\psi + \bar{\psi}\bar{\epsilon} \,,$$
$$\delta\psi = \bar{\epsilon}[-i\dot{x} + \mathcal{W}'(x)] \,,$$
$$\delta\bar{\psi} = \epsilon[i\dot{x} + \mathcal{W}'(x)] \,. \tag{2.32}$$

The classical supercharges are the integrals of motion that correspond to these symmetries, according to the Noether theorem.

For the Landau problem (for simplicity, we restrict ourselves here by the case of constant magnetic field), the classical supercharges and Hamiltonian satisfying the algebra (2.29) are

$$Q \;=\; \sqrt{2}\,\psi \left(\pi - \frac{iBz^*}{2} \right) ,$$
$$\bar{Q} \;=\; \sqrt{2}\,\bar{\psi} \left(\pi^* + \frac{iBz}{2} \right) , \tag{2.33}$$

$$H = \left(\pi - \frac{iBz^*}{2}\right)\left(\pi^* + \frac{iBz}{2}\right) + B\psi\bar{\psi}. \tag{2.34}$$

The corresponding classical Lagrangian reads

$$L = \dot{z}\dot{z}^* + \frac{iB}{2}(\dot{z}z^* - z\dot{z}^*) - i\dot{\psi}\bar{\psi} - B\psi\bar{\psi}. \tag{2.35}$$

It is invariant up to a total derivative under

$$\begin{aligned} \delta z &= \epsilon\psi, & \delta\psi &= -i\bar{\epsilon}\dot{z} \\ \delta z^* &= -\bar{\epsilon}\bar{\psi}, & \delta\bar{\psi} &= i\epsilon\dot{z}^*. \end{aligned} \tag{2.36}$$

In the systems considered so far, there were no ordering ambiguities and natural expressions for the classical supercharges could be easily written. But if the ordering ambiguities are present, the questions:

(1) What particular expressions for the classical supercharges should be written?

(2) Is it always possible to do so in such a way that the quantum supersymmetry algebra would be preserved at the classical level, with the Poisson brackets taking over the role of commutators and anticommutators?

become nontrivial. The answers are also not trivial.

The answer to the second question is "Not always!". The first example of a system where the quantum supersymmetry algebra cannot be preserved at the classical level was constructed in [22]. A simpler example was given in [23]. The latter concerns a supersymmetric system describing the motion of a spinorial particle in Euclidean space of even dimension D equipped by an external gauge field. In this system the Dirac operator $\not{D}$ and also the operator $\not{D}\gamma^{D+1}$ (with γ^{D+1} being the multidimensional counterpart of the familiar to the physicists matrix γ^5) play the role of the quantum supercharges. The operator $\not{D}$ has a classical counterpart, but the operator $\not{D}\gamma^{D+1}$ has not (see Sect. 13.1.1 of the book [4] for detailed discussion).

In quantum field theories, we are used to have quantum anomalies when a classical symmetry cannot be preserved at the quantum level (the chiral anomaly, the conformal anomaly, ...). And here we are dealing with the opposite situation: supersymmetry that is realized at the quantum, but not at the classical level.

The first question in the list above was studied in [24].

We analyzed there different supersymmetric systems where the supercharges and the Hamiltonians include products of canonical coordinates and momenta, and the ordering problem exists. Based on observations performed for various supersymmetric systems involving ordering ambiguities, we formulated the following **conjecture**:

(1) Consider a system enjoying an ordinary or extended classical supersymmetry algebra. Take the classical supercharges and write down the corresponding quantum operators using the Weyl ordering prescription. Then these operators satisfy the quantum supersymmetry algebra.

(2) Inversely: *if* the quantum supersymmetry algebra can be preserved at the classical level, this can always be achieved by choosing the classical supercharges as the Weyl symbols of the quantum ones.

For the benefit of the reader who might not know so well what Weyl ordering and Weyl symbols are, we give here the basic definitions and explain their meaning.

Consider first ordinary systems not involving Grassmann variables.

Definition 2.2. Let $A(p_i, q_i)$ be a function in phase space (an *observable*). Consider its Fourier decomposition

$$A(p_i, q_i) = \int \prod_i d\alpha_i d\beta_i \, h(\alpha_i, \beta_i) e^{i(\alpha_i p_i + \beta_i q_i)} . \tag{2.37}$$

Then the operator

$$\hat{A}(\hat{p}_i, q_i) = \int \prod_i d\alpha_i d\beta_i \, h(\alpha_i, \beta_i) e^{i(\alpha_i \hat{p}_i + \beta_i q_i)} . \tag{2.38}$$

is called the *Weyl-ordered* operator corresponding to the classical function $A(p_i, q_i)$. And the function $A(p_i, q_i)$ is called the *Weyl symbol* of the operator $\hat{A}$.

For the monomials like $p^2 q^2$, this amounts to the symmetric ordering:

$$p^2 q^2 \rightarrow \frac{1}{6}(\hat{p}^2 q^2 + q^2 \hat{p}^2 + \hat{p} q^2 \hat{p} + q \hat{p}^2 q + \hat{p} q \hat{p} q + q \hat{p} q \hat{p}) .$$

This follows from the simple fact that each term in the expansion of the exponential in (2.38) includes such symmetrized products.

To find the Weyl symbol of an arbitrary polynomial operator, one should first represent it as a sum of symmetric combinations using[6] $[\hat{p}, q] = -i\hbar$. For example, the Weyl symbol of $\hat{p}q$ is $pq - i\hbar/2$ and the Weyl symbol of $q\hat{p}$ is $pq + i\hbar/2$.

---

[6]To make the discussion more clear, we are calling off for a while our convention $\hbar = 1$.

Note that the Weyl symbol of the product of two operators does not coincide with the product of their Weyl symbols. Instead of that, one has

$$(\hat{A}\hat{B})_W = \exp\left[-\frac{i\hbar}{2}\left(\frac{\partial^2}{\partial p_i \partial Q_i} - \frac{\partial^2}{\partial P_i \partial q_i}\right)\right] A_W(p_i, q_i) B_W(P_i, Q_i)\bigg|_{p=P,q=Q} \quad (2.39)$$

(the so-called *Grönewold-Moyal product* [25]).

It follows that the Weyl symbol of the commutator of two operators is

$$[\hat{A}, \hat{B}]_W = -2i \sin\left[\frac{\hbar}{2}\left(\frac{\partial^2}{\partial p_i \partial Q_i} - \frac{\partial^2}{\partial P_i \partial q_i}\right)\right] A_W(p_i, q_i) B_W(P_i, Q_i)\bigg|_{p=P,q=Q}$$

$$\stackrel{\text{def}}{=} -i\hbar \{A_W, B_W\}_{GM} \quad (2.40)$$

In the classical limit $\hbar \to 0$, the Grönewold-Moyal bracket $\{A, B\}_{GM}$ reduces to the Poisson bracket. For simple operators relevant for most physical applications, the terms of order $\sim \hbar^3$, etc. in (2.40) vanish, and it is the Poisson bracket that goes over to the commutator under quantization.

Definition 2.2 can be (rather naturally) generalized for systems including Grassmann variables. Then the Weyl-ordered operator corresponding to the product of two Grassmann variables like $\psi\bar{\psi}$ is the antisymmetrized product $(\psi\hat{\bar{\psi}} - \hat{\bar{\psi}}\psi)/2$. The Weyl symbol of the anticommutator of two Grassmann-odd operators is[7] $-i\hbar\{A, B\}_{GM}$ with

$$\{A, B\}_{GM} = 2\sin\left[\frac{\hbar}{2}\left(\frac{\partial^2}{\partial p_i \partial Q_i} - \frac{\partial^2}{\partial P_i \partial q_i}\right) - \frac{i\hbar}{2}\left(\frac{\partial^2}{\partial \psi_\alpha \partial \bar{\Psi}_\alpha} + \frac{\partial^2}{\partial \bar{\psi}_\alpha \partial \Psi_\alpha}\right)\right]$$

$$A(p, q;\, \psi, \bar{\psi}) B(P, Q;\, \Psi, \bar{\Psi})\big|_{p=P,q=Q;\, \psi=\Psi, \bar{\psi}=\bar{\Psi}} . \quad (2.41)$$

Going back to our conjecture, it looks natural. The correspondences (2.40) and (2.41) mean that the quantum algebra is *always* preserved at the classical level if the GM bracket is chosen as the binary operation. It is not counterintuitive to expect that whatever holds for the GM brackets should also hold in most cases for the Poisson brackets. And if this rule of thumb fails, nothing helps. But, of course, it would be interesting to rigourously prove (or disprove) it.

The ordering prescription formulated above will be used extensively in Chapter 5 for supersymmetric sigma models describing the motion over manifolds with nontrivial metric.

[7] This formula demonstrates, among other things, that, for complicated enough theories where the GM bracket of the supercharges includes higher terms of the expansion of (2.41), the Weyl symbol of the quantum Hamiltonian does *not* coincide with the classical Hamiltonian.

2.4   Superspace

Supersymmetric Lagrangians (2.31) and (2.35) are invariant up to a total time derivative under the supersymmetry transformations (2.32) and (2.36). These transformations mix the ordinary and Grassmann variables, which can be compared to rotations mixing different vector components. In the latter case, we know that the description of a vector as a set of its components is possible, but it is much more convenient to use the vector notation and to write a single vector equation instead of several different scalar equations, like $\boldsymbol{B} = \boldsymbol{\nabla} \times \boldsymbol{A}$ instead of $B_x = \partial_y A_z - \partial_z A_y$, etc.

One can ask: Is there an analog of such vector notation in supersymmetry? Can one introduce a single object including the ordinary and Grassmann variables as different components, so that the supersymmetric actions and equations of motion can be written in a more compact and transparent way? The answer is yes, such an object exists, it is called a *superfield*—a function depending on *superspace* coordinates.

At this point one may complain about the universally accepted terminology. Historically, the concept of superspace was first introduced for supersymmetric field theories [107]. But in this chapter we are discussing supersymmetric quantum *mechanics*, and the objects that we are now about to introduce should better be called dynamical *supervariables* depending on *supertime*. However, people still use the terms "superfield" and "superspace" (adding sometimes a specification "one-dimensional") also for mechanical systems. And we will also do so.

The simplest one-dimensional superspace includes on top of time t a single real Grassmann-odd variable $\theta = \bar{\theta}$. The transformations

$$\theta \to \theta + \epsilon\,,$$
$$t \to t + i\epsilon\theta\,. \tag{2.42}$$

with a real Grassmann-odd ϵ play the role of "rotations".

However, the symmetry (2.42) is too poor to bring about interesting dynamics. A dynamical system involving only the symmetry with a single real Grassmann parameter ϵ would have only one real supercharge, while the genuinely supersymmetric systems with double degeneracy of all excited states have at least one complex or two real supercharges $\mathcal{Q}_{1,2}$. Thus, to describe the SQM systems considered above, we need the superspace involving t and a *complex* Grassmann variable θ. The supersymmetry transformations are then defined as

$$\theta \to \theta + \epsilon \,,$$
$$\bar{\theta} \to \bar{\theta} + \bar{\epsilon} \,,$$
$$t \to t + i(\epsilon\bar{\theta} - \theta\bar{\epsilon}) \tag{2.43}$$

with complex Grassmann ϵ.

Consider the composition of two different transformations (2.42). We obtain

$$\theta_1 = \theta + \epsilon_1 \,, \qquad \theta_2 = \theta_1 + \epsilon_2 = \theta + \epsilon_1 + \epsilon_2 \,,$$
$$\bar{\theta}_1 = \bar{\theta} + \bar{\epsilon}_1 \,, \qquad \bar{\theta}_2 = \bar{\theta}_1 + \bar{\epsilon}_2 = \bar{\theta} + \bar{\epsilon}_1 + \bar{\epsilon}_2 \,,$$
$$t_1 = t + i(\epsilon_1\bar{\theta} - \theta\bar{\epsilon}_1) \,, \qquad t_2 = t_1 + i(\epsilon_2\bar{\theta}_1 - \theta_1\bar{\epsilon}_2)$$
$$= t + i[(\epsilon_1 + \epsilon_2)\bar{\theta} - i(\bar{\epsilon}_2 + \bar{\epsilon}_2)\theta] + i(\epsilon_2\bar{\epsilon}_1 - \epsilon_1\bar{\epsilon}_2) \,. \tag{2.44}$$

In other words, the composition of two supertransformations characterized by the complex parameters ϵ_1 and ϵ_2 is a supertransformation with the parameter $\epsilon_1 + \epsilon_2$ supplemented by a shift of time by a real Grassmann-even parameter $\alpha = i(\epsilon_2\bar{\epsilon}_1 - \epsilon_1\bar{\epsilon}_2)$.

Theorem 2.4. *The set of all transformations (2.42) supplemented by time shifts $t \to t + \alpha$ and characterized by Grassmann-even parameters α and Grassmann-odd parameters ϵ form a group.*

Proof. We leave it to the reader. □

Such groups characterized by both Grassmann-even and Grassmann-odd parameters and acting on superspaces are called *supergroups*.

If one uses the notation $g(\alpha, \epsilon)$ for an element of this supergroup, the law (2.44) can be rewritten as $g(0, \epsilon_2)g(0, \epsilon_1) = g[i(\epsilon_2\bar{\epsilon}_1 - \epsilon_1\bar{\epsilon}_2), \epsilon_2 + \epsilon_1]$. We see that this group is not commutative. The group commutator

$$g(0, \epsilon_2)g(0, \epsilon_1)g(0, -\epsilon_2)g(0, -\epsilon_1) = g[2i(\epsilon_2\bar{\epsilon}_1 - \epsilon_1\bar{\epsilon}_2), 0] \tag{2.45}$$

is not trivial and amounts to a pure time shift.

A group element is the exponential of a linear combination of its generators. In our case, we may write

$$g(\alpha, \epsilon) = \exp\{-i\alpha\hat{H} + \epsilon\hat{Q} + \bar{\epsilon}\hat{\bar{Q}}\} \,, \tag{2.46}$$

where

$$\hat{H} = i\partial/\partial t \tag{2.47}$$

is the generator of time shifts, i.e. the Hamiltonian, and

$$\hat{Q} = \frac{\partial}{\partial\theta} + i\bar{\theta}\frac{\partial}{\partial t},$$

$$\hat{\bar{Q}} = \frac{\partial}{\partial\bar{\theta}} + i\theta\frac{\partial}{\partial t} \tag{2.48}$$

are nothing but the *supercharges!* Indeed, it is easy to verify that the generators (2.47), (2.48) obey the algebra (2.1).

To derive in the superspace approach particular Lagrangians (and then the corresponding Hamiltonians and supercharges) as functions of dynamical variables, we have to consider supervariables, alias superfields.

Definition 2.3. A generic bosonic (i.e. representing an even element of the Grassmann algebra) superfield living in the superspace $(t, \theta, \bar{\theta})$ reads

$$Z(t, \theta, \bar{\theta}) = z(t) + \theta\psi(t) + \bar{\theta}\chi(t) + \theta\bar{\theta}F(t), \tag{2.49}$$

where z and F are complex bosonic and ψ and χ are complex fermionic variables. They are called the *components* of the superfield (2.49).

The components z, F, ψ, χ of the superfield Z play the same role as, say, the components $V_{1,2,3}$ of a 3-dimensional vector $\boldsymbol{V}$.

The superfield (2.49) has altogether 4 bosonic and 4 fermionic real degrees of freedom. Their numbers are equal, which actually a property of *any* superfield in SQM and also in supersymmetric field theories that we will meet later.

The transformation (2.43) of superspace coordinates $(t, \theta, \bar{\theta})$ induces the following transformation of the components:

$$\delta z = \epsilon\psi + \bar{\epsilon}\chi,$$
$$\delta\psi = \bar{\epsilon}(F - i\dot{z}),$$
$$\delta\chi = -\epsilon(F + i\dot{z}),$$
$$\delta F = i(\bar{\epsilon}\dot{\chi} - \epsilon\dot{\psi}). \tag{2.50}$$

They can be written in a compact form:

$$\delta Z = (\epsilon\hat{Q} + \bar{\epsilon}\hat{\bar{Q}})Z. \tag{2.51}$$

The set of components $\{z(t), F(t), \psi(t), \chi(t)\}$ and the whole superfield (2.49) realize thus an *infinite-dimensional representation* of the superalgebra (2.1).

But it is not an irreducible representation. Irreducible ones can be constructed by imposing certain constraints on Z. To begin with, we can impose the constraint $Z = \bar{Z}$. This gives a real superfield:

$$X = x + \theta\psi + \bar{\psi}\bar{\theta} + \theta\bar{\theta}F \tag{2.52}$$

with real x, F. A real superfield stays real after supertransformations (2.50), which acquire the form

$$\delta x = \epsilon\psi + \bar{\psi}\bar{\epsilon},$$
$$\delta\psi = \bar{\epsilon}(F - i\dot{x}),$$
$$\delta\bar{\psi} = \epsilon(F + i\dot{x}),$$
$$\delta F = -i(\epsilon\dot{\psi} + \bar{\epsilon}\dot{\bar{\psi}}). \tag{2.53}$$

One can observe the pattern: the variation of a certain component of X involves a higher component in the expansion in θ (ψ and $\bar{\psi}$ for δx, F for $\delta\psi$ and $\delta\bar{\psi}$) and the time derivative of a lower component. This pattern also holds for more complicated superfields to be considered later. In particular, the supersymmetric variation of the highest component of any superfield is a total time derivative.

One can ask: How do we know that (2.52) is irreducible; how to prove that one cannot impose on X an additional constraint compatible with supersymmetry? In fact, this follows immediately from the fact that any superfield representation of the algebra (2.1) must have at least 2 bosonic and 2 fermionic real components. Indeed: take any bosonic component and act on it by two different real supercharges [realized as the generators of the transformations (2.53)] available in the algebra — you obtain two different fermionic components. And starting from a fermion component, one obtains two different bosonic components. The superfield (2.52) saturates this minimum, and the number of components cannot be reduced further.

But (2.52) is not the only irreducible superfield that one can construct. Define the differential operators

$$D = \frac{\partial}{\partial\theta} - i\bar{\theta}\frac{\partial}{\partial t},$$
$$\bar{D} = -\frac{\partial}{\partial\bar{\theta}} + i\theta\frac{\partial}{\partial t}, \tag{2.54}$$

called *supersymmetric covariant derivatives*. They are nilpotent and anticommute with both $\hat{Q}$ and $\hat{\bar{Q}}$ in (2.48).

Definition 2.4. A *holomorphic chiral* superfield Z is a generic superfield (2.49) satisfying the constraint $\bar{D}Z = 0$. An *antiholomorphic chiral* superfield $\bar{Z}$ is a generic superfield (2.49) satisfying the constraint[8] $D\bar{Z} = 0$.

---

[8]In the slang used by the physicists these superfields are called "left chiral" and "right chiral". This terminology came from field theories, where the words "left" and "right" have a physical meaning: see the footnote on p. 15. But the notions of left and right make, of course, little sense for mechanical systems, and we will try not to use here these words, allowing, however, for their traces in the notation t_L and t_R below.

Bearing in mind that $\{\bar{D}, \hat{Q}\} = \{\bar{D}, \hat{\bar{Q}}\} = 0$, one immediately sees that the variation (2.51) is holomorphic chiral if Z is holomorphic chiral. The same is true for antiholomorphic superfields. The conditions $\bar{D}Z = 0$ or $D\bar{Z} = 0$ kill a half of the degrees of freedom and we are left with $2_B + 2_F$ degrees of freedom, meaning that the representation of (2.1) realized by chiral superfields is irreducible.

A generic solution of the constraint $\bar{D}Z = 0$ is[9]

$$Z \ = \ z + \sqrt{2}\,\theta\psi - i\theta\bar{\theta}\dot{z}\,. \tag{2.55}$$

We see that the term $\propto \bar{\theta}$ is now absent and the term $\propto \theta\bar{\theta}$ is not independent any more, but is given by the time derivative of the lowest term.

A convenient technical trick is to introduce a "holomorphic time" $t_L = t - i\theta\bar{\theta}$. It transforms under supersymmetry as $t_L \to t_L - 2i\theta\bar{\epsilon}$ and does not "feel" the presence of $\bar{\theta}$. The covariant derivative $\bar{D}$ reduces in this frame simply to $\partial/\partial\bar{\theta}$ and the solutions to the equation $\bar{D}Z = 0$ have no dependence on $\bar{\theta}$:

$$Z \ = \ Z(t_L, \theta) \ = \ z(t_L) + \sqrt{2}\,\theta\psi(t_L)\,. \tag{2.56}$$

Similarly,

$$\overline{Z} \ = \ z^*(t) - \sqrt{2}\,\bar{\theta}\bar{\psi}(t) + i\theta\bar{\theta}\,\dot{z}^*(t) \ = \ z^*(t_R) - \sqrt{2}\,\bar{\theta}\bar{\psi}(t_R) \tag{2.57}$$

with $t_R = (t_L)^* = t + i\theta\bar{\theta}$.

Definition 2.5. The set of coordinates (t_L, θ) describes the *holomorphic chiral SQM superspace* and the set $(t_R, \bar{\theta})$ describes the *antiholomorphic chiral SQM superspace*.

The superfield $\overline{Z}$ is complex conjugate to Z, but we use here the long bar instead of the star to underline a different *type* of the superfield (2.57). One can prove (though the proof is not so simple, it is actually an essential ingredient in the supersymmetric proof of the so-called Newlander-Nirenberg theorem [27]) that the superspace $(t, \theta, \bar{\theta})$ admits only three types of irreducible superfields: real, holomorphic chiral and antiholomorphic chiral superfields. (The lowest component of a chiral superfield may have bosonic or fermionic nature.)

The transformations of the components of (2.55) induced by transformations of the superspace coordinates read

$$\delta z = \sqrt{2}\,\epsilon\psi\,, \qquad\qquad \delta\psi = -i\sqrt{2}\,\bar{\epsilon}\dot{z} \tag{2.58}$$

---

[9]The factor $\sqrt{2}$ is introduced for further conveniences.

and hence $\delta z^* = -\sqrt{2}\,\bar{\epsilon}\bar{\psi}$, $\delta\bar{\psi} = i\sqrt{2}\,\epsilon\dot{z}^*$. Note that they coincide with (2.36) — the transformations leaving the action of the Landau problem invariant — up to an irrelevant factor $\sqrt{2}$. And that means that the holomorphic superfield (2.55) could be used to construct the action of the Landau problem in the superfield approach!

What is an explicit way to do so? As we mentioned above, the supertransformations of the highest component of any superfield boil down to a total derivative. We have proven the theorem:

Theorem 2.5. *The highest component of any real superfield gives an invariant supersymmetric Lagrangian.*

Thus, the only thing to do is to build up a proper real superfield R out of the chiral superfields Z and $\overline{Z}$ and take its highest component, which, bearing in mind (1.26), amounts to integrating R over the whole superspace,

$$S = \int d\bar{\theta}d\theta dt\, R \quad \text{and} \quad L = \int d\bar{\theta}d\theta\, R \tag{2.59}$$

There are many ways to cook up a real superfield out of the chiral superfield (2.55) that we have at our disposal. One can simply multiply Z by $\overline{Z}$ and the product $Z\overline{Z}$ is obviously real. But to obtain a nontrivial dynamics, we should also get the covariant derivatives D and $\bar{D}$ out of our tool kit. Due to the fact that the covariant derivatives anticommute with $\hat{Q}$ and $\hat{\bar{Q}}$, they make a superfield out of a superfield. But the type of the superfield changes: due to the nilpotency of D and $\bar{D}$, DZ is antiholomorphic and $\bar{D}\overline{Z}$ is holomorphic.

One can be convinced that the action

$$S = \int d\bar{\theta}d\theta dt \left(\frac{1}{4}\bar{D}\overline{Z}\,DZ - \frac{B}{2}Z\overline{Z}\right) \tag{2.60}$$

does the job and the component Lagrangian derived from it coincides up to a total derivative with (2.35). If one replaces $BZ\overline{Z}/2$ in (2.60) by any function $F(\overline{Z}, Z)$, one obtains a system with a non-homogeneous magnetic field

$$B(z, z^*) = 2\frac{\partial^2 F}{\partial z \partial z^*}. \tag{2.61}$$

Eqs. (2.35) and (2.60) are two different ways of representing the same supersymmetric system. In the first case, the Lagrangian is expressed in components, and to show that it is supersymmetric, one has to check the invariance of the action under the transformations (2.36) explicitly. On the other hand, supersymmetry of the action (2.60) is a corollary of Theorem 2.5. The law of transformations of components follows from the transformation law (2.43) of the superspace coordinates. One may but does not need to check anything. The superspace approach gives one, if you will, an "industrial" method to write supersymmetric Lagrangians without much effort, like the calculus deleloped by Newton and Leibniz allowed one to calculate the volume of a ball — the problem first solved by Archimedes by an ingenuous geometric construction — in two lines.

The Lagrangian (2.31) of the Witten model can also be derived using superspace technique. To this end, one has to take the real $\mathcal{N} = 2$ superfield (2.52) and write

$$
\begin{aligned}
L &= \int d\bar{\theta}d\theta \left[\frac{1}{2}\bar{D}XDX - \mathcal{W}(X)\right] \\
&= \frac{1}{2}(\dot{x}^2 + F^2) - i\psi\dot{\bar{\psi}} - \mathcal{W}''(x)\psi\bar{\psi} - F\mathcal{W}'(x).
\end{aligned} \tag{2.62}
$$

This resembles (2.31), but does not coincide with it completely: the Lagrangian (2.62) depends on an extra variable F, which was absent in (2.31). One can notice, however, that F is an *auxiliary* non-dynamical variable: it enters the Lagrangian without derivatives and the corresponding equation of motion is simply

$$
\frac{\partial L}{\partial F} = F - \mathcal{W}'(x) = 0. \tag{2.63}
$$

It has a trivial solution $F = \mathcal{W}'(x)$. If substituting this solution in (2.62), we reproduce (2.31). When doing the same for the supertransformation (2.53), we reproduce (2.32).

One can mention in parentheses at this point that F is auxiliary and can be algebraically excluded for the Lagrangian (2.62), but if one writes a more complicated action including extra time derivative, like

$$
S = \int d\bar{\theta}d\theta dt \, DX\bar{D}\dot{X}, \tag{2.64}
$$

the component Lagrangian *would* depend on $\dot{F}$ [18].

2.5   Extended supersymmetry

The algebra (2.1) involves two different Hermitian supercharges. By the convention adopted now in the literature, such systems are called $\mathcal{N} = 2$

SQM systems, and the superspace and superfields discussed in the preceding section are $\mathcal{N} = 2$ superspace and superfields.

One can also consider also the $\mathcal{N} = 1$ superspace with a single real odd coordinate θ and the law of supertransformations (2.42). As was mentioned above, the $\mathcal{N} = 1$ algebra $\hat{Q}^2 = \hat{H}$ is too poor to entail dynamic consequencies, but in some cases it is convenient to use the $\mathcal{N} = 1$ superfields to describe SQM models with more rich symmetries.

The symmetry can be *richer* that (2.1): a supersymmetric system may involve several pairs of supercharges:

$$\boxed{\begin{aligned} \{\hat{Q}_i, \hat{Q}_j\} &= \{\hat{\bar{Q}}^i, \hat{\bar{Q}}^j\} = 0, \\ \{\hat{Q}_i, \hat{\bar{Q}}^j\} &= 2\,\delta_i^j \hat{H}. \end{aligned}} \qquad i,j = 1,\ldots\mathcal{N}/2 \tag{2.65}$$

The indices i, j distinguishing different complex supercharges and the associated Grassmann variables are written at the different levels to highlight the fact that $\hat{Q}_i$ and the Hermitially conjugated $\hat{\bar{Q}}^i$ belong to the different — fundamental and antifundamental — representations of the $U(\mathcal{N}/2)$ group, a subgroup of the full automorphism group $SO(\mathcal{N})$ of the algebra (2.65), which rotates the complex supercharges and Grassmann variables. In all the extended models considered in this book, this symmetry (called *R symmetry*) will play an important role.

Now, $\mathcal{N} \geq 2$ is an even number. The minimal extension is $\mathcal{N} = 4$. In that case, the R-symmetry group is $U(2)$, its fundamental and the antifundamental representations coincide and the indices may be raised and lowered by the invariant $SU(2)$ tensors $\varepsilon^{ij} = -\varepsilon_{ij}$:

$$X_i = \varepsilon_{ij} X^j \qquad \text{and} \qquad X^i = \varepsilon^{ij} X_j. \tag{2.66}$$

The convention $\varepsilon_{12} = 1$ is chosen.

We give here one example of an $\mathcal{N} = 4$ SQM system. It describes the motion on the complex plane, but, in contrast to the Landau problem, involves two Grassmann variables $\psi_{1,2}$ and the corresponding momenta $\hat{\Pi}_{1,2} = -i\partial/\partial\psi_{1,2} \equiv -i\hat{\bar{\psi}}^{1,2}$.

The supercharges are[10]

$$\hat{Q}_1 = \sqrt{2}\left[\psi_1\hat{\pi} + i(z^*)^2\,\hat{\bar{\psi}}^2\right], \qquad \hat{Q}_2 = \sqrt{2}\left[\psi_2\hat{\pi} - i(z^*)^2\,\hat{\bar{\psi}}^1\right],$$
$$\hat{\bar{Q}}^1 = \sqrt{2}\left[\hat{\bar{\psi}}^1\hat{\pi}^\dagger - iz^2\,\psi_2\right], \qquad \hat{\bar{Q}}^2 = \sqrt{2}\left[\hat{\bar{\psi}}^2\hat{\pi}^\dagger + iz^2\psi_1\right] \tag{2.67}$$

---

[10]Do not confuse these complex $\mathcal{N} = 4$ supercharges with $\mathcal{Q}_{1,2}$ — the real and imaginary parts of the $\mathcal{N} = 2$ supercharge.

with $\hat{\pi} = -i\partial/\partial z$, $\hat{\pi}^\dagger = -i\partial/\partial z^*$. The Hamiltonian

$$\hat{H} = \hat{\pi}^\dagger\hat{\pi} + (z^*z)^2 - 2z\psi_1\psi_2 - 2z^*\hat{\bar{\psi}}^2\hat{\bar{\psi}}^1 \tag{2.68}$$

is not quadratic and the system involves nontrivial interactions. In fact, this SQM system is obtained from the simplest nontrivial 4-dimensional supersymmetric model, the Wess-Zumino model (to be discussed it in the next chapter) by dimensional reduction.

There also exist SQM models with still higher $\mathcal{N}$. The $\mathcal{N} = 8$ models describing the motion on hyper-Kähler manifolds will be presented in Chapter 5 (see Theorem 5.2). The *maximal* $\mathcal{N} = 16$ gauge SQM model will be the subject of Chapter 9.

As we have seen, the excited states in the spectrum of a supersymmetric Hamiltonian form degenerate doublets. For extended SQM models, this assertion can be strengthened.

Theorem 2.6. *For an SQM system satisfying the algebra (2.65), the spectrum of excited states consists of degenerate $2^{\mathcal{N}/2}$ - plets. Half of the states in such a multiplet are bosonic and another half are fermionic.*

Proof. Consider a state Ψ_0 of nonzero energy. The supercharges $\hat{Q}_1$ and $\hat{Q}^1$ satisfy the ordinary supersymmetric algebra (2.1). Then either $\hat{Q}_1\Psi_0$ or $\hat{Q}^1\Psi_0$ does not vanish, giving an eigenstate Ψ_1 with the same energy and opposite fermion parity. Say, $\Psi_1 = \hat{Q}_1\Psi_0$.

Repeating this reasoning for the supercharges $\hat{Q}_2$ and $\hat{Q}^2$, we obtain two more degenerate states — say, $\hat{Q}_2\Psi_0$ and $\hat{Q}_2\Psi_1 = \hat{Q}_1\hat{Q}_2\Psi_0$. The states Ψ_0 and $\hat{Q}_1\hat{Q}_2\Psi_0$ have the same fermion parity, the parity of the states $\hat{Q}_{1,2}\Psi_0$ is opposite.

If a third pair of supercharges is present, we obtain altogether 8 states, etc. $\square$

The $\mathcal{N} = 4$ and $\mathcal{N} = 8$ SQM models may also be described using an extended superspace formalism. But it is beyond the scope of this book. An interested reader may consult Ref. [4].

The algebra (2.65) represents the simplest natural extension of (2.1). But there are SQM systems charaterized by more complicated algebras. For example, one can modify the second line in (2.65) and write

$$\{\hat{Q}_i, \hat{\bar{Q}}_j\} = 2\delta_i^j\hat{H} + \hat{Z}_i^j, \tag{2.69}$$

where $\hat{Z}_i^j$ are *central charges* that commute with the Hamiltonian. In a more simple version, they commute also with the supercharges (we will meet such systems in Chapters 6,7,10), but there are also the algebras where it is not the case. Such is the algebra of *weak supersymmetry* [28] to be discussed in Chapter 11.

Chapter 3

Supersymmetric field theories

3.1 4D algebra and 4D superspace □ Notation

For the half century that passed since the discovery of supersymmetry, theorists have studied supersymmetric field theories in many different dimensions — from $D = 2$ to $D = 10$. But the most attention has been, of course, attracted to 4-dimensional theories, bearing in mind their possible phenomenological relevance. In the following, we use the notations that are close to the notations of Wess and Bagger [29] with the main distinction being the signature of the Minkowski metric, which we choose $(+ - --)$ rather than $(- + ++)$.

The supercharges have fermion nature. In four dimensions, they belong, like the fermion fields, to $(\frac{1}{2}, 0)$ and $(0, \frac{1}{2})$ representations of the Lorentz group. We have thus two doublets $\hat{Q}_\alpha$ and $\hat{\bar{Q}}_{\dot\alpha}$, which are Hermitially conjugated to one another (and the same concerns, as we remember, any spinor η_α and $\bar\eta_{\dot\alpha}$).[1]

The spinor indices are lifted and lowered according to (0.2). It is convenient to introduce the shortcuts

$$\eta\chi = \eta^\alpha\chi_\alpha = \chi^\alpha\eta_\alpha, \qquad \bar\eta\bar\chi = \bar\eta_{\dot\alpha}\bar\chi^{\dot\alpha} = \bar\chi_{\dot\alpha}\bar\eta^{\dot\alpha} \tag{3.1}$$

for the products of two Grassmann spinors. Also $\theta^2 = \theta^\alpha\theta_\alpha$ and $\bar\theta^2 = \bar\theta_{\dot\alpha}\bar\theta^{\dot\alpha}$.

The algebra reads

$$\boxed{\{\hat{Q}_\alpha, \hat{\bar{Q}}_{\dot\alpha}\} = 2(\sigma^\mu)_{\alpha\dot\alpha}\hat{P}_\mu,} \tag{3.2}$$

---

[1]Note the difference in conventions, compared to Eq.(2.65). In four dimensions, it is the presence or absence of the dot rather than the position of the index that indicates that $\hat{Q}_\alpha$ and $\hat{\bar{Q}}_{\dot\alpha}$ belong to different representations of $Spin(3,1)$.

Attention! Later in the book, when we perform dimensional reduction and go down from field theory to quantum mechanics, we will often, though not always, adopt the extended SQM conventions with $\bar{Q}^\alpha = (Q_\alpha)^\dagger$. That will be specially mentioned in every such case.

where $\hat{P}_\mu = i\partial_\mu = (\hat{H}, -\hat{\boldsymbol{P}})$ is the operator of 4-momentum. The anticommutators $\{\hat{Q}_\alpha, \hat{Q}_\beta\}$ and $\{\hat{\bar{Q}}_{\dot\alpha}, \hat{\bar{Q}}_{\dot\beta}\}$ vanish.

The 4-dimensional superspace includes four coordinates x^μ and two doublets of mutually conjugated odd coordinates θ_α and $\bar{\theta}_{\dot\alpha}$. The infinitesimal supertransformations of superspace coordinates are

$$\theta_\alpha \to \theta_\alpha + \epsilon_\alpha, \quad \bar{\theta}_{\dot\alpha} \to \bar{\theta}_{\dot\alpha} + \bar{\epsilon}_{\dot\alpha},$$
$$x^\mu \to x^\mu + i(\sigma^\mu)_{\alpha\dot\alpha}(\epsilon^\alpha\bar{\theta}^{\dot\alpha} - \theta^\alpha\bar{\epsilon}^{\dot\alpha}) \equiv x^\mu + i(\epsilon\sigma^\mu\bar{\theta} - \theta\sigma^\mu\bar{\epsilon}) \quad (3.3)$$

with σ^μ defined in Eq. (0.3). Their generators are

$$\hat{Q}_\alpha = \frac{\partial}{\partial\theta^\alpha} + i(\sigma^\mu)_{\alpha\dot\alpha}\bar{\theta}^{\dot\alpha}\partial_\mu, ,$$
$$\hat{\bar{Q}}_{\dot\alpha} = \frac{\partial}{\partial\bar{\theta}^{\dot\alpha}} + i\theta^\alpha(\sigma^\mu)_{\alpha\dot\alpha}\partial_\mu. \quad (3.4)$$

The covariant derivatives, with the help of which we are going to construct supersymmetric actions, are

$$D_\alpha = \frac{\partial}{\partial\theta^\alpha} - i(\sigma^\mu)_{\alpha\dot\alpha}\bar{\theta}^{\dot\alpha}\partial_\mu, ,$$
$$\bar{D}_{\dot\alpha} = -\frac{\partial}{\partial\bar{\theta}^{\dot\alpha}} + i\theta^\alpha(\sigma^\mu)_{\alpha\dot\alpha}\partial_\mu. \quad (3.5)$$

Their anticommutators are the same as for the supercharges (3.4):

$$\{D_\alpha, D_\beta\} = \{\bar{D}_{\dot\alpha}, \bar{D}_{\dot\beta}\} = 0, \qquad \{D_\alpha, \bar{D}_{\dot\alpha}\} = 2(\sigma^\mu)_{\alpha\dot\alpha}\hat{P}_\mu. \quad (3.6)$$

In addition, they anticommute with $\hat{Q}_\alpha$ and $\hat{\bar{Q}}_{\dot\alpha}$. The properties

$$D_\alpha D_\beta D_\gamma = \bar{D}_{\dot\alpha}\bar{D}_{\dot\beta}\bar{D}_{\dot\gamma} = 0 \quad (3.7)$$

also hold.

As was the case in quantum mechanics, there are three basic kinds of superfields that the superspace $(x^\mu, \theta_\alpha, \bar{\theta}_{\dot\alpha})$ supports: left chiral superfields, right chiral superfields and real superfields (also called *vector superfields* because they include vector gauge fields as the essential physical components). Chiral superfields are simpler. Consider them first.

3.2 Wess-Zumino model

Left and right chiral superfields live in left and right chiral superspaces. The construction is quite analogous to that in quantum mechanics. Introduce left and right coordinates according to

$$x_L^\mu = x^\mu - i\theta\sigma^\mu\bar{\theta}, \quad x_R^\mu = x^\mu + i\theta\sigma^\mu\bar{\theta}. \quad (3.8)$$

Their infinitesimal supertransformations read

$$x_L^\mu \to x_L^\mu - 2i\theta\sigma^\mu\bar\epsilon, \quad x_R^\mu \to x_R^\mu + 2i\epsilon\sigma^\mu\bar\theta \tag{3.9}$$

In other words, the transformation of x_L^μ does not depend on $\bar\theta$ and the transformation of x_R^μ does not depend on θ.

Definition 3.1. The sets of coordinates (x_L^μ, θ_α) and $(x_R^\mu, \bar\theta_{\dot\alpha})$ parametrize *left chiral* and *right chiral* superspaces, correspondingly.

These superspaces are invariant under supersymmetry transformations (3.3).

Definition 3.2. The left chiral superfield is a superfield with the component expansion[2]

$$\Phi = \phi(x_L) + \sqrt{2}\,\theta\psi(x_L) + \theta^2 F(x_L). \tag{3.10}$$

Here $\psi_\alpha(x_L)$ is a Weyl 2-component fermion as in Eq. (1.30). The right chiral superfield is defined as

$$\overline\Phi = \phi^*(x_R) + \sqrt{2}\,\bar\theta\bar\psi(x_R) + \bar\theta^2 F^*(x_R). \tag{3.11}$$

$\overline\Phi$ is conjugate to Φ.

The transformations (3.3) of the superspace coordinates induce the supertransformations of the components:

$$\begin{aligned}
\delta\phi &= \sqrt{2}\,\epsilon\psi, \\
\delta\psi_\alpha &= -i\sqrt{2}\,(\sigma^\mu\bar\epsilon)_\alpha\partial_\mu\phi + \sqrt{2}\,\epsilon_\alpha F, \\
\delta F &= -i\sqrt{2}\,(\sigma^\mu\bar\epsilon)_\alpha\partial_\mu\psi^\alpha
\end{aligned} \tag{3.12}$$

for the components of Φ and

$$\begin{aligned}
\delta\phi^* &= \sqrt{2}\,\bar\epsilon\bar\psi, \\
\delta\bar\psi_{\dot\alpha} &= i\sqrt{2}\,(\epsilon\sigma^\mu)_{\dot\alpha}\partial_\mu\phi^* + \sqrt{2}\,\bar\epsilon_{\dot\alpha} F^*, \\
\delta F^* &= i\sqrt{2}\,(\epsilon\sigma^\mu)_{\dot\alpha}\partial_\mu\bar\psi^{\dot\alpha}
\end{aligned} \tag{3.13}$$

for the components of $\overline\Phi$.

The product $\overline\Phi\Phi$ is a real superfield living in the "large" superspace $(x^\mu, \theta_\alpha, \bar\theta_{\dot\alpha})$. Its highest component is invariant up to total derivative under supersymmetry transformation, and we may take it as a supersymmetric Lagrangian, which can also be expressed as

$$\mathcal{L}_{\text{kin}} = \frac{1}{4}\int d^2\theta d^2\bar\theta\,\overline\Phi\Phi \tag{3.14}$$

---

[2]Here $\theta^2 = \theta\theta = 2\theta_1\theta_2$. It is *not* θ^α with $\alpha = 2$!

with the convention

$$\int d^2\theta\, \theta^2 \;=\; \int d^2\bar\theta\, \bar\theta^2 \;=\; 2\,. \qquad (3.15)$$

Bringing the superfields (3.10) and (3.11) in the large superspace, substituting the obtained expressions in (3.14) and doing the Berezin integral, we derive

$$\mathcal{L}_{\mathrm{kin}} \;=\; \partial_\mu\phi^*\partial^\mu\phi + i\psi\sigma^\mu\partial_\mu\bar\psi + F^*F\,. \qquad (3.16)$$

This Lagrangian describes a free massless scalar field and massless left-handed spinor field [cf. Eqs. (1.1), (1.30)], giving two physical bosonic and two fermionic degrees of freedom. The field F is not dynamical, and one can get rid of it, bearing in mind the "equation of motion" $F = 0$.

It is worth to emphasize here that by "physical degrees of freedom", we mean the eigenstates of the Hamiltonian. Their number is not the same as the number of superfield components (in our case four real bosonic and four real fermionic ones). Recall that, in $\mathcal{N} = 2$ quantum mechanical systems discussed in the previous chapter, we had $2_B + 2_F$ real components in each irreducible superfield and only one bosonic and one fermionic state in each doublet with a given energy. For $4D$ supersymmetric field theories, we have (at least) twice as many real superfield components and twice as many physical states in each degenerate supermultiplet.

The Lagrangian (3.16) is a little bit dull, however. To obtain a nontrivial interacting theory, we have to add to (3.14) something else. The simplest natural choice for "something else" is the *superpotential* term with the action[3]

$$S_{\mathrm{pot}} \;=\; \frac{1}{2}\int d^4x_L d^2\theta\, \mathcal{W}(\Phi) + \frac{1}{2}\int d^4x_R d^2\bar\theta\, \bar{\mathcal{W}}(\bar\Phi), \qquad (3.17)$$

where the superpotential $\mathcal{W}(\Phi)$ is an arbitrary function. The first term involves the integral over d^4x_L, while the second term involves the integral over d^4x_R. However, after doing the integrals over $d^2\theta$ and $d^2\bar\theta$, one may rename the integration variables in the both integrals, $x_L^\mu \to x^\mu$ and $x_r^\mu \to x^\mu$ and add the integrands to obtain the invariant Lagrangian[4] Adding this to (3.16), we obtain the full component Lagrangian of the Wess-Zumino

---

[3]The complex parameters that may be present in $\mathcal{W}(\Phi)$ are conjugated in $\bar{\mathcal{W}}(\bar\Phi)$.

[4]Alternatively, one may not think in terms of the Berezin integral, but simply take the highest components of the chiral superfields $W(\Phi)$ and $\bar{W}(\bar\Phi)$ (which are invariant up to a total derivative under supersymmetry transformations) and add them [29].

model [30]:

$$\mathcal{L}_{WZ} = \partial_\mu \phi^* \partial^\mu \phi + i\psi \sigma^\mu \partial_\mu \bar{\psi} + F^* F$$
$$+ \left\{ F\mathcal{W}'(\phi) - \frac{1}{2}\mathcal{W}''(\phi)\psi^2 + \text{compl. conj.} \right\}. \tag{3.18}$$

By construction, the corresponding action is invariant under the transformations (3.12) and (3.13).

Excluding F and F^*, we finally derive[5]

$$\boxed{\begin{aligned} \mathcal{L}_{WZ} &= \partial_\mu \phi^* \partial^\mu \phi + i\psi \sigma^\mu \partial_\mu \bar{\psi} - \bar{\mathcal{W}}'(\phi^*)\mathcal{W}'(\phi) \\ &\quad - \frac{1}{2}\mathcal{W}''(\phi)\psi^2 - \frac{1}{2}\bar{\mathcal{W}}''(\phi^*)\bar{\psi}^2. \end{aligned}} \tag{3.19}$$

The corresponding action is invariant under *nonlinear* supersymmetry transformations:

$$\begin{aligned} \delta\phi &= \sqrt{2}\,\epsilon\psi, \qquad \delta\phi^* = \sqrt{2}\,\bar{\epsilon}\bar{\psi}, \\ \delta\psi_\alpha &= -i\sqrt{2}\,(\sigma^\mu\bar{\epsilon})_\alpha \partial_\mu\phi - \sqrt{2}\,\epsilon_\alpha \bar{\mathcal{W}}'(\phi^*), \\ \delta\bar{\psi}_{\dot{\alpha}} &= i\sqrt{2}\,(\epsilon\sigma^\mu)_{\dot{\alpha}} \partial_\mu\phi^* - \sqrt{2}\,\bar{\epsilon}_{\dot{\alpha}} \mathcal{W}'(\phi). \end{aligned} \tag{3.20}$$

Choosing *cubic* superpotential,

$$\mathcal{W}(\Phi) = \frac{m}{2}\Phi^2 + \frac{\lambda}{6}\Phi^3, \tag{3.21}$$

we obtain a nice renormalizable theory involving a massive complex scalar, a massive Majorana fermion, the quartic scalar term $\propto (\phi^*\phi)^2$ and the Yukawa terms $\propto \psi^2\phi$ and $\propto \bar{\psi}^2\phi^*$.

3.3 Supersymmetric gauge theories

3.3.1 *Abelian theories*

To describe in superspace formalism the Abelian gauge field A_μ and its fermionic superpartner, consider a real superfield

$$V(x^\mu; \theta_\alpha, \bar{\theta}_{\dot{\alpha}}) = \bar{V}(x^\mu; \theta_\alpha, \bar{\theta}_{\dot{\alpha}}). \tag{3.22}$$

Its expansion in θ_α and $\bar{\theta}_{\dot{\alpha}}$ includes many terms up to the quartic term $\propto \theta^2\bar{\theta}^2$. This gives 8 bosonic and 8 fermionic real components. Obviously, this is too much — A_μ has only 4 components and not all of them are physical; e.g. A_0 is a Lagrange multiplier, as we discussed in Chapter 1.

---

[5]We are slightly deviating here from our conjugation policy and using for $\mathcal{W}$ a bar rather than a star.

That reduction of relevant degrees of freedom was due to gauge invariance of the free photon action under the gauge transformations (1.16). A natural question is whether it is possible to write a supersymmetric action in terms of the superfield (3.22), which would enjoy rich gauge invariance that would allow one to get rid of most degrees of freedom present in (3.22). The answer is positive.

We introduce the superfields[6]

$$W_\alpha = \frac{1}{8}\bar{D}\bar{D}D_\alpha V, \qquad \overline{W}_{\dot\alpha} = \frac{1}{8}DD\bar{D}_{\dot\alpha}V. \qquad (3.23)$$

As follows from (3.7), they are chiral. Noteworthy is their invariance under the gauge transformation

$$V \to V + i(\overline{\Lambda} - \Lambda), \qquad (3.24)$$

where Λ is an arbitrary chiral superfield. This follows from the anticommutators (3.6) and from $\bar{D}_{\dot\beta}\Lambda = D_\beta\overline{\Lambda} = 0$. Consider the supersymmetric action

$$S = \frac{1}{8}\int d^4x_L d^2\theta\, W^\alpha W_\alpha + \frac{1}{8}\int d^4x_R d^2\bar\theta\, \overline{W}_{\dot\alpha}\overline{W}^{\dot\alpha}. \qquad (3.25)$$

It is invariant under (3.24) and hence depends only on the half of components of V. It is convenient, using the gauge freedom (3.24), to get rid of the components $\propto 1, \theta_\alpha, \bar\theta_{\dot\alpha}, \theta\theta$ and $\propto \bar\theta\bar\theta$ and present V in the *Wess-Zumino gauge*:

$$V = -2\theta\sigma^\mu\bar\theta\, A_\mu(x) - 2i(\bar\theta\bar\theta)\theta\lambda(x) + 2i(\theta\theta)\bar\theta\bar\lambda(x) + (\theta\theta)(\bar\theta\bar\theta)D(x).(3.26)$$

Then

$$W_\alpha = i\lambda_\alpha(x_L) + i\theta^\beta F_{\alpha\beta}(x_L) - \theta_\alpha D(x_L) + \theta^2[\sigma^\mu\partial_\mu\bar\lambda(x_L)]_\alpha, \qquad (3.27)$$

where

$$F_{\alpha\beta} = \frac{1}{2}(\sigma^\mu)_{\alpha\dot\alpha}(\sigma^\nu)_{\beta\dot\beta}\varepsilon^{\dot\alpha\dot\beta}F_{\mu\nu} = \varepsilon_{\beta\gamma}\frac{1}{2}(\sigma^\mu\bar\sigma^\nu)_\alpha{}^\gamma F_{\mu\nu} \qquad (3.28)$$

with $F_{\mu\nu} = \partial_\mu A_\nu - \partial_\nu A_\mu$ [the identity (0.4) was used]. The symmetric tensor $F_{\alpha\beta}$ belongs to the representation $(1,0)$ of the Lorentz group.

Likewise,

$$\overline{W}_{\dot\alpha} = -i\bar\lambda_{\dot\alpha}(x_R) - i\theta^{\dot\beta}\bar{F}_{\dot\alpha\dot\beta} - \bar\theta_{\dot\alpha}D(x_R) + \bar\theta^2[\partial_\mu\lambda(x_R)\sigma^\mu]_{\dot\alpha} \qquad (3.29)$$

with

$$\bar{F}_{\dot\alpha\dot\beta} = \frac{1}{2}(\sigma^\mu)_{\alpha\dot\alpha}(\sigma^\nu)_{\beta\dot\beta}\varepsilon^{\alpha\beta}F_{\mu\nu} = \frac{1}{2}\varepsilon_{\dot\beta\dot\gamma}(\bar\sigma^\nu\sigma^\mu)^{\dot\gamma}{}_{\dot\alpha}F_{\mu\nu} \qquad (3.30)$$

---

[6]Nothing to do with the superpotential $\mathcal{W}(\Phi)$ in Eq. (3.17).

lying in the representation $(0, 1)$. Substituting (3.27) and (3.29) in (3.25) and doing the θ and $\bar{\theta}$ integrals, we arrive at the component Lagrangian

$$\mathcal{L} = -\frac{1}{4}F_{\mu\nu}F^{\mu\nu} + i\lambda\sigma^\mu\partial_\mu\bar{\lambda} + \frac{D^2}{2}. \tag{3.31}$$

The field D is *auxiliary*, like the field F in (3.16). The equations of motion following from (3.31) dictate that $D = 0$, and we may, in this case, forget about it.

Then the Lagrangian (3.31) represents a sum of two terms: the Lagrangian (1.11) of free photons and the Lagrangian (1.30) of free massless neutral fermions called *photinos*. The spectrum of the corresponding Hamiltonian involves the photon states (two polarizations for each momentum $\boldsymbol{p}$) and the photino states (a left particle and a right antiparticle). The number of bosonic and fermionic physical states coincides, as it should.

The Wess–Zumino form (3.26) of the superfield V does not fix the gauge completely, and the Lagrangian (3.31) enjoys the remaining one-parameter gauge invariance $A_\mu \to A_\mu + \partial_\mu\chi$.

Free theory (3.31) is not so interesting. Let us see how to include interactions while keeping supersymmetry and gauge symmetry intact. As we saw in Chapter 1, to describe massive charged electrons and positrons, it is necessary to introduce two left-handed fermion fields of opposite electric charge. The Lagrangian of supersymmetric electrodynamics includes besides the gauge superfield V also a positively charged chiral matter superfield

$$S = s(x_L) + \sqrt{2}\,\theta\psi(x_L) + \theta^2 F(x_L)$$

and a negatively charged chiral matter superfield

$$T = t(x_L) + \sqrt{2}\,\theta\xi(x_L) + \theta^2 G(x_L).$$

It reads

$$\boxed{\begin{aligned}\mathcal{L}_{SQED} &= \frac{1}{4}\int d^4\theta(\overline{S}e^{eV}S + \overline{T}e^{-eV}T) \\ &+ \left(\frac{1}{8}\int d^2\theta\,W^\alpha W_\alpha + \frac{m}{2}\int d^2\theta\,ST + \text{c.c.}\right),\end{aligned}} \tag{3.32}$$

where m is the mass of the matter fields and $\pm e$ are their electric charges. This Lagrangian is invariant under supergauge transformations (3.24) supplemented by

$$S \to e^{ie\Lambda}S,\ T \to e^{-ie\Lambda}T, \qquad \overline{S} \to \overline{S}e^{-ie\overline{\Lambda}},\ \overline{T} \to \overline{T}e^{ie\overline{\Lambda}}. \tag{3.33}$$

Substituting in (3.32) the component expressions for the chiral superfields and for the vector superfield V in the Wess-Zumino gauge (3.26) and integrating over θ and $\bar{\theta}$, one derives a comparatively long expression including, besides the physical fields, photons A_μ, photinos λ, matter scalars s, t and matter fermions ψ, ξ, also the auxiliary fields: the real field D and complex fields F, G. After D, F, G, F^*, G^* are excluded, we arrive at the component Lagrangian of supersymmetric electrodynamics:[7]

$$\mathcal{L} = -\frac{1}{4}F_{\mu\nu}F^{\mu\nu} + i\lambda\sigma^\mu\partial_\mu\bar{\lambda} + i\psi\sigma^\mu(\partial_\mu + ieA_\mu)\bar{\psi} + i\xi\sigma^\mu(\partial_\mu - ieA_\mu)\bar{\xi}$$
$$+ |(\partial_\mu - ieA_\mu)s|^2 + |(\partial_\mu + ieA_\mu)t|^2 - m^2(s^*s + t^*t) - m(\psi\xi + \bar{\psi}\bar{\xi})$$
$$+ ie\sqrt{2}[(\psi\lambda)s^* - (\bar{\psi}\bar{\lambda})s - (\xi\lambda)t^* + (\bar{\xi}\bar{\lambda})t] - \frac{e^2}{2}(s^*s - t^*t)^2 . \qquad (3.34)$$

This Lagrangian was first derived in the pioneer paper [16]. The first two lines in (3.34) involve interactions of the matter fields with the gauge field A_μ, while the third line (it is indispensable to make the Lagrangian supersymmetric) includes the Yukawa terms and the quartic scalar potential. The superfied derivation of this Lagrangian and its non-Abelian generalizations was worked out in Ref. [31].

3.3.2 *Non-Abelian theories*

Consider the vector superfield $\hat{V} = V^a t^a$ belonging to the adjoint representation of a Lie group. Define

$$\hat{W}_\alpha = \frac{1}{8g}\bar{D}^2(e^{-g\hat{V}}D_\alpha e^{g\hat{V}}) \qquad (3.35)$$

In these terms, the Lagrangian of the supersymmetric Yang-Mills theory reads

$$\boxed{\mathcal{L}_{\text{SYM}} = \frac{1}{4}\text{Tr}\int d^2\theta\,\hat{W}^\alpha\hat{W}_\alpha + \text{c.c.}} \qquad (3.36)$$

This Lagrangian is invariant under the following supergauge transformations:

$$e^{g\hat{V}} \to e^{i\hat{\bar{\Lambda}}}e^{g\hat{V}}e^{-i\hat{\Lambda}} \quad \text{and hence} \quad \hat{W}_\alpha \to e^{i\hat{\Lambda}}\hat{W}_\alpha e^{-i\hat{\Lambda}} , \qquad (3.37)$$

where $\hat{\Lambda}$ is a chiral superfield belonging, as $\hat{V}$ and $\hat{W}_\alpha$ do, to the adjoint representation of the group.

---

[7]Here e is the charge of the fields $s(x)$ and $\psi(x)$ in our theoretical model, which we have assumed to be positive. It is not the charge of the physical electron in Eq.(1.49), which is negative.

As in the Abelian case, we can choose the Wess-Zumino gauge (3.26) with adjoint component fields $\hat{A}_\mu$, $\hat{\lambda}_\alpha$ and $\hat{D}$. Then

$$\hat{W}_\alpha = i\hat{\lambda}_\alpha(x_L) + i\theta^\beta \hat{F}_{\alpha\beta}(x_L) - \theta_\alpha \hat{D}(x_L) + \theta^2 [\sigma^\mu \nabla_\mu \hat{\bar{\lambda}}(x_L)]_\alpha \quad (3.38)$$

with

$$\nabla_\mu \bar{\lambda}^a = \partial_\mu \bar{\lambda}^a + g f^{abc} A^b_\mu \bar{\lambda}^c . \quad (3.39)$$

The component Lagrangian reads

$$\mathcal{L}_{\text{SYM}} = -\frac{1}{4} F^a_{\mu\nu} F^{a\mu\nu} + i\lambda^a \sigma^\mu \nabla_\mu \bar{\lambda}^a + \frac{D^a D^a}{2} , \quad (3.40)$$

where $F^a_{\mu\nu}$ is the non-Abelian field strength defined in (1.21) and D^a are auxiliary fields that are equal to zero due to equations of motion.

The Lagrangian (3.40) is a supersymmetric generalization of the ordinary Yang-Mills Lagrangian (1.19). In contrast to (3.31), it is a nontrivial interacting theory with rich and interesting dynamics. It resembles QCD, but with the fermions in the adjoint rather than fundamental representation of the group. Similar to QCD, one can expect that the gluons (gauge field excitations) and gluinos (their superpartners) are confined in this theory and the physical states in the imaginary world whose physics is described by the Lagrangian (3.40) would be not massless gluons and gluinos, but massive *glueballs* and *gluinoballs*.

By construction, the action corresponding to the Lagrangian (3.40) is invariant under supertransformations. In the original superfield formulation, these transformations are linear, but they involve many gauge degrees of freedom — the components of $\hat{V}$ that we wish to get rid of by fixing the Wess-Zumino gauge. The Lagrangian (3.40) does not depend on these components and the laws of supertransformations may become nonlinear reflecting the necessity to eliminate gauge degrees of freedom after the transformation is done. This is the case for supersymmetric gluodynamics. For simplicity, we may fix $D^a = 0$, after which one can check[8] that the Lagrangian (3.40) stays invariant up to a total derivative under

---

[8]Use the identities $\sigma^\mu \bar{\sigma}^\nu + \sigma^\nu \bar{\sigma}^\mu = 2\eta^{\mu\nu}$ and

$$\sigma^\mu \bar{\sigma}^\kappa \sigma^\nu - \{\mu \leftrightarrow \nu\} \propto \varepsilon^{\mu\nu\kappa\rho} \sigma_\rho ,$$
$$\bar{\sigma}^\mu \sigma^\kappa \bar{\sigma}^\nu - \{\mu \leftrightarrow \nu\} \propto \varepsilon^{\mu\nu\kappa\rho} \bar{\sigma}_\rho .$$

Then, to show that the variation is the total derivative, use "the second pair of Maxwell's equations" $\varepsilon^{\mu\nu\kappa\rho} (\nabla_\kappa F_{\mu\nu})^a = 0$.

the transformations

$$\delta A_\mu^a = \epsilon \sigma_\mu \bar{\lambda}^a + \lambda^a \sigma_\mu \bar{\epsilon},$$

$$\delta \lambda^{a\alpha} = \frac{i}{2} \epsilon^\beta (\sigma^\mu \bar{\sigma}^\nu)_\beta{}^\alpha F_{\mu\nu}^a,$$

$$\delta \bar{\lambda}^{a\dot{\alpha}} = \frac{i}{2} (\bar{\sigma}^\mu \sigma^\nu)^{\dot{\alpha}}{}_{\dot{\beta}} \bar{\epsilon}^{\dot{\beta}} F_{\mu\nu}^a. \tag{3.41}$$

Similarly to what we saw in the Abelian case, the non-Abelian super-gauge field $\hat{V}$ can be coupled to matter supermultiplets. In the simplest case with the gauge group $SU(N)$, we introduce the chiral superfields S_j and T^j belonging to the fundamental and antifundamental representations of $SU(N)$, respectively. The Lagrangian of the model (called *supersymmetric QCD*) reads:

$$\boxed{\begin{aligned} \mathcal{L}_{\text{SQCD}} &= \mathcal{L}_{\text{SYM}} + \frac{1}{4} \int d^4\theta \, (\overline{S} e^{g\hat{V}} S + T e^{-g\hat{V}} \overline{T}) \\ &+ \frac{m}{2} \left(\int d^2\theta \, T^j S_j + \text{c.c.} \right). \end{aligned}} \tag{3.42}$$

The corresponding action is invariant under the transformations (3.37) supplemented by

$$S \to e^{i\hat{\Lambda}} S, \quad \overline{S} \to \overline{S} e^{-i\hat{\Lambda}}, \quad T \to T e^{-i\hat{\Lambda}}, \quad \overline{T} \to e^{i\hat{\Lambda}} \overline{T}. \tag{3.43}$$

The component fields in the supersymmetric QCD include gluons, gluinos, massive Dirac fermions in the fundamental and antifundamental group representations similar to quarks (though the reader understands that they are not the physical quarks) and "scalar quarks" — complex scalar (anti)fundamentals with the same mass as "quarks". The component Lagrangian is similar to (3.34):

$$\begin{aligned} \mathcal{L} &= -\frac{1}{4} F_{\mu\nu}^a F^{a\mu\nu} + i\lambda^a \sigma^\mu (\partial_\mu \bar{\lambda}^a + g f^{abc} A_\mu^b \bar{\lambda}^c) \\ &- s^* (\partial_\mu - ig\hat{A}_\mu)^2 s - t(\partial_\mu + ig\hat{A}_\mu)^2 t^* - m^2(s^*s + tt^*) + m(\xi\psi + \bar{\psi}\bar{\xi}) \\ &+ i\bar{\psi}\bar{\sigma}^\mu (\partial_\mu - ig\hat{A}_\mu)\psi + i\xi\sigma^\mu(\partial_\mu + ig\hat{A}_\mu)\bar{\xi} \\ &+ ig\sqrt{2}[s^*(\hat{\lambda}\psi) - (\bar{\psi}\hat{\bar{\lambda}})s - (\xi\hat{\lambda})t^* + t(\hat{\bar{\lambda}}\bar{\xi})] - \frac{g^2}{2}(s^* t^a s - t t^a t^*)^2, \end{aligned} \tag{3.44}$$

where $\hat{A}_\mu = A_\mu^a t^a, \hat{\lambda} = \lambda^a t^a$, the fields s, ψ, t^* and $\bar{\xi}$ are fundamental and the fields $s^*, \bar{\psi}, t$ and ξ are antifundamental. The reader will not mix up, of course, the scalar fields t^j and t_j^* with the generators $(t^a)_j^k$, but to make the equation more readable, the group indices are not displayed.

Besides the terms describing gauge interactions of the quarks and scalar quarks, the Lagrangian (3.44) includes Yukawa terms and the quartic scalar self-interaction term.

But one can equally well include in the action the matter multiplets in any other representation of the group. Especially interesting is the theory including besides $\hat{V}$ an *adjoint* massless chiral matter multiplet $\hat{\Phi}$:

$$\mathcal{L} = \mathcal{L}_{\text{SYM}} + \frac{1}{2}\text{Tr}\int d^4\theta\,\hat{\bar{\Phi}}e^{g\hat{V}}\hat{\Phi}e^{-g\hat{V}}. \tag{3.45}$$

The corresponding action is invariant under the gauge transformations (3.37) supplemented by

$$\hat{\Phi} \rightarrow e^{i\hat{\Lambda}}\hat{\Phi}e^{-i\hat{\Lambda}}, \qquad \hat{\bar{\Phi}} \rightarrow e^{i\hat{\bar{\Lambda}}}\hat{\bar{\Phi}}e^{-i\hat{\bar{\Lambda}}}. \tag{3.46}$$

Going to components, fixing the Wess-Zumino gauge and excluding the auxiliary fields, we are left with: *(i)* the gauge field A_μ^a, *(ii)* two adjoint massless fermions — the gluino λ_α^a and the adjoint fermion ψ_α^a present in the expansion of the superfield Φ^a, and *(iii)* the complex adjoint scalar ϕ^a. The component Lagrangian reads

$$\mathcal{L}_{\text{SYM}}^{N=2} = -\frac{1}{4}F_{\mu\nu}^a F^{a\mu\nu} + (\nabla_\mu\phi^a)^*(\nabla^\mu\phi^a) + i\lambda^a\sigma^\mu\nabla_\mu\bar{\lambda}^a + i\psi^a\sigma^\mu\nabla_\mu\bar{\psi}^a$$

$$-g\sqrt{2}f^{abc}[\phi^{*a}(\lambda^b\psi^c) + \phi^a(\bar{\lambda}^b\bar{\psi}^c)] + \frac{g^2}{2}f^{abe}f^{cde}\phi^{*a}\phi^b\phi^{*c}\phi^d, \tag{3.47}$$

The corresponding action is invariant under the following supersymmetry transformations:[9]

$$\delta A_\mu^a = \epsilon_1\sigma_\mu\bar{\lambda}^a + \lambda^a\sigma_\mu\bar{\epsilon}_1,$$

$$\delta\lambda^{a\alpha} = \frac{i}{2}\epsilon_1^\beta(\sigma^\mu\bar{\sigma}^\nu)_\beta{}^\alpha F_{\mu\nu}^a,$$

$$\delta\phi^a = i\sqrt{2}\,\epsilon_1\psi^a,$$

$$\delta\psi_\alpha^a = -\sqrt{2}\,(\sigma^\mu\bar{\epsilon}_1)_\alpha\partial_\mu\phi^a, \tag{3.48}$$

which mix A_μ^a with λ^a and ϕ^a with ψ_α^a. But one can notice that the Lagrangian (3.47) enjoys an extra discrete symmetry:

$$\lambda \leftrightarrow \psi, \qquad \phi \rightarrow -\phi. \tag{3.49}$$

---

[9]The third and the fourth line in (3.47) follow from (3.20) if one takes into account the absence of the superpotential and introduces the extra factors $\pm i$, which ensure the invariance of the gauge-matter interaction terms.

The superposition of (3.48) and (3.49) brings about *another* supersymmetry:

$$\delta A_\mu^a = \epsilon_2 \sigma_\mu \bar\psi^a + \psi^a \sigma_\mu \bar\epsilon_2 ,$$

$$\delta\psi^{a\alpha} = \frac{i}{2}\epsilon_2^\beta (\sigma^\mu \bar\sigma^\nu)_\beta{}^\alpha F_{\mu\nu}^a ,$$

$$\delta\phi^a = -i\sqrt{2}\,\epsilon_2 \lambda^a ,$$

$$\delta\lambda_\alpha^a = \sqrt{2}\,(\sigma^\mu \bar\epsilon_2)_\alpha \partial_\mu \phi^a , \tag{3.50}$$

which mixes A_μ^a with ψ^a and ϕ^a with λ_α^a.

As is not difficult to see, the supertransformations of two different types commute, which means (bearing in mind the Grassmann nature of the transformation parameters) that their generators, the supercharges, anticommute. The full algebra reads

$$\{\hat Q_\alpha^j, \hat Q_\beta^k\} = \{\hat{\bar Q}_{\dot\alpha}^j, \hat{\bar Q}_{\dot\beta}^k\} = 0,$$

$$\{\hat Q_\alpha^j, \hat{\bar Q}_{\dot\alpha}^k\} = 2(\sigma^\mu)_{\alpha\dot\alpha} \hat P_\mu\, \delta^{jk} , \tag{3.51}$$

$j, k = 1, 2$. This is the algebra of extended $\mathcal{N} = 2$ four-dimensional supersymmetry.

The physics of the imaginary world described by the Lagrangian (3.45) is very rich and interesting. In fact, in this case, the Lagrangian does not completely define the physics. The spectrum of the corresponding Hamiltonian has *infinity* of the degenerate zero energy vacuum states associated with the "valley" (*vacuum valley*) where the quartic term $\sim |f^{abe}\phi^a\phi^{*b}|^2$ in the potential vanishes. The physics, in particular the properties of the excitations, depend(s) on the vacuum choice. We will not go into details here, instead directing the reader to the classical paper [32] and the vast literature that it inspired, but just note that in this case it has become possible to obtain many *exact* results: the richer is the symmetry of the problem (and $\mathcal{N} = 2$ supersymmetry is not so poor), the more tractable the problem is.

Another interesting theory includes *three* different massless adjoint chiral multiplets $\Phi_{f=1,2,3}^a$. The Lagrangian reads

$$\boxed{\begin{aligned} \mathcal{L} &= \mathcal{L}_{\text{SYM}} + \frac{1}{2}\text{Tr}\int d^4\theta\, \hat{\bar\Phi}_f e^{g\hat V}\hat\Phi_f e^{-g\hat V} \\ &\quad + \frac{g}{6\sqrt{2}}\varepsilon_{fgh}f^{abc}\left(\int d^2\theta\, \Phi_f^a \Phi_g^b \Phi_h^c + \text{c.c.}\right). \end{aligned}} \tag{3.52}$$

The physical component fields include the gauge field, four adjoint fermions λ_α^a and $\psi_{\alpha f}^a$ and three complex adjoint scalars ϕ_f^a. Note the presence of the superpotential term. It brings about the Yukawa terms $\sim \varepsilon_{fgh} f^{abc} \phi_f^a (\psi_g^b \psi_h^c) + \text{c.c.}$ The full component Lagrangian reads

$$\mathcal{L}_{\text{SYM}}^{N=4} = -\frac{1}{4} F_{\mu\nu}^a F^{a\mu\nu} + (\nabla_\mu \phi_f^a)^* (\nabla^\mu \phi_f^a) + i\lambda^a \sigma^\mu \nabla_\mu \bar{\lambda}^a + i\psi_f^a \sigma^\mu \nabla_\mu \bar{\psi}_f^a$$

$$-g\sqrt{2} f^{abc} \left[\phi_f^{*a} (\lambda^b \psi_f^c) + \frac{1}{2} \varepsilon_{fgh} \phi_f^a (\psi_g^b \psi_h^c) + \phi_f^a (\bar{\lambda}^b \bar{\psi}_f^c) + \frac{1}{2} \varepsilon_{fgh} \phi_f^{*a} (\bar{\psi}_g^b \bar{\psi}_h^c) \right]$$

$$-\frac{g^2}{2} \left(f^{abe} f^{adc} + f^{ade} f^{abc} \right) \phi_g^b \phi_g^{*d} \phi_h^c \phi_h^{*e}. \tag{3.53}$$

The scalar potential in the last line is obtained as a sum of two contributions: *(i)* the D-term contribution of the same origin as the potential in (3.47) and *(ii)* the F-term contribution coming from the last superpotential term in (3.52).

This Lagrangian enjoys three discrete symmetries:

$$\mathbf{a}: \quad \lambda \leftrightarrow \psi_1, \ \psi_2 \leftrightarrow \psi_3, \ \phi_1 \to -\phi_1, \ \phi_2 \to -\phi_2^*, \ \phi_3 \to \phi_3^*;$$

$$\mathbf{b}: \quad \lambda \leftrightarrow \psi_2, \ \psi_3 \leftrightarrow \psi_1, \ \phi_2 \to -\phi_2, \ \phi_3 \to -\phi_3^*, \ \phi_1 \to \phi_1^*;$$

$$\mathbf{c}: \quad \lambda \leftrightarrow \psi_3, \ \psi_1 \leftrightarrow \psi_2, \ \phi_3 \to -\phi_3, \ \phi_1 \to -\phi_1^*, \ \phi_2 \to \phi_2^*. \tag{3.54}$$

The algebra

$$\mathbf{a}^2 = \mathbf{b}^2 = \mathbf{c}^2 = 1, \quad \mathbf{ab} = \mathbf{ba} = \mathbf{c}, \quad \mathbf{bc} = \mathbf{cb} = \mathbf{a}, \quad \mathbf{ac} = \mathbf{ca} = \mathbf{b}$$

holds. We are dealing with the so-called *Klein four-group* isomorphic to $\mathbb{Z}_2 \times \mathbb{Z}_2$.

The superpositions of the discrete symmetries (3.54) and the manifest supersymmetry following from the superfield Lagrangian (3.52) brings about three extra supersymmetries. The full algebra has the form (3.51) with $j, k = 1, 2, 3, 4$. This is the algebra of extended $\mathcal{N} = 4$ four-dimensional supersymmetry. Vast symmetry of the theory allows one to establish its many remarkable features.

First of all, it is a *conformal* theory. At the classical level, it is a trivial statement — the coupling constant carries no dimension and the Lagrangian has no other dimensional parameters. But this alone does not guarantee conformal symmetry at the quantum level. The ordinary Yang-Mills theory also does not seem to have dimensional parameters in the Lagrangian, but it is not conformal. The matter is that the coupling constant may *run*. To define quantum field theory, one has to regularize it in the ultraviolet, introduce the ultraviolet cutoff Λ_{UV} and define a bare coupling at that scale. The physical dimensional scale of the theory is the scale where the running coupling constant grows up to the value of order 1.

The same phenomenon of *dimensional transmutation* occurs in $\mathcal{N} = 1$ and $\mathcal{N} = 2$ SYM theories. But in $\mathcal{N} = 4$ SYM the beta function vanishes, the coupling constant does not run and the conformal symmetry is exact [33]! It has also become possible to obtain many exact results concerning the correlators in this theory.[10]

3.4 Supersymmetry and phenomenology

We do not see supersymmetry around us. There is no massless photino, there are no scalar particles with the same mass as electrons. Thus, one can get an impression that supersymmetric field theories describe some imaginary worlds, not the universe we live in. This may be true. But many physicists still believe that the *fundamental* theory of the universe is supersymmetric. The supersymmetry does not show up at the energy scale that we can explore in accelerator experiments because it is *broken* at some larger energy scale. There are several serious arguments in favor of this hypothesis and also some arguments against. Let us briefly discuss them.

3.4.1 *Arguments in favor*

Supersymmetry may help to resolve the so-called *hierarchy problem*. Basically, this is the problem of quadratic ultraviolet divergences in the mass of *Higgs particle*[11] coming from graphs like the one depicted in Fig. 3.1.

One can, of course, always renormalize the theory and *subtract* these divergences, but it does not look natural. In addition, people believe that the Standard Model [describing strong interactions with the gauge group $SU(3)$ and electroweak interactions with two building blocks $SU(2) \times U(1)$] is not a fundamental theory, but the latter represents a *unified* theory with a large gauge group $\mathcal{G}$ including $SU(3) \times SU(2) \times U(1)$ as a subgroup [the minimal option is $\mathcal{G} = SU(5)$] and only one coupling constant. As we will soon explain, the unification is expected to occur at the scale $M_{\text{GUT}} \sim 10^{15}$–$10^{16}$ GeV, and in this case it is very difficult to protect "our" Higgs boson

---

[10]One can often hear a statement that the $\mathcal{N} = 4$ SYM is *integrable*, but it is not so. For an integrable theory, *all* classical trajectories wind around the tori in a regular way. But for any $\mathcal{N}$, the SYM theories include the purely bosonic sector with chaotic trajectories [34].

[11]A necessary ingredient of the modern theory of electroweak interactions. In 2012, its existence was confirmed in accelerator experiment at CERN [35].

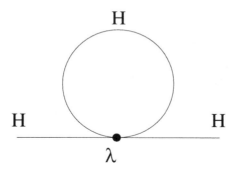

Fig. 3.1: A divergent contribution in the Higgs mass.

from being much lighter than M_{GUT}.

In supersymmetric versions of the Standard Model, there are additional graphs contributing to the Higgs mass. If supersymmetry were exact, these graphs would exactly cancel the contributions from the ordinary Standard Model graphs. If supersymmetry is broken at the scale Λ_{SUSY}, one could expect the mass to be of order $m_H \sim \Lambda_{\mathrm{SUSY}}$, but not $\sim M_{\mathrm{GUT}}$.

Another argument in favor of underlying fundamental supersymmetry is also related to the *Grand Unification* scenario. This hypothesis is supported by the following observation. The Standard Model has three coupling constants: $g_{SU(3)}$, $g_{SU(2)}$ and $g_{U(1)}$. From experiment, we know their values at the scale ~ 100 GeV. But these values "run" with energy. Non-Abelian couplings decrease with the energy increase due to *asymptotic freedom* and the Abelian constant $g_{U(1)}$ grows (as it does in conventional QED). As a result, the constants approach one another as is shown in Fig. 3.2.

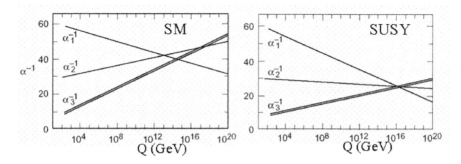

Fig. 3.2: Running couplings

Now the reader is invited first to close his right eye and look at the left-hand plot. It shows how the effective couplings

$$\alpha_1^{-1} = \frac{4\pi}{g_{U(1)}^2}, \qquad \alpha_2^{-1} = \frac{4\pi}{g_{SU(2)}^2}, \qquad \alpha_3^{-1} = \frac{4\pi}{g_{SU(3)}^2} \qquad (3.55)$$

depend on the logarithm of the characteristic energy Q in the Standard Model. We see that the three constants, being rather different at $Q \sim$ 100 GeV, approach one another as the energy grows and become *roughly* equal at $Q \sim M_{\text{GUT}} \sim 10^{15}$ GeV. This is the characteristic scale of Grand Unification.

Grand unified theories are characterized by the presence of the so-called X bosons — the vector particles that convert quarks to leptons and the other way round (see Fig. 3.3). Their mass is of order M_{GUT}. As a result, the proton is no longer stable, it can decay, say, into a positron and π^0. As m_X is pretty large, the probabilty of the proton decay is pretty small.

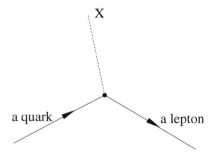

Fig. 3.3: Quark-lepton transitions

But everything is perceived in comparison. One can evaluate this probability and derive that nonsupersymmetric GUTs contradict experiment. The estimate for the proton lifetime appears to be smaller that the experimental limit ($\tau_p > 10^{34}$ years) by 6 orders of magnitude. In addition, the three lines in the left-hand plot do not intersect at *exactly* the same point, as the GUTs predict. We should conclude that the simple-minded nonsupersymmetric GUT models do not describe Nature.

And now the reader is allowed to open the second eye and look at the right-hand plot. S/he will see that, in the minimal supersymmetric extension of the Standard Model, the runnings of the couplings are modified such that the lines cross at the same point! The scale of the unification is somewhat larger than in the supersymmetric models,

$$M_{\text{SUSY GUT}} \approx 10^{16}\text{GeV}. \tag{3.56}$$

The larger unification scale means larger masses of the X bosons. Because of this and for some other reasons, the probability of proton decay is further suppressed, and there is no longer any contradiction with experiment!

To reiterate, there are at least three problems of the Standard Model that supersymmetry may help to solve:

• hierarchy problem,

• not exact matching of coupling constants at the unification scale and

• the absence (as far as the current experiment is concerned) of the proton decay.

Further arguments come from astrophysics:

• We now know that the main part of gravitating mass in galaxies dwells not in stars or gas, but in the so-called *dark matter*. We do not know what is its nature, but, according to the most popular hypothesis, dark matter consists of some electrically neutral weakly interacting massive particles (*WIMPs*). If the fundamental supersymmetry is really there and if it is broken at the scale of several hundred GeV, such WIMPs should be in abundance starting from photino that acquired mass due to supersymmetry breaking...

• The last argument may better be qualified as a *hint*. There is an old and unresolved mystery, probably the most dark and troublesome mystery in modern physics: the vacuum energy density that we see from cosmological observations (it is called *cosmological term*) is very very small, and that is difficult to explain theoretically. In the ordinary quantum field theory framework, the vacuum energy density diverges as a fourth power of the ultraviolet cutoff, and the only known way to make it finite is to consider a supersymmetric theory, where the bosonic zero-point fluctuations cancel with the fermionic ones, so that the vacuum energy is zero.

This would be fine, but unfortunately once the supersymmetry is broken, the vacuum energy should reappear, with a natural scale for its density being of order $\sim \Lambda^4_{\text{SUSY}}$. And this exceeds the observed value of the cosmological term by ~ 85 orders!

3.4.2 *Arguments against*

When physicists contemplated and then realized the construction of the Large Hadron Collider at CERN, they hoped to find there something *really* new, and the discovery of supersymmetry was probably their main aspiration. However, nothing (except the Higgs particle, about which existence nobody had any doubt) was found, and that was a very big disappointment.

In particular, the present experimental lower limit for the mass of scalar superpartner of the heaviest quark is around 1200 GeV [36]. Then a natural estimate for the mass of the Higgs particle that appears due to the absence of exact cancellation between different diagrams contributing in m_H is of order ~ 2 TeV, to be compared with the observed value of the Higgs mass $m_H \approx 125$ GeV. Strictly speaking, this is not a *contradiction* (small mass of Higgs particle still can be explained by adjusting parameters, by *fine tuning*), but is definitely a troublesome indication that something is wrong in our understanding and that at the end of the day supersymmetry might not be there.

Finally, there is an aesthetic argument against supersymmetric modifications of the Standard Model. The Standard Model has 26 free parameters. In the search for fundamental theory, it is natural to try to reduce this number, to make the theory *simpler*. However, introducing supersymmetry does not serve this purpose. Rather, it works in the opposite direction, bringing about at least a hundred *new* parameters...

3.5 Other dimensions

3.5.1 *Gauge quantum mechanics*

Our world is 4-dimensional, but supersymmetric field theories can also be constructed for $D \neq 4$. It is easy to go down in dimension. For this purpose, it is sufficient to take any 4-dimensional supersymmetric Lagrangian and assume that the fields do not depend on one of the spatial coordinates (to obtain a 3-dimensional theory), on two spatial coordinates (to obtain a 2-dimensional theory) or on all of them to obtain a supersymmetric classical or quantum mechanics. This procedure is called *dimensional reduction*. The reduction of a minimal $\mathcal{N} = 1$ 4-dimensional theory involving the supercharges Q_α and $\bar{Q}_{\dot\alpha}$ results in an extended $\mathcal{N} = 4$ supersymmetric system with four real or two complex supercharges.

In particular, one can take the Lagrangian (3.19) of the Wess-Zumino model, suppress there the spatial derivatives and assume that ϕ and ψ depend only on time. We arrive then at supersymmetric mechanical system. The Legendre transformation gives the Hamiltonian. Its quantum version coincides with (2.68).

Interesting systems are obtained after dimensional reduction of super-symmetric gauge theories. Take the Lagrangian (3.40). In the mechanical limit, we derive

$$L = \frac{1}{2}(\dot{A}_j^a + gf^{abc}A_0^b A_j^c)^2 - \frac{g^2}{4}(f^{abc}A_j^b A_k^c)^2$$
$$+ i\lambda^a(\dot{\bar{\lambda}}^a + gf^{abc}A_0^b \bar{\lambda}^c) - igf^{abc}\lambda^a \sigma_j A_j^b \bar{\lambda}^c. \qquad (3.57)$$

The variables A_0^a enter without time derivatives, the variation over them gives *gauge constraints* G^a. The variables A_j^a are dynamical. Their canon-ical momenta are

$$E_j^a = \dot{A}_j^a + gf^{abc}A_0^b A_j^c. \qquad (3.58)$$

The quantum Hamiltonian and quantum constraints read [37]

$$\hat{H} = \frac{1}{2}(\hat{E}_j^a)^2 + \frac{g^2}{4}(f^{abc}A_j^b A_k^c)^2 + igf^{abc}\lambda^{a\alpha}(\sigma_j)_{\alpha\beta}\hat{\bar{\lambda}}^{c\beta}A_j^b \qquad (3.59)$$

and

$$\hat{G}^a = f^{abc}(A_j^b \hat{E}_j^c - i\lambda^{ba}\hat{\bar{\lambda}}^{ca}) \qquad (3.60)$$

with $\hat{E}_j^a = -i\partial/\partial A_j^a$, $\hat{\bar{\lambda}}^{a\alpha} = \partial/\partial\lambda^{a\alpha}$.

The operator $\hat{G}^a$ is nothing but the generator of gauge transformations. The commutator $[\hat{H}, \hat{G}^a]$ vanishes, as it should. To find the spectrum, we have to solve the Schrödinger equation $\hat{H}\Psi(A,\lambda) = E\Psi(A,\lambda)$ in the Hilbert space involving the gauge-invariant wave functions satisfying the constraints

$$\hat{G}^a\Psi = 0. \qquad (3.61)$$

The Hermitially conjugated supercharges are

$$\hat{Q}_\alpha = (\lambda^a\sigma_j)_\alpha(\hat{E}_j^a + iB_j^a), \qquad \hat{\bar{Q}}_\alpha = (\sigma_j\hat{\bar{\lambda}}^a)_\alpha(\hat{E}_j^a - iB_j^a), \qquad (3.62)$$

where $B_j^a = \frac{g}{2}\varepsilon_{jkl}f^{abc}A_k^b A_l^c$. The anticommutators $\{\hat{Q}_\alpha, \hat{Q}_\beta\}$ and $\{\hat{\bar{Q}}_\alpha, \hat{\bar{Q}}_\beta\}$ vanish, while

$$\{\hat{\bar{Q}}_\alpha, \hat{Q}_\beta\} = 2\delta_{\alpha\beta}\hat{H} - 2g(\sigma_j)_{\alpha\beta}A_j^a\hat{G}^a. \qquad (3.63)$$

In addition, $[\hat{Q}_\alpha, \hat{G}^a] = [\hat{\bar{Q}}_\alpha, \hat{G}^a] = 0$, which means that, acting with the supercharges on a gauge invariant function, the result is also gauge invari-ant.

The second term in the anticommutator (3.63) gives zero, when acting on gauge invariant wave functions, and, in the restricted Hilbert space in-cluding the wave functions satisfying the constraint (3.61), we reproduce the algebra (2.65) of the extended supersymmetric $\mathcal{N} = 4$ quantum mechanics.

We can also perform similar dimensional reductions for the Lagrangians (3.47) and (3.53) describing extended supersymmetric Yang-Mills theories. As a result, we obtain gauge SQM systems involving 8 or 16 Hermitian supercharges. The $\mathcal{N} = 16$ SQM system obtained by the reduction of the $\mathcal{N} = 4, D = 4$ SYM theory (3.53) is especially interesting. The Hamiltonian, Hermitian supercharges and the gauge constraints of this theory have the form

$$\hat{H} = \frac{1}{2}(\hat{E}_I^a)^2 + \frac{g^2}{4}(f^{abc}A_I^a A_J^b)^2 + \frac{ig}{2}f^{abc}\hat{\lambda}_\alpha^a(\Gamma_I)_{\alpha\beta}\hat{\lambda}_\beta^b A_I^c , \qquad (3.64)$$

$$\hat{Q}_\alpha = \left[(\Gamma_I)_{\alpha\gamma}\hat{E}_I^a + \frac{g}{2}(\Gamma_I\Gamma_J)_{\alpha\gamma}f^{abc}A_I^b A_J^c\right]\hat{\lambda}_\gamma^a , \qquad (3.65)$$

and

$$\hat{G}^a = f^{abc}\left(A_I^b \hat{E}_I^c - \frac{i}{2}\hat{\lambda}_\alpha^b \hat{\lambda}_\alpha^c\right) . \qquad (3.66)$$

In these expressions, $I, J = 1, \ldots, 9$ and $\alpha = 1, \ldots, 16$. Now $\hat{\lambda}_\alpha$ are the Majorana spinors belonging to the 16-plet of[12] $SO(9)$. They are quantum operators satisfying the Clifford algebra, $\{\hat{\lambda}_\alpha^a, \hat{\lambda}_\beta^b\} = \delta^{ab}\delta_{\alpha\beta}$, and Γ_I are the 9-dimensional real and symmetric gamma martices, which also satisfy the Clifford algebra $\{\Gamma_I, \Gamma_J\} = 2\delta_{IJ}$. Note the identity

$$\text{Tr}\{\Gamma_I\Gamma_J\Gamma_K\Gamma_L\} = 16(\delta_{IJ}\delta_{KL} + \delta_{IL}\delta_{JK} - \delta_{IK}\delta_{JL}).$$

The algebra

$$\{\hat{Q}_\alpha, \hat{Q}_\beta\} = 2\delta_{\alpha\beta}\hat{H} + 2g(\Gamma_I)_{\alpha\beta}A_I^a\hat{G}^a \qquad (3.67)$$

holds.

The operators (3.64), (3.65) and (3.66) act on the wave functions depending on A_I^a and *holomorphic* fermion variables $\mu_{1,\ldots,8}^a$, which can be chosen as

$$\begin{cases} \mu_1^a = (\lambda_1^a + i\lambda_9^a)/\sqrt{2} \\ \qquad \cdots \\ \mu_8^a = (\lambda_8^a + i\lambda_{16}^a)/\sqrt{2} \end{cases} \qquad (3.68)$$

We will discuss the physics (i.e. the spectrum) of this model in Chapter 9.

---

[12]The reader might be perplexed at this point — why $SO(9)$ popped up here — but we will explain it pretty soon.

3.5.2 $D = 6$ *and* $D = 10$

One may go down in dimension, but one may also go *up*.

Consider the bosonic part of the Lagrangian (3.47) and trade the complex scalar field ϕ^a for a couple of real adjoint scalars, which we fancy to denote as A_4^a and A_5^a: $\phi^a = (A_4^a + iA_5^a)/\sqrt{2}$. In these terms,

$$
\mathcal{L}_{\text{bos}}^{N=2 \text{ SYM}} = -\frac{1}{4}F_{\mu\nu}^a F^{a\mu\nu} + \frac{1}{2}(\partial_\mu A_4^a + gf^{abc}A_\mu^b A_4^c)^2
$$
$$
+ \frac{1}{2}(\partial_\mu A_5^a + gf^{abc}A_\mu^b A_5^c)^2 - \frac{g^2}{2}(f^{abc}A_4^b A_5^c)^2 . \quad (3.69)
$$

And this can be rewritten as

$$
\mathcal{L}_{\text{bos}}^{N=2 \text{ SYM}} = -\frac{1}{4}F_{MN}^a F^{aMN}, \quad F_{MN}^a = \partial_M A_N^a - \partial_N A_M^a + gf^{abc}A_M^b A_N^c,
$$
$$
M, N = 0, 1, \ldots, 5 \quad (3.70)
$$

with a restriction that the fields A_M^a depend only of four coordinates t, x_1, x_2, x_3. The expression (3.70) coincides with the Lagrangian of the 6-dimensional Yang-Mills theory. In other words, the bosonic part of the $\mathcal{N} = 2$ supersymmetric Yang-Mills Lagrangian coincides with the 6-dimensional Yang-Mills Lagrangian dimensionally reduced from six to four dimensions. Having observed this, we may consider the full 6-dimensional Yang-Mills Lagrangian with the fields A_M^a depending on six coordinates $x_{M=0,1,\ldots,5}$.

A similar analysis can be performed for the fermion part. Two 4-dimensional adjoint Weyl fermions $\lambda^{a,\alpha=1,2}$ and $\psi^{a,\alpha=1,2}$ can be interpreted as different components of a 6-dimensional adjoint fermion $\chi^{a,\alpha=1,2,3,4}$ belonging to one of the two spinor represesanions of $SO(5,1)$.

It is convenient to define, instead of the complex 4-component spinor χ^α, a couple of spinors $\chi_{j=1,2}^\alpha$ obeying the pseudoreality condition

$$
\overline{\chi_j^a} \equiv -C_\beta^\alpha(\chi_j^\beta)^* = \chi^{j\alpha} \equiv \varepsilon^{jk}\chi_k^\alpha , \quad (3.71)
$$

where C is the charge conjugation matrix with the properties $C^T = -C, C^2 = -\mathbb{1}$. Then the full 6-dimensional SYM Lagrangian can be written in the form [38–40] [13]

---

[13] In six dimensions, it is convenient to change our conventions and include the coupling constant f in the common factor in front of the Lagrangian. This constant carried dimension, $[f] = m^{-1}$ to make the action $\int d^6x \, \mathcal{L}$ dimensionless. And the field density tensor and the covariant derivative are now defined as

$$
F_{MN}^a = \partial_M A_N^a - \partial_N A_M^a + f^{abc}A_M^b A_N^c, \quad \nabla_M \chi_k^a = \partial_M \chi_k^a + f^{abc}A_M^b \chi_k^b .
$$

$$\mathcal{L}^{D=6\ SYM} = -\frac{1}{4f^2}F^a_{MN}F^{aMN} + \frac{i}{2f^2}\chi^{ka}\gamma^M\nabla_M\chi^a_k, \qquad (3.72)$$

where $(\gamma^M)_{\alpha\beta}$ (a 6-dimensional analog of σ_μ) are real skew-symmetric matrices satisfying

$$\gamma^M\tilde\gamma^N + \gamma^N\tilde\gamma^M = -2\eta^{MN} \qquad (3.73)$$

with

$$(\tilde\gamma^M)^{\alpha\beta} = \frac{1}{2}\varepsilon^{\alpha\beta\gamma\delta}(\gamma^M)_{\gamma\delta}. \qquad (3.74)$$

The 6-dimensional model (3.72) may represent an interest on its own, irrespectively of the fact that its dimensional reduction gives (3.47). Its Lagrangian can also be expressed in terms of superfields.[14]

The presence of the dimensionful coupling constant makes this model not renormalizable. One can also construct a renormalizable $D = 6$ SYM theory involving no dimensional constants [39], but one will need to pay a price: its Lagrangian involves higher derivatives bringing about ghosts...

In a similar way, the $\mathcal{N} = 4$ 4-dimensional supersymmetric YM theory can be obtained by dimensional reduction of a 10-dimensional SYM theory. The Lagrangian (3.53) includes three complex adjoint scalars, which can be traded for six extra components of the vector potential A_M. Four 4-dimensional adjoint fermions λ^a and ψ^a_f can be united in a single 10-dimensional Majorana-Weyl fermion with 16 real components. Further reducing it in $(0+1)$ dimensions gives the system (3.64).

We can understand now the origin of $O(9)$ symmetry which the Hamiltonian (3.64) enjoys. It comes from the reduced spatial coordinates of the 10-dimensional theory. As was also the case for (3.72), the $D = 10$ SYM theory is interesting on its own, being related to the effective Lagrangian for open superstrings [42]. But these fascinating issues are beyond the scope of our book...

3.5.3 $D = 3$: *Supersymmetric Yang-Mills-Chern-Simons theory*

Consider the following 3-dimensional Lagrangian [43]:

$$\mathcal{L} = \frac{1}{g^2}\left[-\frac{1}{4}F^a_{\mu\nu}F^{a\mu\nu} + \frac{i}{2}\lambda^{a\alpha}(\gamma^\mu)_{\alpha\beta}\nabla_\mu\lambda^{a\beta}\right]$$
$$+ \frac{\kappa}{2}\left[\varepsilon^{\mu\nu\rho}\left(A^a_\mu\partial_\nu A^a_\rho + \frac{1}{3}f^{abc}A^a_\mu A^b_\nu A^c_\rho\right) + i\lambda^{a\alpha}\lambda^a_\alpha\right], \qquad (3.75)$$

---

[14]This can be done using the *harmonic superspace* formalism [41].

where $\mu = 0, 1, 2$, $\varepsilon^{012} = 1$, λ^a_α is a two-component real adjoint spinor and $(\gamma^\mu)_{\alpha\beta} = (\mathbb{1}, \sigma^1, \sigma^3)_{\alpha\beta}$ are $3D$ counterparts of the matrices σ^μ in (0.3). Note the identity

$$(\gamma^\mu)_\alpha{}^\beta (\gamma^\nu)_\beta{}^\gamma = -g^{\mu\nu}\delta^\gamma_\alpha - \varepsilon^{\mu\nu\rho}(\gamma_\rho)_\alpha{}^\gamma, \qquad (3.76)$$

where the spinor indices are still raised and lowered as in (0.2).

The coupling constant g^2 carries the dimension of mass in three dimensions, while κ is dimensionless. g is included here in the definition of A so that $F^a_{\mu\nu} = \partial_\mu A^a_\nu - \partial_\nu A^a_\mu + f^{abc} A^b_\mu A^c_\nu$. The canonical dimensions of the fields are $[A] = m$ and $[\lambda] = m^{3/2}$, as in four dimensions.

The Lagrangian (3.75) is supersymmetric, being invariant up to a total derivative under the transformations

$$\delta A^a_\mu = -i\lambda^{a\alpha}(\gamma_\mu)_{\alpha\beta}\epsilon^\beta, \qquad \delta\lambda^a_\alpha = \frac{1}{2}\varepsilon^{\mu\nu\rho}F^a_{\mu\nu}(\gamma_\rho)_{\alpha\beta}\epsilon^\beta, \qquad (3.77)$$

as one can check using the identity (3.76).

The first line in (3.75) is the Lagrangian of $3D$ supersymmetric Yang-Mills theory — it can be obtained by dimensional reduction from (3.40) when only the real part of the gluino field, $\mathrm{Re}[\lambda^{4D}] \to \lambda^{3D}/\sqrt{2}$, is kept (we have only traded σ^2 for σ^3 to make all the gamma matrices real and symmetric), while the second line involves the *Chern-Simons* term $\propto \mathrm{Tr}\int(\hat{A}d\hat{A} - 2i\hat{A}^3/3)$ and, to supersymmetrize it, the term $i\lambda^{a\alpha}\lambda^a_\alpha$ is added. The second line, in contrast to the first one, is odd under the parity transformations. It is especially clear for the Chern-Simons term. When the direction of *one* of the spatial axes is reversed, it changes sign.[15]

The physical boson and fermion degrees of freedom in this theory are massive[16]

$$m = \kappa g^2. \qquad (3.78)$$

In contrast to the gauge theories considered so far, the Lagrangian (3.75) is not quite invariant under gauge transformations: the variation of the Chern-Simons term may bring about a total derivative in the Lagrangian. For local gauge transformations, it does not change the action, but there also exist so-called *large* gauge transformations, which may change the Chern-Simons number,

$$\boxed{N_{CS} = \frac{1}{16\pi^2}\varepsilon^{\mu\nu\rho}\int d^3x \left(A^a_\mu\partial_\nu A^a_\rho + \frac{1}{3}f^{abc}A^a_\mu A^b_\nu A^c_\rho \right),} \qquad (3.79)$$

---

[15] For a nice review of the Chern-Simons dynamics, see Ref. [44].

[16] We are talking now about the excited particle states in the spectrum. The model has also rich vacuum dynamics, which will be discussed in detail in Chapter 10.

by an integer. The action is also modified. For consistency, we must require that the functional integral involving the factor e^{iS} stays invariant and that seems to bring about the quantization condition

$$\kappa = \frac{k}{4\pi} \tag{3.80}$$

with integer k.

However, this reasoning does not take into account the loop effects and the exact statement (we will prove it in Chapter 10) is that $4\pi\kappa$ for the gauge groups $SU(N)$ must be integer when N is even and half-integer when N is odd [45, 46].

Chapter 4

Witten index: generalities

4.1 Definition

The title and the main subject of our book is Witten index. The time has come to *define* what it is and explain the importance of this notion.

Let us recall our discussion of the simplest SQM model, Witten's model, in Chapter 2. The spectrum of the Hamiltonian (2.8) involves two towers of states: the bosonic states $\Psi^B(x, \psi) = a_0(x)$ and fermionic states $\Psi^F(x, \psi) = \psi a_1(x)$. For nonzero energies, these states are all paired in accordance with Theorem 2.2 and are related by the action of the supercharges (2.7) as is shown in Fig. 2.1. But on top of these degenerate doublets, the spectrum may involve zero-energy states, which are annihilated by the action of both $\hat{Q}$ and $\hat{\bar{Q}}$ and are not paired. In Theorem 2.3, we proved that for a polynomial $\mathcal{W}(x)$ with an even highest power, the spectrum involves one zero-energy state, which is bosonic if $\lim_{x \to \pm \infty} \mathcal{W}(x) = \infty$ and fermionic if $\lim_{x \to \pm \infty} \mathcal{W}(x) = -\infty$. If the highest power of $\mathcal{W}(x)$ is odd, the zero-energy states are absent.

This pattern is quite general. Theorem 2.2 dictates that, for any SQM system with discrete spectrum, the excited states are paired into the degenerate doublets, with each doublet including a bosonic and a fermionic state, and there may exist also unpaired zero-energy states, which might be bosonic or fermionic.

If a Hamiltonian enjoys a certain symmetry, but the ground state(s) in its spectrum is (are) not annihilated by the action of the symmetry generators, physicists say that the symmetry is *spontaneously broken*. A familiar physical example is a ferromagnet (of small enough size such that we are dealing with a single domain). Its Hamiltonian has rotational symmetry, but the ground state is characterized by a nonzero magnetization $\boldsymbol{B}$. The vector $\boldsymbol{B}$ and hence the ground state of the Hamiltonian are not invariant under all rotations and the symmetry $O(3)$ is broken spontaneously down to $O(2)$.

Similarly, for an SQM system, we say that supersymmetry is not spontanously broken if the ground state has zero energy and is annihilated by the action of the supercharges and it *is* broken if the ground states have nonzero energy and are not annihilated either by the action of $\hat{Q}$ or by the action of $\hat{\bar{Q}}$.

Definition 4.1. The Witten index is defined as the difference

$$\boxed{I_W \;=\; n_B^{(0)} - n_F^{(0)}\,,} \tag{4.1}$$

where $n_B^{(0)}$ is the number of bosonic zero-energy states and $n_F^{(0)}$ is the number of fermionic zero-energy states.[1]

If $I_W \neq 0$, supersymmetry is not spontaneously broken. If $I_W = 0$, then in most cases both $n_B^{(0)}$ and $n_F^{(0)}$ are equal to zero and supersymmetry is broken. [If not and zero-energy states still exist, there should be a special reason for that — some other index, more sophisticated than the index (4.1) might be different from zero. See p. 120 to get acquainted with a system where such a special reason *exists*.]

The fundamental property of the Witten index is its invariance under smooth deformations of the Hamiltonian. Indeed, the Witten index is an integer, and it cannot change abruptly under a smooth parameter change. One can imagine, e.g., that the energy of the states belonging to a degenerate supersymmetric doublet and having initially nonzero value goes to zero after a certain deformation, but this does not affect the *difference* (4.1).

That is why the notion of the Witten index is interesting and important for physical applications. Suppose that we have a complicated supersymmetric system (not necessarily SQM, it can be a field theory system) and we want to know whether supersymmetry is broken spontaneously there or not, i.e. whether there exist or not the zero-energy states annihilated by the action of the supercharges. To answer this question by explicitly solving the Schrödinger equation of the equations

$$\hat{Q}\Psi = \hat{\bar{Q}}\Psi \;=\; 0 \tag{4.2}$$

might be a very difficult task.

---

[1] For an arbitrary SQM system, an excited state may be called bosonic if it is annihilated by $\hat{\bar{Q}}$, but not by $\hat{Q}$. An excited state may be called fermionic if it is annihilated by $\hat{Q}$, but not by $\hat{\bar{Q}}$. To attribute the labels "bosonic" and "fermionic" also to the zero-energy states, one should introduce the notion of fermion charge $\hat{F}$ or rather the operator $(-1)^{\hat{F}}$ — see the discussions around Eq.(2.10) and after Eq.(4.50).

However, the invariance of I_W allows one to *simplify* the system by a smooth deformation such that an analysis of the simplified system may be already feasible. And the values of I_W for the simplified system and the initial system in interest must be the same. If supersymmetry is not broken for the simplified system, we can also be sure that it is not broken for the initial one.

If the latter is a field theory system, we first put it in a large finite spatial box to regularize it in the infrared and make the spectrum discrete, and then we may send the size of the box to zero so that only zero Fourier modes of the fields are left and the problem is reduced to a SQM problem [47].

4.2 Atiyah-Singer theorem

The mathematical interest and importance of the Witten index consists in its relationship to the Atiyah-Singer theorem and Atiyah-Singer index [2]. Supersymmetry allows one to give a rather transparent reformulation of this theorem and significantly simplify its proof [48, 49].

The AS theorem is a very general statement concerning the properties of certain elliptic differential operators. Some special cases of this statement (e.g. the Riemann-Roch theorem) have been known to mathematicians since the 19th century, but the general statement was first conjectured by Gelfand in 1960 [50]. It was announced as a theorem by Atiyah and Singer in 1963 and the complete proof was published by them in a series of papers in 1968 and 1971 [2].

Let us explain what is this theorem about. To please our reader-mathematician, we will use a moderately fancy mathematical language — the same as in the original papers.

Let X be a compact closed manifold.[2] Let F and G be two *vector bundles* over X, let $\Gamma(F)$ and $\Gamma(G)$ be the spaces of smooth sections of these bundles and let $\hat{D}$ be an elliptic differential operator acting on the sections in $\Gamma(F)$ in such a way that the results of this action belong to $\Gamma(G)$ and the action of $\hat{D}$ on any section in $\Gamma(G)$ gives zero. Correspondingly, the conjugate operator $\hat{D}^\dagger$ acts on the sections in $\Gamma(G)$ to produce the

---

[2]A generalization to manifolds with boundaries also exists [51]. The requirement for the manifold to be compact is also not absolutely necessary. In Chapter 5 and especially in Chapter 9, we will discuss the SQM systems living in flat Euclidean spaces, which are not compact, but the Atiyah-Singer index (or Witten index) can still be defined there. But to make things simpler, we begin by imposing this requirement.

sections in $\Gamma(F)$ and annihilates the latter. Obviously, both $\hat{D}$ and $\hat{D}^\dagger$ are nilpotent.

Definition 4.2. *The* index *of $\hat{D}$ is the difference between the number of the independent solutions of the equation $\hat{D}f = 0$ and the number of the independent solutions of the equation $\hat{D}^\dagger g = 0$ with $f \in \Gamma(F)$ and $g \in \Gamma(G)$:*

In mathematical notation,

$$I \ = \ \dim \ker \hat{D} - \dim \ker \hat{D}^\dagger . \tag{4.3}$$

The AS theorem says that *this index is determined by the topology of the manifold and of its structures.* What particular topology and what particular structures—depends on $\hat{D}$.

We may now observe that the operators $\hat{D}$ and $\hat{D}^\dagger$ in the formulation of the theorem can be interpreted as supercharges $\hat{Q}$ and $\hat{\bar{Q}}$. Indeed, being nilpotent, they satisfy the algebra (2.1), with the operator $\{\hat{D}, \hat{D}^\dagger\}$ playing the role of the Hamiltonian. $\Gamma(F)$ may be associated with the Hilbert space of the bosonic wave functions and $\Gamma(G)$ with the Hilbert space of the fermionic wave functions. Then the index (4.3) is none other than the Witten index (4.1) ! We noticed above that the Witten index stays invariant under smooth deformations of the parameters of the model. But this is tantamout to saying that the index has topological nature.

Table 4.1: AS-SUSY correspondences.

AS theorem	Supersymmetry
$\Gamma(F)$	bosonic states
$\Gamma(G)$	fermionic states
$\hat{D}, \hat{D}^\dagger$	$\hat{Q}, \hat{\bar{Q}}$
$\{\hat{D}, \hat{D}^\dagger\}$	Hamiltonian
AS index	Witten index

Like Mr. Jourdain back in 1670 and like Dr. Landau in 1930, Prof. Atiyah and Prof. Singer spoke in 1963 supersymmetric prose without realizing this!

To make the last statement absolutely clear, we have given a small dictionary of correspondences in Table 4.1 .

4.3 Path integral representation

How to *evaluate* the index? Is there any other way than solving the equations (4.2), as we did in Chapter 2 for the simplest SQM models, and counting their bosonic and fermionic solutions?

The answer is positive. In many cases the Witten index can be found by evaluating the *supersymmetric partition function* given by a proper *path integral.*

4.3.1 *Path integrals in quantum mechanics*

Classical mechanics has two basic ways of description — the Lagrange formalism and the Hamilton formalism. The same concerns quantum mechanics. Historically, it was first formulated in the framework of the quantum Hamiltonian formalism. The quantum counterpart of the Lagrange formalism is the path integral approach developed by Feynman. We will remind here its salient features, addressing the reader for details to the book [52] and many textbooks that have been written afterwards.

Consider the time-dependent Schrödinger equation.

$$\hat{H}\Psi = i\hbar\frac{\partial}{\partial t}\Psi \, . \tag{4.4}$$

To focus on its mathematical structure, we set $\hbar = 1$. To make things even simpler, we assume the presence of only one dynamical real variable $q \in (-\infty, \infty)$.

A formal solution to (4.4) reads

$$\Psi(q, t) = \hat{U}(t - t_{\rm in})\Psi(q, t_{\rm in}) \, , \tag{4.5}$$

where

$$\hat{U}(t - t_{\rm in}) = \exp\left\{-i(t - t_{\rm in})\hat{H}\right\} \tag{4.6}$$

is the *evolution operator.* We now consider the *kernel* of the evolution operator, that is, the matrix element

$$\mathcal{K}(q_{\rm f}, q_{\rm in}; t_{\rm f} - t_{\rm in}) = \langle q_{\rm f}|\hat{U}(t_{\rm f} - t_{\rm in})|q_{\rm in}\rangle \equiv \langle q_{\rm f}, t_{\rm f}|q_{\rm in}, t_{\rm in}\rangle \tag{4.7}$$

$(t_{\rm f} - t_{\rm in} \equiv \Delta t > 0)$. It describes the probability amplitude that the system will find itself at the position $q_{\rm f}$ at $t = t_{\rm f}$ provided it was located at $q = q_{\rm in}$

at $t = t_{\text{in}}$. In terms of (4.7), the solution (4.5) is expressed as

$$\Psi(q_{\text{f}}, t_{\text{f}}) = \int dq_{\text{in}} \, \mathcal{K}(q_{\text{f}}, q_{\text{in}}; t_{\text{f}} - t_{\text{in}}) \, \Psi(q_{\text{in}}, t_{\text{in}}) . \tag{4.8}$$

The function $\mathcal{K}$ is a *fundamental solution* to the Schrödinger equation. And it plays a fundamental role in the whole approach. The following *spectral decomposition* holds:

$$\boxed{\mathcal{K}(q_{\text{f}}, q_{\text{in}}; \Delta t) = \sum_k \Psi_k(q_{\text{f}}) \overline{\Psi_k(q_{\text{in}})} e^{-iE_k \Delta t}} , \tag{4.9}$$

where $\Psi_k(q)$ are the eigenstates of $\hat{H}$ with eigenvalues E_k. (It follows from (4.8), from orthogonality, $\langle k|l \rangle = \delta_{kl}$, and from the fact that the time evolution of an eigenfunction $\Psi_k(q, t)$ amounts to a multiplication by the phase factor $e^{-iE_k \Delta t}$.)

For simple systems, the kernel can be found explicitly. For a free particle of unit mass moving along the line,

$$\mathcal{K}(q_f, q_{\text{in}}; \Delta t) = \frac{1}{\sqrt{2\pi i \Delta t}} \exp \left\{ \frac{i(q_f - q_{\text{in}})^2}{2\Delta t} \right\} . \tag{4.10}$$

An exact analytical expression for the kernel can also be derived for the Hamiltonian of the harmonic oscillator [52]. But how do we solve the problem in the general case?

A crucial observation is that $\mathcal{K}$ satisfies the completeness relation

$$\langle q_{\text{f}}, t_{\text{f}} | q_{\text{in}}, t_{\text{in}} \rangle = \int_{-\infty}^{\infty} dq_* \, \langle q_{\text{f}}, t_{\text{f}} \, | \, q_*, t_* \rangle \langle q_*, t_* \, | \, q_{\text{in}}, t_{\text{in}} \rangle , \tag{4.11}$$

where t_* is any time moment between t_{in} and t_{f}. The relation (4.11) has a transparent physical meaning: the probability amplitude for the particle to go from the point q_{in} at the initial moment t_{in} to the point q_{f} at the final moment t_{f} is given by the convolution of the probability amplitudes to go first to an intermediate point q_* at t_* and then to the final point.

We now divide the time interval Δt in a large number n of equal tiny time slices and write

$$\langle q_{\text{f}}, t_{\text{f}} | q_{\text{in}}, t_{\text{in}} \rangle =$$
$$\left(\prod_{j=1}^{n-1} \int_{-\infty}^{\infty} dq_j \right) \langle q_{\text{f}}, t_{\text{f}} | q_{n-1}, t_{\text{f}} - \epsilon \rangle \cdots \langle q_1, t_{\text{in}} + \epsilon | q_{\text{in}}, t_{\text{in}} \rangle \tag{4.12}$$

with $\epsilon = \Delta t / n$.

The integrand involves the product of a large number of factors. To evaluate each such factor, an *infinitesimal* evolution kernel, we use (4.6) and (4.7). For example,

$$\langle q_1, t_{\text{in}} + \epsilon | q_{\text{in}}, t_{\text{in}} \rangle = \langle q_1 | \exp\{-i\epsilon\hat{H}\} | q_{\text{in}} \rangle =$$

$$= \int \frac{dp}{2\pi} \langle q_1 | p \rangle \langle p | e^{-i\epsilon\hat{H}} | q_{\text{in}} \rangle = \int \frac{dp}{2\pi} e^{ipq_1} \langle p | e^{-i\epsilon\hat{H}} | q_{\text{in}} \rangle, \quad (4.13)$$

where we used the completeness relation again, inserting the plane wave states $|p\rangle$ that form a complete Hilbert space basis. The factor $1/2\pi$ in the measure provides the correct normalization:

$$\langle q_1 | q_{\text{in}} \rangle = \int \frac{dp}{2\pi} e^{ip(q_1 - q_{\text{in}})} = \delta(q_1 - q_{\text{in}}). \quad (4.14)$$

Suppose that the Hamiltonian has a simple form

$$\hat{H} = \frac{\hat{p}^2}{2} + V(q). \quad (4.15)$$

We may then write for small ϵ

$$\langle p | e^{-i\epsilon\hat{H}} | q_{\text{in}} \rangle \approx e^{-i\epsilon H(p,q_{\text{in}})} \langle p | q_{\text{in}} \rangle = e^{-ipq_{\text{in}} - i\epsilon H(p,q_{\text{in}})}, \quad (4.16)$$

where $H(p,q)$ is the classical Hamiltonian. Plugging it in (4.13), we derive

$$\langle q_1, t_{\text{in}} + \epsilon | q_{\text{in}}, t_{\text{in}} \rangle \approx \int \frac{dp}{2\pi} \exp\{ip(q_1 - q_{\text{in}}) - iH(p,q_{\text{in}})\epsilon\}. \quad (4.17)$$

For a generic Hamiltonian mixing the coordinates and momenta, the situation is more complicated. As we discussed in Chapter 2, the order in which $\hat{p}$ and q enter $\hat{H}$ matters. This ordering ambiguity matches the ambiguity in the path integral definition. Generically, it matters whether we write $H(p, q_{\text{in}})$, $H(p, q_1)$ or maybe $H\left(p, \frac{q_1 + q_{\text{in}}}{2}\right)$ in the integrand in Eq. (4.17). For example, the relation (4.16) holds for $\hat{H} = \hat{p}^2 q^2$, but not for $\hat{H} = q^2 \hat{p}^2$. We direct the reader to Ref. [53] where this question is thouroughly studied and pedagogically explained. We will see later, however, that this ordering uncertainty is irrelevant for the final result we are interested in — the integral representation of the Witten index.

Doing the analogous transformation in all other factors in (4.12) and sending n to infinity, we obtain the evolution kernel at finite time interval in the following form:

$$\mathcal{K}(q_{\text{f}}, q_{\text{in}}; t_{\text{f}} - t_{\text{in}}) = \lim_{n \to \infty} \int \exp\{ip_n(q_{\text{f}} - q_{n-1}) + \ldots + ip_1(q_1 - q_{\text{in}}) -$$

$$i\epsilon[H(p_n, q_{n-1}) + \ldots + H(p_1, q_{\text{in}})]\} \frac{dp_n}{2\pi} \frac{dp_{n-1}dq_{n-1}}{2\pi} \cdots \frac{dp_1 dq_1}{2\pi} \quad (4.18)$$

with $\epsilon = (t_f - t_i)/n$. In the limit $n \to \infty$, we obtain an integral over a *continuous* number of variables, which can be formally written as

$$\mathcal{K}(q_f, q_{\text{in}}; t_f - t_{\text{in}}) = \int \exp\left\{ i \int_{t_{\text{in}}}^{t_f} [p\dot{q} - H(p, q)]dt \right\} \prod_t \frac{dp(t)dq(t)}{2\pi} \quad (4.19)$$

with the boundary conditions

$$q(t_{\text{in}}) = q_{\text{in}}, \qquad q(t_f) = q_f, \quad (4.20)$$

while the function $p(t)$ is quite arbitrary.

We may also integrate over momenta[3] and express the evolution kernel in the form

$$\mathcal{K}(q_f, q_{\text{in}}; t_f - t_{\text{in}}) = \mathcal{N} \int \prod_t dq(t) \exp\left\{ i \int_{t_{\text{in}}}^{t_f} L[\dot{q}(t), q(t)]dt \right\}, \quad (4.21)$$

where $L[\dot{q}(t), q(t)]$ is the Lagrangian of the system. An infinite normalization factor $\mathcal{N}$ plays no role in practical applications.

In other words, to calculate the evolution kernel, one has to resum the contributions of all possible paths connecting the points q_{in} at $t = t_{\text{in}}$ and q_f at $t = t_f$ with the weight e^{iS}, where S is the classical action on this path. The classical limit corresponds to the case where all such actions are large (in the units of $\hbar$ that we do not display). Then the main contributions come from the paths close to the classical path where the action has an extremum.

The path integral has now been used in physics for three quarters of a century. But not in mathematics — mathematicians still do not like it and consider it not to be a rigorously defined object. To be more precise, they have nothing against the *Euclidean* path integral where the time is considered to be imaginary, $t = -i\tau$. In that case, the function to be integrated in (4.21) goes over to e^{-S_E}, where S_E is the positive definite Euclidean action. For example, when the Hamiltonian has a simple form (4.15) with positive definite $V(q)$, the Euclidean action on the interval $0 \leq \tau \leq \beta$ is

$$S_E = \int_0^{\beta} d\tau \left[\frac{1}{2}\left(\frac{dq}{d\tau}\right)^2 + V(q) \right] > 0.$$

The Euclidean evolution operator is then defined as

$$\mathcal{K}(q_f, q_{\text{in}}; -i\beta) \propto \int \prod_{\tau} dq(\tau)e^{-S_E} \quad (4.22)$$

with the boundary conditions $q(0) = q_{\text{in}}, q(\beta) = q_f$.

---

[3]This amounts to Legendre transformation and is especially simple for the Hamiltonian (4.15) where the integral is Gaussian.

For this integral, the contribution of the trajectories with wildly deviating $q(\tau)$ is exponentially suppressed. Mathematicians call such integrals the integrals with *Wiener measure* and can work with them. A good news for a reader-mathematician is that we will also only work with Euclidean path integrals in this book!

Another name for the Euclidean evolution operator $\mathcal{K}(q_f, q_{in}; -i\beta)$ is the *heat kernel*. It is called so because it represents the fundamental solution of the *diffusion* equation or *heat* equation in the same way as $\mathcal{K}(q_f, q_{in}; \Delta t)$ represents the fundamental solution of the Schrödinger equation.

The object of principal interest for us in this book is the *partition function* and especially its supersymmetric generalization, to be discussed soon. In statistical physics, the partition function is defined as the sum

$$Z(\beta) = \sum_k e^{-E_k/T}, \qquad (4.23)$$

where T is the temperature of the system. The point is [54] that this quantity can be represented as the integral of the Euclidean evolution operator,

$$Z = \int dq \, \mathcal{K}(q, q; -i\beta) \qquad (4.24)$$

with $\beta = 1/T$. [The identity (4.24) follows immediately from the spectral decomposition (4.9) with Δt replaced by $-i\beta$.] In other words, the partition function is expressed as the integral,

$$Z \propto \int \prod_\tau dq(\tau) e^{-S_E}, \qquad (4.25)$$

over all Euclidean trajectories $q(\tau)$ satisfying the *periodic* boundary conditions,

$$q(\beta) = q(0). \qquad (4.26)$$

All the formulas above were written for a QM system with one single degree of freedom q. But the whole analysis can be easily generalized to the systems with many degrees of freedom and to quantum field theories with a continuous number of degrees of freedom.

4.3.2 *Grassmann evolution kernel*

We now go over to the systems including Grassmann dynamical variables. As a toy model, consider first the Grassmann oscillator discussed earlier on p. 10 in Chapter 1. It is described by a single holomorphic Grassmann variable $\xi \to \psi$. The classical Hamiltonian is

$$H = \omega \psi \bar{\psi}. \qquad (4.27)$$

The corresponding Lagrangian related to (4.27) by the Legendre transformation can be chosen as

$$L = -i\dot{\psi}\bar{\psi} + \omega\bar{\psi}\psi. \tag{4.28}$$

The quantum Hamiltonian reads

$$\hat{H} = \omega\psi\frac{\partial}{\partial\psi}. \tag{4.29}$$

The Hilbert space includes only two states: the state $|1\rangle$ with $\Psi(\psi) = 1$ and the state $|\psi\rangle$ with $\Psi(\psi) = \psi$. To proceed, we need to define the inner product in this space. We should have $\langle 1|1\rangle = \langle\psi|\psi\rangle = 1$ and $\langle 1|\psi\rangle = \langle\psi|1\rangle = 0$. This is enforced by the following definition:[4]

$$\langle f|g\rangle = \int d\psi d\bar{\psi}\, e^{-\psi\bar{\psi}}\, \overline{f(\psi)}g(\psi). \tag{4.30}$$

For the systems including many Grassmann dynamical variables ψ_α, the fermion measure is

$$d\mu = \prod_\alpha d\psi_\alpha d\bar{\psi}_\alpha\, e^{-\psi_\alpha\bar{\psi}_\alpha}. \tag{4.31}$$

The states $|1\rangle$ and $|\psi\rangle$ are the eigenstates of the Hamiltonian (4.29) : $\hat{H}|1\rangle = 0$ and $\hat{H}|\psi\rangle = \omega|\psi\rangle$. Now the kernel of the evolution operator, the Grassmann analog of (4.7), has the form

$$\langle f|e^{-i\hat{H}\Delta t}|i\rangle = \mathcal{K}(\psi_f, \bar{\psi}_{in}; \Delta t). \tag{4.32}$$

Then the wave function at a later moment t_f is given by the convolution

$$\Psi(\psi_f, t_f) = \int d\psi_{in}d\bar{\psi}_{in}\, e^{-\psi_{in}\bar{\psi}_{in}}\, \mathcal{K}(\psi_f, \bar{\psi}_{in}; \Delta t)\, \Psi(\psi_{in}, t_{in}). \tag{4.33}$$

It involves the integral over the initial phase space with the measure prescribed in (4.30).

For the oscillator or any quantum problem where the spectrum is explicitly known, an explicit expression for the evolution kernel can be written. This is given by the spectral decomposition that has the form

$$\mathcal{K}(\psi_f, \bar{\psi}_{in}; \Delta t) = \sum_k \Psi_k(\psi_f)\overline{\Psi_k(\psi_{in})}e^{-iE_k\Delta t}. \tag{4.34}$$

The Euclidean evolution operator reads

$$\mathcal{K}(\psi_f, \bar{\psi}_{in}; -i\beta) = 1 + \psi_f\bar{\psi}_{in}e^{-\omega\beta}. \tag{4.35}$$

---

[4] As we see, the measure includes now the factor $e^{-\psi\bar{\psi}}$. Its appearance is not so surprising—an exponential weight e^{-aa^*} in the measure arises also for the ordinary oscillator in the holomorphic representation when the phase space variables q, p are traded for a, a^* written in Eq. (1.37).

Consider a natural generalization of (4.24):

$$\tilde{Z}(\beta) \;=\; \int d\psi d\bar{\psi}\, e^{-\psi\bar{\psi}}\, \mathcal{K}(\psi, \bar{\psi}; -i\beta)\,, \tag{4.36}$$

where the measure (4.30) was used again. Doing the integral, we obtain

$$\tilde{Z}(\beta) \;=\; 1 - e^{-\beta\omega}\,. \tag{4.37}$$

This resembles the partition function (4.23), but the second term, the contribution of the excited oscillator state, enters with the negative sign!

Remember this remarkable fact and look again at the kernel (4.35). For $\beta = 0$, it is reduced to $\mathcal{K}(\psi_f, \bar{\psi}_{\rm in}; 0) = 1 + \psi_f\bar{\psi}_{\rm in} = e^{\psi_f\bar{\psi}_{\rm in}}$. For small $\beta \ll \omega^{-1}$, the kernel can be written as

$$\mathcal{K}(\psi_{\rm f}, \bar{\psi}_{\rm in}; -i\beta) \;\approx\; e^{\psi_{\rm f}\bar{\psi}_{\rm in} - \beta H(\psi_{\rm f}, \bar{\psi}_{\rm in})}\,, \tag{4.38}$$

where $H(\psi, \bar{\psi})$ is the classical Hamiltonian (4.27).

The sign $\approx$ in Eq. (4.38) means the following. For small β, we keep in $\mathcal{K}(\psi_{\rm f}, \bar{\psi}_{\rm in}; -i\beta)$ the terms of order $\sim \beta\psi\bar{\psi}$, but neglect the terms of order $\sim \beta^{n \geq 2}\psi\bar{\psi}$ (and would neglect the terms of order $\sim \beta^{n \geq 1}$ not involving the factor $\psi\bar{\psi}$, were such terms present). The meaning of this selective neglection will be clarified somewhat later.

Representing the kernel (4.32) as a convolution of a large number of infinitesimal kernels, we may derive, in analogy with (4.22), the representation

$$\tilde{Z}(\beta) \;\propto\; \int \prod_{\tau} d\psi(\tau)d\bar{\psi}(\tau) \exp\left\{ -\int_0^\beta L_E[\psi(\tau), \bar{\psi}(\tau)]d\tau \right\}, \tag{4.39}$$

with $\psi(\tau)$ satisfying the periodic boundary conditions $\psi(\beta) = \psi(0)$.

This analysis can be generalized for any quantum system involving an arbitrary number of bosonic q_j and fermionic ψ_α degrees of freedom. The Euclidean kernel $\mathcal{K}(q_j, \psi_\alpha; q_j, \bar{\psi}_\alpha; -i\beta)$ allows a spectral decomposition

$$\mathcal{K}(q_j, \psi_\alpha; q_j, \bar{\psi}_\alpha; -i\beta) \;=\; \sum_k \Psi_k(q_j, \psi_\alpha)\overline{\Psi_k(q_j, \psi_\alpha)}e^{-\beta E_k}\,. \tag{4.40}$$

Consider the quantity

$$\tilde{Z}(\beta) \;=\; \int \prod_j dq_j \prod_\alpha d\psi_\alpha d\bar{\psi}_\alpha\, e^{-\psi_\alpha\bar{\psi}_\alpha}\, \mathcal{K}(q_j, \psi_\alpha; q_j, \bar{\psi}_\alpha; -i\beta)\,. \tag{4.41}$$

Substituting (4.40) for $\mathcal{K}$, we derive

$$\boxed{\tilde{Z}(\beta) \;=\; \sum_k \eta_k e^{-\beta E_k}\,,} \tag{4.42}$$

where $\eta_k = 1$ for the bosonic states with Grassmann-even wave functions Ψ_k and $\eta_k = -1$ for the fermionic states with Grassmann-odd wave functions.

Representing the evolution kernel as a convolution of many infinitesimal kernels, one can derive in analogy with (4.25) the representation

$$\tilde{Z}(\beta) \propto \int \prod_{j\tau} dq_j(\tau) \prod_{\alpha\tau} d\psi_\alpha(\tau) d\bar{\psi}_\alpha(\tau)$$

$$\exp\left\{-\int_0^\beta L_E[q_j(\tau), \psi_\alpha(\tau), \bar{\psi}_\alpha(\tau)] d\tau\right\}, \qquad (4.43)$$

where both the ordinary and Grassmann variables satisfy the periodic boundary conditions:

$$q_j(\beta) = q_j(0); \quad \psi_\alpha(\beta) = \psi_\alpha(0); \quad \bar{\psi}_\alpha(\beta) = \bar{\psi}_\alpha(0). \qquad (4.44)$$

One can ask whether it is possible to represent in a similar way the *conventional* partition function where all the states, both bosonic and fermionic, contribute in the sum with the positive sign. Yes, one can. One just has to write

$$Z(\beta) = \int \prod_j dq_j \prod_\alpha d\psi_\alpha d\bar{\psi}_\alpha\, e^{-\psi_\alpha\bar{\psi}_\alpha}\, \mathcal{K}(q_j, \psi_\alpha; q_j, -\bar{\psi}_\alpha; -i\beta). \ (4.45)$$

The minus before the second Grassmann argument in $\mathcal{K}$ brings about an extra minus in the spectral sum for Grassmann-odd wave functions, which compensates the factor $\eta_k = -1$ in (4.42). The partition function (4.45) may also be represented as a path integral (4.43) with the periodic boundary conditions imposed on the bosonic variables and *antiperiodic* boundary conditions imposed on the fermionic variables.

The conventional partition function is interesting for a physicist when s/he studies thermal properties of the system. For a supersymmetric system at finite temperature, one observes that the symmetry between the bosons and fermions, i.e. supersymmetry, is broken due to different boundary conditions in the Matsubara imaginary time formalism.[5]

4.3.3   *The index*

But in our book we are not going to break supersymmetry in such a brutal way. We are interested not in the conventional partition function, but in

---

[5]This is also seen in the operator real time approach, where one can observe that the breaking is not explicit but *spontaneous*, bringing about a massless goldstino mode in the spectrum of collective excitations, a supersymmetric analog of the sound [55].

the graded sum (4.42) for supersymmetric dynamic systems. This graded sum, called the *supersymmetric partition function*, is nothing but the Witten index (4.1). Indeed, only the zero-energy states contribute in (4.42); the contribution of excited states cancels out due to the boson-fermion degeneracy in the spectrum, a distinguishing feature of supersymmetric systems.

We can now capitalize on the fact that $\tilde{Z}(\beta)$, given by the sum (4.42), does not depend on β. Then one may calculate it using the integral representation (4.41) for any β, and it is convenient to do so when β is small. In this case, we can substitute for $\mathcal{K}$ in the integrand its approximate infinitesimal expression. This expression reads[6]

$$\mathcal{K}_{\text{inf}}(q_j, \psi_\alpha; q_j, \bar{\psi}_\alpha; -i\beta) \;\approx\; e^{\psi_\alpha \bar{\psi}_\alpha} \int \prod_j \frac{dp_j}{2\pi} e^{-\beta H(p_j, q_j; \psi_\alpha, \bar{\psi}_\alpha)}, \quad (4.46)$$

where $H(p_j, q_j; \psi_\alpha, \bar{\psi}_\alpha)$ is the classical Hamiltonian. Calculating the integral in Eq. (4.41), we obtain [56]

$$\boxed{I_W \;=\; \lim_{\beta \to 0} \int \prod_j \frac{dp_j dq_j}{2\pi} \prod_\alpha d\psi_\alpha d\bar{\psi}_\alpha \, e^{-\beta H(p_j, q_j; \bar{\psi}_\alpha, \psi_\alpha)}.} \quad (4.47)$$

This is an *ordinary* integral, and it can be evaluated analytically for many SQM systems.

Alternatively, this result may be derived in the Lagrangian language from the representation (4.43) of the supersymmetric partition function. If β is small and the boundary conditions are periodic, the path integral is saturated for most systems by *constant* field configurations (only the constant modes in the Fourier expansions of $q_j(\tau)$, $\psi_\alpha(\tau)$ and $\bar{\psi}_\alpha(\tau)$ should be taken into consideration). An ordinary integral thus obtained coincides with what one obtains doing momenta integrations in (4.47).

One can also represent the result (4.47) in an alternative form when the Grassmann dynamical variables are not introduced, but the supersymmetric Hamiltonian has matrix nature. For example, for the Witten model, we may write

$$\hat{H} \;=\; \frac{P^2 + [\mathcal{W}'(x)]^2}{2} + \frac{1}{2}\mathcal{W}''(x)\sigma^3. \quad (4.48)$$

Here the hat over H only signifies that $\hat{H}$ is a matrix, but it is not the full-scale quantum Hamiltonian representing a differential operator. The momentum P is a classical phase space variable.

---

[6]This is a combination of the purely bosonic formula (4.17) with $\epsilon \to -i\beta$ (it multidimensional generalization) and the purely fermionic formula (4.38).

In this case, the Euclidean evolution kernel also has matrix nature. The functional trace of this kernel including the ordinary matrix trace and the integration over coordinates, as in (4.45), gives us the partition function of the system:

$$Z(\beta) = \int \prod_j dq_j \, \text{Tr}\{\hat{\mathcal{K}}(q_j; q_j; -i\beta)\} \,. \tag{4.49}$$

To find the supersymmetric partition function, we have to recall that the spectrum of a supersymmetric Hamiltonian includes two towers of degenerate states, as in Fig. 2.2, bosonic and fermionic states. We introduce then the operator $(-1)^{\hat{F}}$ which commutes with the Hamiltonian and whose eigenvalues are $+1$ for the bosonic states and -1 for the fermionic states.

To understand the commonly used notation $(-1)^{\hat{F}}$ for this operator, one should go back to the Grassmann representation. Then $\hat{F}$ is the operator of the fermion charge,

$$\hat{F} = \psi_j \frac{\partial}{\partial \psi_j} \,. \tag{4.50}$$

It simply counts the number of the Grassmann factors in the wave function. The operator $\hat{F}$ may commute with the Hamiltonian, and in most cases it does, but this is not *always* the case. In Chapter 9, we will meet a system where it does not.

The property $[(-1)^{\hat{F}}, \hat{H}]$ always holds, however. Indeed, the supercharge operators always include an odd number of Grassmann factors ψ or $\hat{\psi}$ and they transform the states with even F to the states with odd F and vice versa. Thus, in the Grassmann language, the two towers of states mentioned above are the states of even and odd fermionic charge. It is natural to call "bosonic" the states with even F and "fermionic" the states with odd F.

The supersymmetric partition function can thus be represented as

$$\tilde{Z}(\beta) = \int \prod_j dq_j \, \text{Tr}\{(-1)^{\hat{F}} \hat{\mathcal{K}}(q_j; q_j; -i\beta)\} \,. \tag{4.51}$$

The formula (4.47) acquires the form

$$I_W = \lim_{\beta \to 0} \int \prod_j \frac{dp_j dq_j}{2\pi} \, \text{Tr}\left\{(-1)^{\hat{F}} e^{-\beta \hat{H}(p_j, q_j)}\right\} \,. \tag{4.52}$$

As with any method, the Cecotti-Girardello (CG) method has its limits of applicability.

- To begin, the very notion of the Witten index, as defined above, applies only to the systems with discrete spectrum. If the spectrum is continuous and the states are not normalizable, most methods of the functional analysis do not work. And even a physicist who is brave enough to consider systems with continuous spectrum (like the free motion along a line), always has in mind a finite-volume regularization making the spectrum discrete. For the systems involving bands of continuous spectrum not separated from zero by a gap (the continuous spectrum associated with the motion of massive excitations in field theory does not represent a problem), the integrals (4.47) often have fractional values, which does not have an immediate sense.[7]

- Even for the systems with discrete spectrum, in some cases, the integral (4.47) gives wrong fractional numbers. The reason is that the approximation (4.46) does not work in *all* the cases.

  First of all, it concerns certain SQM models living on curved manifolds that will be treated in Chapter 5. The curvature effects invalidate in these cases the simple CG procedure when the higher harmonics in the Fourier expansion of $q_j(t)$ and $\psi_\alpha(t)$ are ignored. Instead, one should take them into account and carefully integrate over them in the Gaussian approximation. This gives a correct answer.

  The second class of problematic models are chiral supersymmetric 4-dimensional gauge theories considered in Chapter 8. To make the spectrum discrete, we put them in a finite box, but then we see that the effective Hamiltonian has dangerous singularities in the corners of the box, and these singularities may invalidate the CG procedure. In fact, they do so for non-Abelian chiral theories where the the integral (4.47) has a fractional value. However, it still gives the correct result for the index in chiral SQED and in a certain toy SQM model involving such singularities (see Chapter 8 for more details).

4.3.4 *Examples*

We will now show how this general machinery works for the simple SQM models discussed in Chapter 2. Consider first the Witten model with the

---

[7]In Chapter 9, we will consider, however, a certain rather complicated supersymmetric *gauge* SQM models where the spectrum is continuous and this gives a fractional contribution to the integral (4.47), but on top of that, ground zero-energy normalizable states exist and Witten index can still be defined [57–61].

classical Hamiltonian

$$H = \frac{P^2 + [\mathcal{W}'(x)]^2}{2} + \mathcal{W}''(x)\psi\bar\psi. \tag{4.53}$$

Substitute this into (4.47). Integrating over $d\psi d\bar\psi$ and over momenta, we derive

$$\boxed{I_W = \lim_{\beta\to 0}\sqrt{\frac{\beta}{2\pi}}\int_{-\infty}^{\infty}dx\,\mathcal{W}''(x)\exp\left\{-\frac{[\mathcal{W}'(x)]^2}{2}\right\}.} \tag{4.54}$$

Now it has become clear why we had to keep $\propto \beta\psi\bar\psi$ in (4.38) and (4.39). Otherwise the fermion integral would vanish and the result for the index would be wrong.

Changing the variable $y = \mathcal{W}'(x)$, we obtain

$$I_W = \lim_{\beta\to 0}\sqrt{\frac{\beta}{2\pi}}\int_{\mathcal{W}'(-\infty)}^{\mathcal{W}'(\infty)}dy\,\exp\left\{-\frac{\beta y^2}{2}\right\}. \tag{4.55}$$

We are mostly interested in the systems where the potential $V(x) = [\mathcal{W}'(x)]^2/2$ grows at infinity. Otherwise, the spectrum of the Hamiltonian would not be discrete and the notion of the index would not be easily defined.[8] Let $\mathcal{W}(x)$ be a polynomial. There are three possibilities.

(1) The highest power in $\mathcal{W}(x)$ is odd. Then $\mathcal{W}'(\infty) = \mathcal{W}'(-\infty)$, the limits in the integral (4.55) coincide and $I_W = 0$. This signalizes the spontaneous supersymmetry breaking, which agrees with the explicit analysis done in Chapter 2.

(2) The highest power in $\mathcal{W}(x)$ is even and the coefficient in front is positive. Then $\mathcal{W}'(\pm\infty) = \pm\infty$ and the integral in (4.55) does not depend

---

[8]Consider as an example the *superconformal* quantum mechanics [62] with $\mathcal{W}(x) = \ln x$ and the Hamiltonian

$$H = \frac{P^2}{2} + \frac{1}{2x^2} - \frac{\psi\bar\psi}{x^2}. \tag{4.56}$$

The spectral problem is defined on a half-line $x \in (0, \infty)$. There is a band of continuous spectrum starting from $E = 0$. And the integral (4.54) for the index reduces in this case to

$$I_W = \lim_{\beta\to 0}\sqrt{\frac{\beta}{2\pi}}\int_{0}^{\infty}dx\,\exp\left\{-\frac{\beta}{2x^2}\right\} = \frac{1}{2}. \tag{4.57}$$

In Chapter 9, we will show how one can still deal with such systems and define the index that takes into account only *normalizable* zero-energy states. For the system (4.56), such states are absent.

on β, giving $I_W = 1$ — there is one bosonic zero-energy eigenstate in the spectrum[9] in agreement with the explicit analysis in Sect. 2.1.

(3) The higher power in $\mathcal{W}(x)$ is even and the coefficient in front is negative. Then $\mathcal{W}'(\pm\infty) = \mp\infty$ and the integral (4.55) gives $I_W = -1$ and there is one fermionic zero-energy eigenstate in the spectrum. Again, this agrees with Sect. 2.1.

Consider now the generalized Landau problem, with the magnetic field having the form

$$\boldsymbol{B} = [0, 0, B(x, y)]. \tag{4.58}$$

In the Grassmann variables description, the classical Hamiltonian reads

$$H = \frac{1}{2}(P_x + A_x)^2 + \frac{1}{2}(P_y + A_y)^2 + B(x, y)\psi\bar{\psi}, \tag{4.59}$$

where $A_{x,y}$ are the components of the vector potential. We have

$$I_W = \lim_{\beta \to 0} \int \frac{dP_x \, dx}{2\pi} \frac{dP_y \, dy}{2\pi} d\psi d\bar{\psi}$$

$$\exp\left\{-\frac{\beta}{2}[(P_x + A_x)^2 + (P_y + A_y)^2] - \beta B(x, y)\psi\bar{\psi}\right\}. \tag{4.60}$$

Doing the integral over $dP_x dP_y d\psi d\bar{\psi}$, we obtain

$$\boxed{I_W = \frac{1}{2\pi} \int dx dy \, B(x, y) = \frac{\Phi}{2\pi},} \tag{4.61}$$

where Φ is the magnetic flux.

By definition, the index I_W is an integer. It follows that the flux Φ should be quantized, $\Phi = 2\pi n$. The sign of the index coincides with the sign of the flux, which aligns with the analysis of Sect. 2.2: if $B > 0$, the zero-energy states are spin-down in the spinorial language and bosonic in the Grassmannian language.

In principle, one can also consider the quantum systems with a fractional magnetic flux [63]. In contrast to what is usually assumed, in this case one can define a spectral problem, but the Hamiltonian is not supersymmetric anymore and is hence not interesting for us. In the following, we will

---

[9]Strictly speaking, we can only say that $n_B^{E=0} - n_F^{E=0} = 1$, which does not exclude the possibility for the system to have e.g. two bosonic and one fermionic vacua. But one can make here the same comment that we did on p. 66 when discussing the case $I_W = 0$ for a generic system: there should be a special reason for the existence of extra vacuum states not dictated by the index considerations. For the Witten model, there is no such special reason.

assume, when talking about the gauge fields, that the magnetic flux (2.27) as well as its generalizations for more complicated systems *are* quantized.

The reader might have noticed that, in the examples considered above, the integral did not depend on β and taking the limit $\beta \to 0$ was superfluous. Actually, this is due to the clever choice of the classical Hamiltonians (4.53) and (4.59). These Hamiltonians coincide with the Poisson brackets of the classical supercharges (2.33) and (2.7) (with removed hats), so that the classical supersymmetry algebra (2.29) holds. The latter condition fixes the ordering ambiguity, which is in principle there even for the simple systems: we could e.g. choose as a classical counterpart of the operator $(\psi\hat{\bar{\psi}} - \hat{\bar{\psi}}\psi)/2$ entering the quantum Hamiltonian (2.8) not just $\psi\bar{\psi}$, but add to this expression an arbirtrary constant. Then the integral in (4.47) would depend on β, but it would still give the correct value of the Witten index in the limit $\beta \to 0$.

In the Witten and Landau models, an ordering ambiguity is seen for the Hamiltonian, but not for the supercharges. In more complicated systems to be considered in the next chapter, such ambiguity also manifests in the supercharges, but, as was mentioned in Chapter 2, we always resolve it choosing classical supercharges as the Weyl symbols of the quantum ones. At this point, we can formulate a **conjecture**:

Let $\hat{Q}, \hat{\bar{Q}}$ and $\hat{H}$ be quantum operators satisfying the supersymmetry algebra (2.1). Let the spectrum of $\hat{H}$ be discrete. Let Q and $\bar{Q}$ be the Weyl symbols of $\hat{Q}$ and $\hat{\bar{Q}}$ and let the Poisson bracket $\{Q, Q\}_P$ and hence $\{\bar{Q}, \bar{Q}\}_P$ vanish.[10] Define the classical Hamiltonian as $H = -\frac{i}{2}\{Q, \bar{Q}\}_P$. Then the integral in (4.47) does not depend on β.

We do not know a general proof of this statement, but it holds in all the SQM systems that we know.

---

[10]This reservation is necessary to exclude the systems suffering from the classical supersymmetry anomaly, see p. 27.

Chapter 5

Classical complexes and their SQM description

In Chapters 2 and 4, we discussed the simplest SQM models and calculated the Witten indices there. Now we proceed with the analysis of more complicated models. In this chapter, we will discuss supersymmetric *sigma models* and will show how their analysis with the tools displayed in the preceding chapters allows one to derive in a rather transparent way many classical results of differential geometry.

Definition 5.1. An ordinary mechanical sigma model is a dynamical system including D dynamical variables $x^M(t)$ with the Lagrangian

$$L = \frac{1}{2} g_{MN}(x)\, \dot{x}^M \dot{x}^N \,. \tag{5.1}$$

Here x^M has the meaning of the coordinates of a D-dimensional manifold and $g_{MN}(x)$ is its metric tensor. The equations of motion derived from (5.1) read

$$\frac{d^2 x^M}{dt^2} = -\Gamma^M_{NP} \frac{dx^N}{dt} \frac{dx^P}{dt} \,, \tag{5.2}$$

where

$$\Gamma^M_{NP} = \frac{1}{2} g^{MQ} \left(\partial_N g_{PQ} + \partial_P g_{NQ} - \partial_Q g_{NP} \right) \tag{5.3}$$

are the Christoffel symbols. The equations (5.2) describe the motion along the geodesic lines.

Supersymmetric sigma models include, besides $x^M(t)$, Grassmann dynamical variables. There are many different models of this type [4], but in this chapter we will only be interested with two basic models: the model involving the real superfields $X^M(t, \theta, \bar{\theta})$, which describes the *de Rham* complex and the model involving the chiral superfields $Z^m(t_L, \theta)$, $\overline{Z}^{\bar{m}}(t_R, \bar{\theta})$, which describes the *Dolbeault* complex.

5.1   De Rham complex

Take a set of D real superfields (2.52) and write the Lagrangian [64]

$$\boxed{L \ = \ \frac{1}{2}\int d\bar{\theta}d\theta \, g_{MN}(X)\bar{D}X^{M}DX^{N}} \tag{5.4}$$

with $\mathcal{N} = 2$ covariant derivatives (2.54).

In components, this gives

$$L \ = \ \tfrac{1}{2}g_{MN}(x)\left[\dot{x}^{M}\dot{x}^{N} + F^{M}F^{N} + i\big(\bar{\psi}^{M}\boldsymbol{\nabla}\psi^{N} - \boldsymbol{\nabla}\bar{\psi}^{M}\psi^{N}\big)\right]$$
$$+ \Gamma_{M,PQ}F^{M}\psi^{P}\bar{\psi}^{Q} + \tfrac{1}{2}(\partial_{P}\partial_{Q}g_{MN})\,\bar{\psi}^{P}\bar{\psi}^{M}\psi^{Q}\psi^{N}\,, \tag{5.5}$$

where

$$\boldsymbol{\nabla}\Psi^{M} = \dot{\Psi}^{M} + \Gamma^{M}_{PS}\dot{x}^{P}\Psi^{S}\,. \tag{5.6}$$

The dynamical bosonic part of (5.5) is the same as in (5.1) and describes the motion over a manifold endowed with the metric g_{MN}. We are going to calculate later the Witten index of the model (5.4), which, with some reservations (see the footnote on p. 79), is well defined only for finite motions, and hence we will mostly consider *compact* manifolds in this chapter.

The auxiliary variables F^{M} enter the Lagrangian without derivatives and the corresponding equations of motion $\partial L/\partial F^{M} = 0$ are purely algebraic and can be resolved for F^{M}. The solution reads

$$F^{M} \ = \ -\Gamma^{M}_{PQ}\psi^{P}\bar{\psi}^{Q}\,. \tag{5.7}$$

Plugging this into (5.5), we derive

$$L \ = \ \frac{1}{2}g_{MN}\big[\dot{x}^{M}\dot{x}^{N} + i\big(\bar{\psi}^{M}\boldsymbol{\nabla}\psi^{N} - \boldsymbol{\nabla}\bar{\psi}^{M}\psi^{N}\big)\big]$$
$$- \frac{1}{4}R_{PMQN}\,\bar{\psi}^{P}\bar{\psi}^{M}\psi^{Q}\psi^{N}\,, \tag{5.8}$$

where R_{PMQN} is the Riemann tensor.

The Lagrangian (5.8) is invariant under the following nonlinear supersymmetry transformations

$$\delta x^{M} = \epsilon\psi^{M} + \bar{\psi}^{M}\bar{\epsilon},$$
$$\delta\psi^{M} = -\bar{\epsilon}\big(i\dot{x}^{M} + \Gamma^{M}_{PQ}\psi^{P}\bar{\psi}^{Q}\big), \quad \delta\bar{\psi}^{M} = \epsilon\big(i\dot{x}^{M} - \Gamma^{M}_{PQ}\psi^{P}\bar{\psi}^{Q}\big). \tag{5.9}$$

Noether's theorem allows us to find the classical supercharges:

$$Q = \psi^{M}\left[\Pi_{M} + \frac{i}{2}\Gamma_{M,NP}\,\psi^{N}\bar{\psi}^{P}\right] = \psi^{M}\left[\Pi_{M} - \frac{i}{2}\partial_{M}g_{NP}\,\psi^{N}\bar{\psi}^{P}\right],$$

$$\bar{Q} = \bar{\psi}^{M}\left[\Pi_{M} - \frac{i}{2}\Gamma_{M,NP}\,\psi^{N}\bar{\psi}^{P}\right] = \bar{\psi}^{M}\left[\Pi_{M} + \frac{i}{2}\partial_{M}g_{NP}\,\psi^{N}\bar{\psi}^{P}\right],$$

$$\tag{5.10}$$

where

$$\Pi_M = \frac{\partial L}{\partial \dot{x}^M} = g_{MN}\dot{x}^N + \frac{i}{2}\left(\partial_Q g_{PM} - \partial_P g_{QM}\right)\psi^P \bar{\psi}^Q \quad (5.11)$$

is the canonical momentum conjugate to x^M.

For some purposes (in particular, for quantization, which is our next task), it is more convenient to express the classical supercharges and the Hamiltonian in terms of the tangent space fermion variables $\psi_A = e_{AM}\psi^M$, where e_{AM} are the *vielbeins*, $g_{MN} = e_{AM}e_{AN}$. Then the supercharges acquire the following form [37, 65]:

$$Q = \psi_C\, e_C^M\left(P_M - i\omega_{AB,M}\psi_A\bar{\psi}_B\right),$$
$$\bar{Q} = \bar{\psi}_C\, e_C^M\left(P_M - i\omega_{AB,M}\psi_A\bar{\psi}_B\right), \quad (5.12)$$

where

$$\omega_{AB,M} = e_{AN}(\partial_M e_B^N + \Gamma_{MP}^N e_B^P) \quad (5.13)$$

is the *spin connection*.

We want to make two comments at this juncture.

(1) First, a technical comment. When comparing (5.10) and (5.12), it is necessary to keep in mind that the canonical momenta Π_M and P_M do *not* coincide. Π_M is given by the variation of the Lagrangian with respect to $\dot{x}^M$ when ψ^M are kept fixed, while P_M is this variation taken with fixed ψ_A.

(2) Second, an historical comment. The notion and the term "spin connection" came from the studies of general relativity, of curved 4-dimensional spaces with Minkowski signature. This object is a necessary tool to describe interactions of the particles of spin 1/2 (like electrons) with gravitational field. But spin connection can also be defined on Euclidean manifolds of any dimension. It has a transparent *geometric* meaning: it can be considered as a gauge field associated with the group $SO(D)$ of tangent space rotations.

A related notion are the *spinors*, which we will deal with in the second half of this chapter.

The only nonzero Poisson brackets in the basis $(x^M, P_M; \psi_A, \bar{\psi}_A)$ are

$$\{P_N, x^M\}_P = \delta_N^M, \qquad \{\psi_A, \bar{\psi}_B\}_P = i\delta_{AB}. \quad (5.14)$$

The Poisson brackets $\{Q, Q\}_P$ and $\{\bar{Q}, \bar{Q}\}_P$ vanish, while $-i\{Q, \bar{Q}\}_P/2$ gives the classical Hamiltonian,

$$H = \frac{1}{2} g^{MN} (P_M - i\omega_{AB,M} \psi_A \bar{\psi}_B)(P_N - i\omega_{CD,N} \psi_C \bar{\psi}_D)$$

$$+ \frac{1}{4} R_{ABCD} \, \bar{\psi}_A \bar{\psi}_B \psi_C \psi_D \qquad (5.15)$$

with $R_{ABCD} = e_A^M e_B^N e_C^P e_D^Q R_{MNPQ}$.

To quantize this system, we take the classical supercharges (5.12) and replace $P_M \to \hat{P}_M = -i\partial/\partial x^M$ and $\bar{\psi}_A \to \hat{\bar{\psi}}_A = \partial/\partial\psi_A$, ordering the operators following the Weyl prescription (see the discussion in the second half of Sect. 2.3 and at the end of Chapter 4). After some calculation, we obtain

$$\hat{Q} = -i\,\psi^M \left[\frac{\partial}{\partial x^M} + \omega_{AB,M} \, \psi_A \frac{\partial}{\partial\psi_B} - \frac{1}{4} \partial_M \ln g \right],$$

$$\hat{\bar{Q}} = -i\,e_C^M \frac{\partial}{\partial\psi_C} \left[\frac{\partial}{\partial x^M} + \omega_{AB,M} \, \psi_A \frac{\partial}{\partial\psi_B} - \frac{1}{4} \partial_M \ln g \right]. \qquad (5.16)$$

These supercharges are Hermitian conjugate to each other in the Hilbert space involving the wave functions

$$\Psi(x^M, \psi_A) = \Psi_0(x^M) + \Psi_A(x^M)\psi_A + \ldots \qquad (5.17)$$

with the complex-valued coefficients and endowed with the inner product with the flat measure

$$d\mu_{\text{flat}} \sim \prod_M dx^M \prod_A d\psi_A d\bar{\psi}_A \, e^{-\psi_A \bar{\psi}_A} \qquad (5.18)$$

[cf. Eq. (4.31)]. If we want to make a contact with geometry, we have to derive the expressions for the *covariant* operators acting in the Hilbert space with the measure involving the extra factor $\sqrt{g}$. This is achieved by a similarity transformation

$$(\hat{Q}^{\text{cov}}, \hat{\bar{Q}}^{\text{cov}}) = g^{-1/4} (\hat{Q}^{\text{flat}}, \hat{\bar{Q}}^{\text{flat}}) g^{1/4}. \qquad (5.19)$$

This transformation kills the last terms in (5.16) and we finally derive

$$\boxed{\begin{aligned} \hat{Q}^{\text{cov}} &= -i\,\psi^M \left[\frac{\partial}{\partial x^M} + \omega_{AB,M} \, \psi_A \frac{\partial}{\partial\psi_B} \right], \\ \hat{\bar{Q}}^{\text{cov}} &= -i\,e_C^M \frac{\partial}{\partial\psi_C} \left[\frac{\partial}{\partial x^M} + \omega_{AB,M} \, \psi_A \frac{\partial}{\partial\psi_B} \right]. \end{aligned}} \qquad (5.20)$$

The operators (5.20) are nilpotent. The anticommutator $\{\hat{Q}^{\text{cov}}, \hat{\bar{Q}}^{\text{cov}}\}$ gives the covariant quantum Hamiltonian, which has the form[1]

$$\hat{H}^{\text{cov}} = -\frac{1}{2\sqrt{g}} \mathcal{D}_M \sqrt{g}\, g^{MN} \mathcal{D}_N - \frac{1}{2} R_{ABCD}\, \hat{\bar{\psi}}_A \psi_B \hat{\bar{\psi}}_C \psi_D \qquad (5.21)$$

with the covariant derivative

$$\mathcal{D}_M = \frac{\partial}{\partial x^M} + \omega_{AB,M} \psi_A \frac{\partial}{\partial \psi_B}. \qquad (5.22)$$

In the first term in (5.21) with suppressed Grassmann variables, the reader may recognize the Laplace-Beltrami operator.

Now the key observation is the following [3].

The Hilbert space including the wave functions (5.17) is isomorphic to the de Rham complex representing the set of all p-forms,

$$\alpha^{(p)} = \alpha_{M_1 \ldots M_p}(x^N)\, dx^{M_1} \wedge \cdots \wedge dx^{M_p}, \qquad (5.23)$$

$p = 0, \ldots, D$. *The anticommuting fermion variables ψ^M are isomorphic to the anticommuting differentials dx^M.*

One can then show (see e.g. Theorem 8.1 in the book [4]) that *the action of the quantum supercharges $\hat{Q}$ and $\hat{\bar{Q}}$ written in Eq. (5.20) is isomorphic (up to the factors $-i$ and i) to the action of the operator of exterior derivative d,*

$$d\alpha = \partial_N \alpha_{M_1 \ldots M_p}\, dx^N \wedge dx^{M_1} \wedge \cdots \wedge dx^{M_p}, \qquad (5.24)$$

and its Hermitian conjugate $d^\dagger$ in the de Rham complex.

All the statements above are valid for a generic Riemann manifold. But if the manifold has some special features, these statements can be strengthened.

Definition 5.2.

A *Kähler manifold* is an even-dimensional manifold admitting the tensor field I_{MN} satisfying the properties:

$$I_{MN} = -I_{NM}, \qquad I_M{}^N I_N{}^P = -\delta_M^P, \qquad \nabla_P I_M{}^N = 0, \qquad (5.25)$$

where ∇_P is the covariant Levi-Civita derivative. The tensor I_{MN} is called *complex structure.*

---

[1] To see that the classical limit of (5.21) coincides with (5.15), use the identity

$$R_{ABCD} + R_{ACDB} + R_{ADBC} = 0.$$

Kähler manifolds are complex, i.e. they can be parametrized by holomorphic coordinates $z^{k=1,\ldots,D/2}$ which can be found as solutions to the equation system

$$\frac{\partial z^k}{\partial x^M} - iI_M{}^N(x)\frac{\partial z^k}{\partial x^N} = 0. \tag{5.26}$$

In the complex basis $x^N \equiv (z^n, z^{*\bar{n}})$, the complex structure acquires a simple form

$$I_m{}^n = -i\delta_m^n, \quad I_{\bar{m}}{}^{\bar{n}} = i\delta_{\bar{m}}^{\bar{n}}, \quad I_m{}^{\bar{n}} = I_{\bar{m}}{}^n = 0. \tag{5.27}$$

Definition 5.3. A *hyper-Kähler manifold* is a Kähler manifold admitting three different complex structures I^a satisfying the quaternionic algebra,

$$I^a I^b = -\delta^{ab} + \varepsilon^{abc}I^c. \tag{5.28}$$

One can prove the following theorems:[2]

Theorem 5.1. *A Kähler manifold admits two pairs of covariant supercharges:*

$$\begin{aligned}
\hat{Q}_0 &= -i\psi^M \mathcal{D}_M, & \hat{Q}_1 &= -iI_N{}^M \psi^N \mathcal{D}_M, \\
\hat{\bar{Q}}_0 &= -i\bar{\psi}^M \mathcal{D}_M, & \hat{\bar{Q}}_1 &= -iI_N{}^M \bar{\psi}^N \mathcal{D}_M.
\end{aligned} \tag{5.29}$$

These supercharges satisfy the extended $\mathcal{N} = 4$ *supersymmetry algebra (2.65).*

This algebra is isomorphic to what mathematicians call *Kähler-de Rham complex*.

Theorem 5.2. *A hyper-Kähler manifold admits four pairs of covariant supercharges:*

$$\begin{aligned}
\hat{Q}_0 &= -i\psi^M \mathcal{D}_M, & \hat{Q}_a &= -i(I^a)_N{}^M \psi^N \mathcal{D}_M, \\
\hat{\bar{Q}}_0 &= -i\bar{\psi}^M \mathcal{D}_M, & \hat{\bar{Q}}_a &= -i(I^a)_N{}^M \bar{\psi}^N \mathcal{D}_M.
\end{aligned} \tag{5.30}$$

These supercharges satisfy the extended $\mathcal{N} = 8$ *supersymmetry algebra (2.65).*

---

[2]These statements are well known, are a little bit tangential to the main subject of this book (which is the Witten index), and we will not give their proofs here. The proofs are given in many different textbooks including the book [4].

5.1.1 Euler characteristic and Morse theory

This isomorphism allows one to prove, using the supersymmetric language, many pure mathematical theorems in the realm of differential geometry. For example, the famous *Hodge decomposition* theorem saying that any differential form α can be uniquely presented as a sum of an exact, a coexact and a harmonic form,

$$\alpha = d\beta + d^\dagger \gamma + h, \tag{5.31}$$

is an almost trivial corollary of the supersymmetry algebra illistrated in Fig. 2.2: the Hilbert space of a SQM system includes the states annihilated by the action of $\hat{Q}$ but not of $\hat{\bar{Q}}$ (which are isomorphic to the exact forms $d\beta$), the states annihilated by $\hat{\bar{Q}}$ but not by $\hat{Q}$ (the coexact forms) and zero modes annihilated both by $\hat{Q}$ and $\hat{\bar{Q}}$ (the harmonic forms).

Another classical mathematical notion that shines differently, when looking through the supersymmetric glasses, is the *de Rham cohomology*. It is the space of all closed forms, i.e. the forms satisfying the condition $d\alpha_p = 0$, factorized over the space of exact forms that can be represented as $\alpha_p = d\beta_{p-1}$. For a given degree p, the coset *closed/exact* represents a vector space $H_{\mathrm{dR}}^p(\mathcal{M})$.

Definition 5.4. The *Betti number* b_p is the dimension of H_{dR}^p.

Definition 5.5. The *Euler characteristic* of the manifold is the alternating sum

$$\chi(\mathcal{M}) = \sum_p (-1)^p \, b_p(\mathcal{M}). \tag{5.32}$$

The following remarkable result holds.

Theorem 5.3. *The Euler characteristic of a smooth manifold coincides with the Witten index of the supersymmetric sigma model (5.4).*

Proof. A look at Fig. 2.2 makes it evident that any eigenstate Ψ of an SQM Hamiltonian that is annihilated by $\hat{Q}$ but is not representable as $\Psi = \hat{Q}\Psi'$ does not belong to a supersymmetric doublet of states carrying nonzero energy. Thus, bearing in mind the isomorphism established above, b_p counts the number of zero-energy states of the Hamiltonian (5.21) having fermion charge p. If p is even, they are bosonic and, if p is odd, they are fermionic. Then the sum (5.32) coincides with the supersymmetric partition function (4.42) which in turn coincides with the Witten index (4.1) for this system. $\qquad\square$

In Chapter 4, we learned that the Witten index is expressed via a functional integral, which in many cases can be reduced to the phase space integral (4.47). Let us calculate the latter for the Hamiltonian (5.15). Integrating over momenta, we obtain

$$
\boxed{
\begin{aligned}
I_W = \chi = \\
\frac{1}{(2\pi\beta)^{D/2}} \int d^D x \sqrt{g} \int \prod_{A=1}^{D} d\psi_A d\bar{\psi}_A \exp\left\{-\frac{\beta}{4} R_{ABCD}\, \bar{\psi}_A \bar{\psi}_B \psi_C \psi_D \right\}.
\end{aligned}
}
$$

$$(5.33)$$

The first immediate observation is that the fermion integral vanishes when D is odd. And, indeed, the Euler characteristic of any odd-dimensional manifold is zero, as follows from the definition (5.32) and from the identity $b_p = b_{D-p}$ stemming from the fact that p-forms and $(D - p)$-forms are interrelated by the duality transformation.

If D is even and $R_{ABCD} \neq 0$, the fermion integral does not generically vanish. A nonzero contribution stems from the $D/2$-th term in the expansion of the exponential. The factor $\beta^{D/2}$ cancels the factor $\beta^{-D/2}$ in front of the integral in (5.33) that came from the momentum integrations. In the simplest 2-dimensional case, we obtain

$$
\chi_2 = \frac{1}{2\pi} \int d^2 x \sqrt{g}\, \mathbb{R}, \tag{5.34}
$$

where $\mathbb{R} = R_{1212}$ is the Gaussian curvature. This is none other than the very well known *Gauss-Bonnet formula* for the Euler characteristic of a 2-dimensional manifold. For an arbitrary $D = 2n$ we derive the generalized Gauss-Bonnet formula,

$$
\chi_{2n} = \frac{1}{(8\pi)^n n!} \int d^D x \sqrt{g}\, \varepsilon_{A_1 B_1 \ldots A_n B_n} \varepsilon_{C_1 D_1 \ldots C_n D_n} \prod_{k=1}^{n} R_{A_k B_k C_k D_k}. \tag{5.35}
$$

In four dimensions, this can be presented as

$$
\chi_4 = \frac{1}{32\pi^2} \int d^4 x \sqrt{g}\, R_{ABCD} \tilde{R}_{ABCD}, \tag{5.36}
$$

where

$$
\tilde{R}_{ABCD} = \frac{1}{4} \varepsilon_{ABEF} \varepsilon_{CDGH} R_{EFGH} \tag{5.37}
$$

is the dual Riemann tensor.

As was mentioned, the Euler characteristic of a manifold is represented as an alternating sum (5.32) of its Betti numbers. There is another way to represent the Euler characteristicits known to mathematicians as *Morse theory*. A relationship of Morse theory to supersymmetry was unravelled in [3].

As we know from the preceding chapter, the Witten index is invariant under smooth deformations that keep supersymmetry. Let us deform the Lagrangian (5.4) by adding a superpotential term

$$L_{\text{pot}} = -\int d\bar{\theta}d\theta\, \mathcal{W}(X^M)\,. \tag{5.38}$$

The standard procedure, the same as for the undeformed Lagrangian (5.4), gives the following expressions for the quantum supercharges:

$$\hat{Q}_W = \hat{Q}_0 - i\,\partial_M \mathcal{W}\,\psi^M = e^{-\mathcal{W}}\hat{Q}_0\,e^{\mathcal{W}}\,,$$
$$\hat{\bar{Q}}_W = \hat{\bar{Q}}_0 + i\,e^M_A(\partial_M \mathcal{W})\frac{\partial}{\partial\psi_A} = e^{\mathcal{W}}\hat{\bar{Q}}_0\,e^{-\mathcal{W}}\,, \tag{5.39}$$

where $\hat{Q}_0$ and $\hat{\bar{Q}}_0$ are the undeformed covariant quantum supercharges (5.20). The anticommutator $\{\hat{Q}_W, \hat{\bar{Q}}_W\}$ gives the quantum Hamiltonian:

$$\hat{H}_W = \hat{H}_0 + \frac{1}{2}\partial_M \mathcal{W}\,\partial^M \mathcal{W} + \frac{1}{2}\partial_M\partial_N \mathcal{W}(\psi^M\hat{\bar{\psi}}^N - \hat{\bar{\psi}}^N\psi^M)\,. \tag{5.40}$$

Similarly, the Poisson bracket of the classical counterparts of (5.39) gives the classical Hamiltonian:

$$H_W = H_0 + \frac{1}{2}\partial_M \mathcal{W}\,\partial^M \mathcal{W} + \psi^M\bar{\psi}^N\,\partial_M\partial_N \mathcal{W}\,. \tag{5.41}$$

The Witten index of the Hamiltonian (5.40) is given by the integral (4.47) with H_W in the exponent.

Capitalizing on the invariance of the index, we can calculate it for any smooth function $\mathcal{W}(X^M)$: it is the same for all of them and the same as for $\hat{H}_0$. As is seen from Eq. (5.41), the Hamiltonian acquires after the deformation the potential term $\propto (\partial_M \mathcal{W})^2$. This potential turns to zero at the *critical points* of the function $\mathcal{W}$. For a compact manifold and for a smooth function $\mathcal{W}$, such critical points are always there.[3] With our smooth deformation, we can scale up $\mathcal{W}$ so that the integrand in (4.47) would be exponentially suppressed everywhere except the neighbourhood of the critical points. We can thus evaluate the contribution of each critical point separately and then add them up.

---

[3]To give a Greek-style proof of this assertion, we invite the reader to *look* at Fig. 5.1.

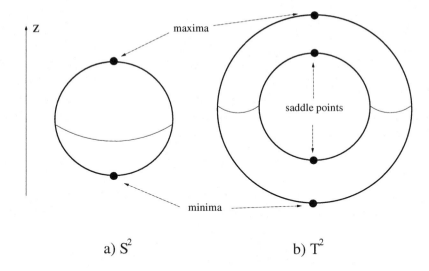

a) S^2 b) T^2

Fig. 5.1: Critical points of the Morse function $z(\mathcal{M})$. The Morse indices are $f_P = 0$ at the minima, $f_P = 2$ at the maxima, and $f_P = 1$ at the saddle points.

Substituting (5.41) in (4.47) and integrating over the momenta, we obtain

$$I_W = \frac{1}{(2\pi\beta)^{D/2}} \int d^D x \sqrt{g} \prod_A d\psi_A d\bar{\psi}_A$$

$$\exp\left\{-\beta\left[\frac{1}{2}\partial_M \mathcal{W}\partial_N \mathcal{W}\, g^{MN} + \psi^M \bar{\psi}^N\, \partial_M\partial_N \mathcal{W}\right.\right.$$

$$\left.\left. + \frac{1}{4}R_{ABCD}\,\bar{\psi}_A\bar{\psi}_B\psi_C\psi_D\right]\right\}. \tag{5.42}$$

At the vicinity of each critical point, one can expand

$$\mathcal{W}(x) = \mathcal{W}_0^{(P)} + \frac{1}{2}b_{MN}^{(P)}x^M x^N + \dots \tag{5.43}$$

If $b_{MN}^{(P)}$ are chosen to be large enough, one can neglect in this expansion the higher-order terms and neclect also the term $\propto R_{ABCD}$ in the exponential. The fermion integral can be done using (1.29), and we can write

$$I_W^{(P)} = \left(\frac{\beta}{2\pi}\right)^{D/2} \int d^D x \sqrt{g}\,\det[e_A^M e_B^N b_{MN}^{(P)}]\exp\left\{-\frac{\beta}{2}b_{MQ}^{(P)}b_{NR}^{(P)}\,x^Q x^R g^{MN}\right\}$$

$$= \frac{\det(b^{(P)})}{|\det(b^{(P)})|} = (-1)^{f_P}, \tag{5.44}$$

where f_P is the number of negative eigenvalues in the matrix $b_{MN}^{(P)}$. We have derived the formula representing the Euler characteristic of a manifold as a sum over the critical points of a generic smooth function $\mathcal{M} \overset{\mathcal{W}}{\to} \mathbb{R}$ (mathematicians call it the *Morse function*):

$$\chi(\mathcal{M}) = \sum_P (-1)^{f_P} . \qquad (5.45)$$

This result is illustrated in Fig. 5.1. We embed S^2 and T^2 in $\mathbb{R}^3$. Let $\mathcal{W}$ be the coordinate z of a manifold point. For the sphere, $\mathcal{W}$ has two critical points—a maximum and a minimum. The contribution of each point in the sum (5.45) is $+1$ and we derive $\chi(S^2) = 2$ — a well-known result for the Euler characteristic of the sphere. For the torus, $\mathcal{W}$ has 4 critical points: a maximum, a minimum and two saddle points. The contribution of each extremum is $+1$ and the contribution of each saddle point is -1. We derive that $\chi(T^2) = 0$, which agrees again with what is well known.

The *Morse index* f_P also has a physical interpretation. Choose locally flat coordinates that diagonalize the matrix b in the vicinity of the critical point P and are rescaled so that $g_{MN}(P) = \delta_{MN}$. The dominant contribution to the quantum Hamiltonian is $\hat{H} = \sum_P \hat{H}^{(P)}$ with

$$\hat{H}^{(P)} = \frac{1}{2} \sum_M \left[\hat{P}_M^2 + \omega_M^2 (x^M)^2 + \omega_M (\psi^M \hat{\bar{\psi}}^M - \hat{\bar{\psi}}^M \psi^M) \right] , \qquad (5.46)$$

where ω_M are the eigenvalues of b_{MN}. Now, $\hat{H}^{(P)}$ represents the sum of D Hamiltonians (2.9) describing supersymmetric oscillators. We have seen that the ground state of each term in the sum (2.9) has zero energy and is bosonic if $\omega_M > 0$ and fermionic if $\omega_M < 0$. If the matrix b has f_P negative eigenvalues, the ground state involves f_P Grassmann factors and has fermion charge f_P.

However, we constructed only *approximate* ground states for the original Hamiltonian (5.40). The properties of the *true* ground states may not be the same. For example, for S^2, two supersymmetric bosonic states of (5.46) go over to two supersymmetric ground states of the full Hamiltonian because $I_{\mathcal{W}} = 2$ in this case. On the other hand, $I_{\mathcal{W}} = 0$ for the torus and the exact Hamiltonian has no supersymmetric ground states. Four zero-energy ground states of $\hat{H}$ are shifted from zero if one takes into account the neglected terms in the Hamiltonian. Instead of four unpaired zero-energy ground states, one obtains two supersymmetric doublets with small (if the torus is large), but nonzero energies.

5.2 Dolbeault and Dirac complexes

5.2.1 *Dolbeault complex*

Take a set of d chiral superfields

$$Z^m(t_L, \theta) = z^m(t_L) + \sqrt{2}\theta\psi^m(t_L),$$
$$\overline{Z}^{\bar{n}}(t_R, \bar{\theta}) = z^{*\bar{n}}(t_R) - \sqrt{2}\bar{\theta}\bar{\psi}^{\bar{n}}(t_R) \tag{5.47}$$

with $m, \bar{n} = 1, \ldots, d$.

Consider a supersymmetric Lagrangian

$$\boxed{L = \int d\bar{\theta}d\theta \left[\frac{1}{4}h_{m\bar{n}}(Z, \overline{Z})\bar{D}\overline{Z}^{\bar{n}}DZ^m + W(Z, \overline{Z})\right]} \tag{5.48}$$

with Hermitian $h_{m\bar{n}}$ and real W. We will see soon that the term $\propto W$ describes a gauge field, like it does in the Landau system (2.60).

But consider first the system with $W = 0$. The component Lagrangian takes the form [66]

$$L = h_{m\bar{n}}\left[\dot{z}^m\dot{z}^{*\bar{n}} + \frac{i}{2}\left(\psi^m\dot{\bar{\psi}}^{\bar{n}} - \dot{\psi}^m\bar{\psi}^{\bar{n}}\right)\right]$$
$$- \frac{i}{2}\left[(2\partial_m h_{k\bar{n}} - \partial_k h_{m\bar{n}})\dot{z}^k - (2\partial_{\bar{n}}h_{m\bar{k}} - \partial_{\bar{k}}h_{m\bar{n}})\dot{z}^{*\bar{k}}\right]\psi^m\bar{\psi}^{\bar{n}}$$
$$+ (\partial_k\partial_{\bar{q}}h_{m\bar{n}})\psi^k\psi^m\bar{\psi}^{\bar{q}}\bar{\psi}^{\bar{n}}. \tag{5.49}$$

The bosonic part of this Lagrangian describes the motion over a complex manifold of complex dimension d with the coordinates $z^m, z^{*\bar{m}}$ and Hermitian metric $h_{m\bar{n}}$. To make the motion finite, we assume that the manifold is compact, as we did in the previous section. The Lagrangian (5.49) includes a 4-fermion term, but one can observe that, for Kähler manifolds where the metric satisfies the condition

$$\partial_l h_{k\bar{m}} - \partial_k h_{l\bar{m}} = 0, \tag{5.50}$$

the 4-fermion term vanishes. For the most part of this section, we will discuss the Kähler manifolds, making only occasional remarks on the generic complex case.

We introduce the complex vielbeins e_m^a and their complex conjugates $e_{\bar{m}}^{\bar{a}}$ with the properties

$$e_m^a e_{\bar{n}}^{\bar{a}} = h_{m\bar{n}}, \qquad e_{\bar{a}}^{\bar{n}}e_a^m = h^{\bar{n}m},$$
$$e_m^a e_a^n = \delta_m^n, \quad e_{\bar{m}}^{\bar{a}}e_{\bar{a}}^{\bar{n}} = \delta_{\bar{m}}^{\bar{n}}, \quad e_a^m e_m^b = \delta_a^b, \quad e_{\bar{a}}^{\bar{m}}e_{\bar{m}}^{\bar{b}} = \delta_{\bar{a}}^{\bar{b}}. \tag{5.51}$$

If the manifold is Kähler, the only nonzero components of the Christoffel symbols are holomorphic and antiholomorphic:[4]

$$\Gamma^p_{mn} = h^{\bar{q}p}\partial_m h_{n\bar{q}}, \qquad \Gamma^{\bar{p}}_{\bar{m}\bar{n}} = h^{\bar{p}q}\partial_{\bar{m}} h_{q\bar{n}}. \qquad (5.52)$$

The only nonzero components of the Kähler spin connections are

$$\omega_{\bar{b}a,m} = -\omega_{a\bar{b},m} = e^{\bar{n}}_{\bar{b}}\partial_m e^a_{\bar{n}}, \qquad \omega_{b\bar{a},\bar{m}} = -\omega_{\bar{a}b,\bar{m}} = e^n_b\partial_{\bar{m}} e^{\bar{a}}_n. \quad (5.53)$$

The only nonzero components of the Kähler Riemann tensor are

$$R_{m\bar{p}n\bar{q}} = -R_{\bar{p}mn\bar{q}} = -R_{m\bar{p}\bar{q}n} = R_{\bar{p}m\bar{q}n} =$$
$$\partial_m\partial_{\bar{p}} h_{n\bar{q}} - h^{\bar{s}t}(\partial_m h_{n\bar{s}})(\partial_{\bar{p}} h_{t\bar{q}}). \qquad (5.54)$$

Like in the de Rham case, it is convenient to introduce the fermion variables with tangent space indices, $\psi^a = e^a_m\psi^m$ and $\bar{\psi}^{\bar{a}} = e^{\bar{a}}_{\bar{m}}\bar{\psi}^{\bar{m}}$. In these terms, the canonical Kähler classical Hamiltonian that corresponds to the Lagrangian (5.49) acquires a very compact form:

$$H = h^{\bar{n}m}\mathcal{P}_m\mathcal{P}^*_{\bar{n}}, \qquad (5.55)$$

where

$$\mathcal{P}_m = p_m - i\omega_{a\bar{b},m}\,\psi^a\bar{\psi}^{\bar{b}}, \qquad \mathcal{P}^*_{\bar{n}} = p^*_{\bar{n}} - i\omega_{a\bar{b},\bar{n}}\,\psi^a\bar{\psi}^{\bar{b}} \qquad (5.56)$$

and the canonical momenta p_m and $p^*_{\bar{m}}$ are given by the derivatives of the Lagrangian over the canonical velocities $\dot{z}^m, \dot{z}^{*\bar{m}}$ with fixed ψ^c and $\bar{\psi}^{\bar{c}}$.

The action of the model is invariant under the transformations

$$\delta z^m = \epsilon\psi^m, \qquad \delta\psi^m = -i\bar{\epsilon}\dot{z}^m,$$
$$\delta z^{*m} = -\bar{\epsilon}\bar{\psi}^{\bar{m}}, \qquad \delta\bar{\psi}^{\bar{m}} = i\epsilon\dot{z}^{*\bar{m}} \qquad (5.57)$$

[cf. Eq. (2.58)]. The corresponding conserved classical supercharges read

$$Q = \sqrt{2}\,\psi^m\mathcal{P}_m, \qquad \bar{Q} = \sqrt{2}\,\bar{\psi}^{\bar{m}}\mathcal{P}^*_{\bar{m}}. \qquad (5.58)$$

The classical algebra

$$\{Q,Q\}_P = \{\bar{Q},\bar{Q}\}_P = 0, \qquad \{Q,\bar{Q}\}_P = 2iH \qquad (5.59)$$

holds.

Now we turn our attention to the second term in the action (5.48). The component Lagrangian acquires the extra term

$$\Delta L_W = i(\dot{z}^{*\bar{m}}\partial_{\bar{m}}W - \dot{z}^m\partial_m W) + 2\,\partial_m\partial_{\bar{n}}W\psi^m\bar{\psi}^{\bar{n}}. \qquad (5.60)$$

---

[4]This means, in particular, that the holonomy group of a Kähler manifold is $U(d)$ rather than $SO(2d)$.

This brings about the extra terms in the supercharges:

$$\Delta Q_W = i\sqrt{2}\,\psi^m \partial_m W\,, \qquad \Delta \bar{Q}_W = -i\sqrt{2}\,\bar{\psi}^{\bar{m}} \partial_{\bar{m}} W\,. \qquad (5.61)$$

The classical Hamiltonian now reads

$$H_W = h^{\bar{n}m}\left(\mathcal{P}_m + i\partial_m W\right)\left(\mathcal{P}^*_{\bar{n}} - i\partial_{\bar{n}} W\right) - 2\,\partial_m \partial_{\bar{n}} W \psi^m \bar{\psi}^{\bar{n}}\,. \qquad (5.62)$$

Compare this expression with the Pauli Hamiltonian (2.20). We see that the vector

$$A_M = (i\partial_m W, -i\partial_{\bar{m}} W) \qquad (5.63)$$

has the meaning of the *gauge vector potential*.

Note that the function $W(Z, \overline{Z})$ has nothing to do with the superpotential $\mathcal{W}(X)$ of the preceding section and plays a completely different role.[5] In the de Rham case, the superpotential $\mathcal{W}(X^M)$ in (5.38) provided a *smooth* deformation of the model (5.4) — $\mathcal{W}$ could be arbitrarily small (and arbitrarily large). And now W describes a gauge field, which cannot be made arbitrarily small because the magnetic flux (4.61) and its multidimensional generalizations (we will discuss them later in this chapter) are quantized. The model (5.48) with a nonzero W *does* not represent a smooth deformation of the model without W and the value of the Witten index *depends* on W.

Well, quantization of fluxes is a feature of *quantum* rather than classical theories. Thus, we proceed with quantization. As in the previous section, we resolve the ordering ambiguities as prescribed in [24], i.e. we use the symmetric Weyl ordering for the supercharges supplemented by a similarity transformation (5.19). We obtain the following expressions for the covariant quantum supercharges:[6]

$$\hat{Q} = \sqrt{2}\,\psi^m \left[\hat{p}_m - \frac{i}{4}\partial_m(\ln \det h) + i\partial_m W - i\,\omega_{a\bar{b},m}\,\psi^a \hat{\bar{\psi}}^{\bar{b}}\right],$$

$$\hat{\bar{Q}} = \sqrt{2}\,\hat{\bar{\psi}}^{\bar{m}} \left[\hat{p}^*_{\bar{m}} - \frac{i}{4}\partial_{\bar{m}}(\ln \det h) - i\partial_{\bar{m}} W - i\,\omega_{a\bar{b},\bar{m}}\,\psi^a \hat{\bar{\psi}}^{\bar{b}}\right], \qquad (5.64)$$

where $\hat{p}_m, \hat{p}^*_{\bar{m}}$ and $\hat{\bar{\psi}}^{\bar{a}}$ are the operators $-i\partial/\partial z^m$, $-i\partial/\partial z^{*\bar{m}}$ and $\partial/\partial\psi^a$. These expressions almost coincide by form with the classical expressions (5.58), but note the presence of the extra terms $\propto \partial_m \ln \det h$ and $\propto$

---

[5] Unfortunately, the number of letters in the Latin alphabet is finite.

[6] To avoid unnecessary complications, we assumed that the complex vielbeins e^a_m are chosen so that their determinant is real: $\det e = \det \bar{e} = \sqrt{\det h}$.

$\partial_{\bar{m}} \ln \det h$ in the quantum supercharges, which are added to the gauge fields.

Consider first the case $W = (1/4) \ln \det h$ when the terms $\propto \partial_m W$ and $\propto \partial_m \ln \det h$ in the supercharge $\hat{Q}$ (but not the similar terms in the supercharge $\hat{\bar{Q}}$!) exactly cancel. A known fact is (see e.g. Theorem 9.1 in the book [4]) that in this case *the action of $\hat{Q}$ and $\hat{\bar{Q}}$ on the wave functions $\Psi(z^m, z^{*\bar{m}}, \psi^a)$ is isomorphic to the action of the holomorphic exterior derivative operator ∂ and its Hermitian conjugate $\partial^\dagger$ in the Dolbeault complex.*

It is worth recalling here what *is* the Dolbeault complex.

Consider *holomorphic* (p,0) forms living in our complex manifold:

$$\alpha_{p,0} = \alpha_{m_1,\ldots,m_p} dz^{m_1} \wedge \cdots \wedge dz^{m_p} , \qquad (5.65)$$

$p \leq d$. The operator of exterior holomorphic derivative is defined as

$$\partial \alpha = \partial_n \alpha_{m_1 \ldots m_p} dz^n \wedge dz^{m_1} \wedge \cdots \wedge dz^{m_p} . \qquad (5.66)$$

By the same token as the operator d of the de Rham complex defined in Eq. (5.24), the operator ∂ is nilpotent. Consider also the Hermitian conjugate operator $\partial^\dagger$. Then the anticommutator $-\{\partial, \partial^\dagger\}$ is a Hermitian operator called the *Dolbeault laplacian*.

Definition 5.6. The set of all holomorphic forms (5.65) equiped by the operators ∂ and $\partial^\dagger$ is called the Dolbeault complex.

One can also define the anti-Dolbeault complex involving the antiholomorphic forms

$$\bar{\alpha} = \alpha_{\bar{m}_1 \ldots \bar{m}_p} dz^{*\bar{m}_1} \wedge \cdots \wedge dz^{*\bar{m}_p} \qquad (5.67)$$

and the nilpotent operators $\bar{\partial}$ and $\bar{\partial}^\dagger$. Generically, the anti-Dolbeault laplacian equal to $-\{\bar{\partial}, \bar{\partial}^\dagger\}$ does not coincide with the Dolbeault one, but for the Kähler manifolds it does!

The fact that the supercharges (5.64) were isomorphic to the Dolbeault rather than anti-Dolbeault complex was due to our choice of W. Had we chosen $W = -(1/4) \ln \det h$, the extra derivative terms would cancel in $\hat{\bar{Q}}$ rather than in $\hat{Q}$, and in this case *the operators $\hat{\bar{Q}}$ and $\hat{Q}$ would be isomorphic to the antiholomorphic exterior derivative $\bar{\partial}$ and its conjugate $\bar{\partial}^\dagger$, respectively.* (To see that, one should write the wave functions in the antiholomorphic representation: $\Psi(z, z^*, \bar{\psi})$ — see the footnote on p. 20).

For an arbitrary W (modulo the requirement that the associated topological charge is integer), we are dealing with the *twisted* Dolbeault complex. The following theorem holds.

Theorem 5.4. *The action of a generic* $\hat{Q}_W$ *on the wave functions* $\Psi(z^m, z^{*\bar{m}}; \psi^a)$ *maps to the action of the nilpotent operator*

$$\partial_W = \partial - \partial \left(W - \frac{1}{4} \ln \det h \right) \wedge \qquad (5.68)$$

in the twisted Dolbeault complex. The action of a generic $\hat{\bar{Q}}_W$ *on the wave functions* $\Psi(z^m, z^{*\bar{m}}, \bar{\psi}^a)$ *maps to the action of the operator*

$$\bar{\partial}_W = \bar{\partial} + \bar{\partial} \left(W + \frac{1}{4} \ln \det h \right) \wedge \qquad (5.69)$$

in the twisted anti-Dolbeault complex.

5.2.2   Dirac operator, Dirac complex and Dolbeault-Dirac isomorphism

There is another remarkable isomorphism. For Kähler manifolds, the SQM system described by the Lagrangian (5.48) and the Dolbeault complex, to which it is isomorphic, are in turn isomorphic to the *Dirac complex* including the Dirac operator living on a manifold and its eigenfunctions.[7]

We have in fact already met the Dirac operator when we talked about 4-dimensional field theories, supersymmetric and not, in Chapters 1 and 3. It was the first order matrix differential operator $\gamma^\mu(\partial_\mu + ieA_\mu)$, the structure that entered the Lagrangian (1.59) of QED. A non-Abelian version of the Dirac operator enters the Lagrangian of QCD (1.61). The Abelian or non-Abelian Dirac operators also enter the Lagrangian of supersymmetric gauge theories discussed in Chapter 3.

But those were the operators defined in physical 4-dimentional Minkowski space. In this chapter, we are interested in the Dirac operators living in an arbitrary even-dimensional *Euclidean* manifold. The simplest example of such operator is the flat three-dimensional operator $\boldsymbol{\sigma} \cdot (\hat{\boldsymbol{P}} + \boldsymbol{A})$ projected on a 2-sphere.

Let us establish some definitions. To begin, we recall what *spinor* is from the mathematical viewpoint (for a physicist, the spinor is a wave function of a spin 1/2 particle).

First, consider first a flat D-dimensional space. The group $SO(D)$ [or rather $Spin(D)$, see below] involves spinor representations. For odd D, there is a unique irreducible spinor representation, but, to make contact with supersymmetry, we will only be interested in *even* D, where there are two irreducible representations S_L and S_R — left and right spinors.

---

[7]To the best of our knowledge, this fact was first noticed by N. Hitchin [67].

Consider a set of complex functions $\Phi_\alpha(x)$ belonging to $S_L \oplus S_R$ (the *bispinor* representation). The index α takes $N_D = 2^{D/2}$ different values.

We now introduce a convenient object:

Definition 5.7. Hermitian matrices $\gamma^{A=1,\dots,D}$ of dimension $N_D \times N_D$ satisfying the Clifford algebra

$$\gamma^A\gamma^B + \gamma^B\gamma^A = 2\delta^{AB} \tag{5.70}$$

are called Euclidean multidimensional *gamma matrices*.[8]

Explicit choices for γ^A may be different, and we need not specify it.

In these terms, the law of infinitesimal global rotations for the bispinor Φ_α acquires a nice form:

$$\delta\Phi_\alpha = \frac{1}{4}\theta_{AB}\left(\gamma^A\gamma^B\right)_{\alpha\beta}\Phi_\beta, \tag{5.71}$$

where $\theta_{AB} = -\theta_{BA}$ is a small rotation angle in the plane $[AB]$. The reader can be convinced that the generators $t^{AB} \sim i\gamma^{[A}\gamma^{B]}$ commute as they should in $so(D)$ and that the norm $\Phi_\alpha^*\Phi_\alpha$ is invariant under the rotations (5.71).

The group $Spin(D)$ is the set of all matrices

$$g = \exp\left\{\frac{1}{4}\theta_{AB}\,\gamma^A\gamma^B\right\}.$$

One can note that the matrix $\exp\left\{\pi\gamma^A\gamma^B\right\}$ of the finite rotation by 2π in the plane $[AB]$ coincides with $-\mathbb{1}$, so that $Spin(D)$ represents a two-fold covering of $SO(D)$. But we need not plunge deeper into these group theory details.

The Dirac operator $\slashed{\nabla}$ on a curved manifold in the presence of an Abelian gauge field is defined as

$$\boxed{\slashed{\nabla} = e_A^M\gamma^A\left(\partial_M + \frac{1}{4}\omega_{BC,M}\,\gamma^B\gamma^C + iA_M\right).} \tag{5.72}$$

It acts on the bispinors Φ_α. One can show that $(\slashed{\nabla}\Phi)_\alpha$ transforms in the same way as Φ_α under both gauge rotations,

$$\Phi_\alpha \to e^{-i\chi(x)}\Phi_\alpha, \qquad A_M \to A_M + \partial_M\chi, \tag{5.73}$$

---

[8]They were first introduced a century ago by Dirac in Minkowski 4-dimensional space. In this case, Dirac matrices satisfy the relation $\{\gamma^\mu, \gamma^\nu\} = 2\eta^{\mu\nu} = 2\,\mathrm{diag}(1, -1, -1, -1)$.

and orthogonal rotations in the tangent plane with x-dependent parameters $\theta_{AB}(x)$.

Eq. (5.72) is a local expression. A nontrivial question is whether the Dirac operator can be defined *globally* on the whole manifold. To do so, one has to define the *spinor bundle*, i.e. divide the manifold in several charts and define the bispinor field in each chart in such a way that, in the intersection regions, the bispinor fields from two different charts U_α and U_β were interrelated by a transformation $\Omega_{\alpha\beta} \in Spin(D)$. If non-empty overlap regions $U_\alpha \cap U_\beta \cap U_\gamma$ of three different charts exist, the group-valued functions $\Omega_{\alpha\beta}$ should satisfy there the consistency condition $\Omega_{\alpha\beta}\Omega_{\beta\gamma} = \Omega_{\alpha\gamma}$.

Definition 5.8. People say that the manifolds admitting spinor bundles satisfying the conditions above admit a *spin structure*.

For such manifolds, one can globally define a "purely gravitational" Dirac operator. It has the form (5.72) with $A_M = 0$ in each chart. In the overlapping regions, the bispinor fields are related by $\Omega_{\alpha\beta}$, while the sets of vielbeins e_A^M defined in each chart are related by the $SO(D)$ rotations that correspond to $\Omega_{\alpha\beta}$.

Not all manifolds have this property. For example, $\mathbb{CP}^{2k}$ do not admit spin structure and the Dirac operator not including an extra gauge field A_M is not well defined there. However, we will see soon that the so-called *spin*$^\mathbb{C}$ *structure* on $\mathbb{CP}^{2k}$ is admissible: if an Abelian gauge field A_M of a special form is present, the Dirac operator (5.72) is globally well defined. Such A_M represents a *principal fiber bundle* with the fiber $U(1)$ (or else, *line bundle*) on $\mathbb{CP}^{2k}$: the functions $A_M(x)$ in the overlapping regions are related by a gauge transformation,

$$A_M^\alpha = A_M^\beta + \partial_M \chi^{\alpha\beta}. \tag{5.74}$$

The bispinor fields in the overlapping charts are then related by $\Omega_{\alpha\beta} \in Spin(D) \times U(1)$.

Now define the matrix

$$\gamma^{D+1} = \frac{(-i)^{D/2}}{D!} \varepsilon_{A_1,\dots,A_D} \gamma^{A_1} \cdots \gamma^{A_D}, \tag{5.75}$$

the multidimensional analog of the γ^5 matrix widely used in physical applications. If D is even, the set of matrices $\gamma^{A=1,\dots,D}, \gamma^{D+1}$ also obey the Clifford algebra, i.e. γ^{D+1} anticommutes with all $\gamma^{A=1,\dots,D}$ and $(\gamma^{D+1})^2 = \mathbb{1}$ (if D is odd, the RHS of Eq. (5.75) is proportional to unity).

It follows that γ^{D+1} can be diagonalized with the eigenvalues ± 1. In fact, a half of these eigenvalues are $+1$ and a half -1.

Definition 5.9. A bispinor satisfying the condition $\gamma^{D+1}\Psi = \Psi$ is called *right-handed* and a bispinor satisfying the condition $\gamma^{D+1}\Psi = -\Psi$ is called *left-handed.*

As γ^{D+1} anticommutes with $\slashed{\nabla}$, the eigenstates of the latter can be divided in two classes: right-handed and left-handed ones.

One can then make the following crucial observation:

Theorem 5.5. *The operators*

$$\hat{Q}_D = -\frac{i}{2}\slashed{\nabla}\left(1 - \gamma^{D+1}\right), \qquad \hat{Q}_D^\dagger = -\frac{i}{2}\slashed{\nabla}\left(1 + \gamma^{D+1}\right), \qquad (5.76)$$

the chiral projections of $\slashed{\nabla}$ that annihilate either right-handed or left-handed bispinors, obey the supersymmetry algebra (2.1) with the Hamiltonian

$$\hat{H} = -\frac{\slashed{\nabla}^2}{2}. \qquad (5.77)$$

The bispinor Hilbert space together with the chiral projections of $\slashed{\nabla}$ defined above constitute the Dirac complex.

Now, the idea is to divide the set of operators γ^A into two subsets, $\gamma^A = (\gamma^a, \gamma^{\bar{a}})$ with the (anti)holomorphic indices $a, \bar{a} = 1, \ldots, d$, try to identify γ^a with the holomorphic fermion variables ψ^a [rather $\sqrt{2}\psi^a$ bearing in mind the factor 2 in (5.70)] entering the supercharges (5.64), $\gamma^{\bar{a}}$ with the operators $\sqrt{2}\,\partial/\partial\psi^a$, and see whether we can identify the supercharges (5.76) with the supercharges (5.64) and represent the Dirac operator (5.72) as the sum

$$\slashed{\nabla} = i(\hat{Q} + \hat{\bar{Q}}). \qquad (5.78)$$

For Kähler manifolds, where the only components of the spin connections $\omega_{AB,M}$ have the form (5.53), we can do so! The matrix γ^{D+1} then maps to the fermion parity operator $(-1)^{\hat{F}}$, the left-handed spinors map to the bosonic states and the right-handed spinors to the fermionic states.

For a generic complex manifold, this does not quite work because the spin connection $\omega_{AB,M}$ involves also the components $\omega_{ab,\bar{m}}$ and $\omega_{\bar{a}\bar{b},m}$, but one can consider a *deformed* Dirac operator

$$\slashed{\nabla}^{\text{tors}} = e_A^M \gamma^A \left(\partial_M + \frac{1}{4}\omega_{BC,M}\,\gamma^B\gamma^C - \frac{1}{24}C_{MKP}\gamma^K\gamma^P + iA_M\right) \qquad (5.79)$$

that involves besides the spin connections also the completely antisymmetric *torsion* tensor C_{MKP} with nonzero components

$$C_{mk\bar{p}} = \partial_k h_{m\bar{p}} - \partial_m h_{k\bar{p}}, \qquad C_{\bar{m}\bar{k}p} = \overline{C_{mk\bar{p}}} = \partial_{\bar{k}}h_{p\bar{m}} - \partial_{\bar{m}}h_{p\bar{k}}, \qquad (5.80)$$

the components $C_{m\bar{k}p}$ etc. being restored by antisymmetry. The troublesome components $\omega_{ab,\bar{m}}$, $\omega_{\bar{a}\bar{b},m}$ cancel in the operator ∇^{tors}, so that the latter can be represented as in (5.78).

The most known and widely studied family of complex manifolds are the complex projective spaces $\mathbb{CP}^n$. A $\mathbb{CP}^n$ manifold is defined as a set of complex $(n+1)$-tuples $(w^0, \ldots, w^n)$ identified under the multiplication by a nonzero complex number λ: $(w^0, \ldots, w^n) \equiv (\lambda w^0, \ldots, \lambda w^n)$. A naturally chosen "round" [symmetric under $SU(n+1)$ rotations] metric on this space reads[9]

$$ds^2 = 2\frac{w^{*\alpha}w^\alpha\, dw^{*\beta}dw^\beta - w^{*\alpha}w^\beta\, dw^{*\beta}dw^\alpha}{(w^{*\alpha}w^\alpha)^2} \qquad (5.81)$$

($\alpha = 0, \ldots, n$). $\mathbb{CP}^n$ is topologically nontrivial; it can be described by an atlas including $n+1$ charts with the topology $\mathbb{C}^n$. One of these charts (call it $\mathcal{C}_0$) excludes the points with $w^0 = 0$ (the set of all such points has the topology $\mathbb{CP}^{n-1}$). The complex coordinates uniquely describing the points on this chart can be chosen as $z^j = w^j/w^0$, $j = 1, \ldots, n$. Then the metric (5.81) reduces to the *Fubini-Study* form: $ds^2 = 2h_{j\bar{k}}dz^j dz^{*k}$ with

$$h_{j\bar{k}} = \partial_j\partial_{\bar{k}}\ln(1 + z^*z) = \frac{1}{1 + z^*z}\left(\delta_{j\bar{k}} - \frac{z^k z^{*j}}{1 + z^*z}\right), \qquad (5.82)$$

the shorthand $z^*z \equiv z^{*l}z^l$ being used. This metric satisfies the conditions (5.50) and is Kähler. A similar construction can be done for all other charts excluding the points with $w^1 = 0$ etc.

The simplest case is $n = 1$. The manifold $\mathbb{CP}^1$ is nothing but S^2. It is described by a single complex coordinate z and has the metric

$$ds^2 = 2\frac{dz^*dz}{(1 + z^*z)^2}. \qquad (5.83)$$

Keeping the conventions of the book [4], we assume that z is expressed via the real coordinates x, y as $z = (x+iy)/\sqrt{2}$, so that the area of our sphere is

$$\mathcal{A} = \int h\, dz^*dz = \int dxdy\, \frac{1}{\left(1 + \frac{x^2+y^2}{2}\right)^2} = 2\pi \qquad (5.84)$$

and its radius is equal to $1/\sqrt{2}$.

---

[9]To avoid confusion, we are not putting bars on the indices of antiholomorphic coordinates in Eq. (5.81) and some further formulas. In the expressions like $w^{*\alpha}w^\alpha$, the first index is, strictly speaking, antiholomorphic, but is has the same numerical value as the second (holomorphic) index, and summation over α is assumed.

We consider now the supersymmetric action (5.48), where besides the first term involving the metric, the second term describing the Abelian gauge field (5.63) is also present. For $\mathbb{CP}^1$, we may take

$$W = \frac{q}{4} \ln h = -\frac{q}{2} \ln(1 + z^*z), \qquad (5.85)$$

which gives

$$A = -\frac{iqz^*}{2(1 + z^*z)}, \qquad A^* = \frac{iqz}{2(1 + z^*z)} \qquad (5.86)$$

It is nothing but the gauge field of the *Dirac monopole* of magnetic charge q placed at the origin and projected on the surrounding sphere.

As is well-known, only integer values of the monopole charge are admissible. That follows from the physical requirement that the angular momentum of the electromagnetic field proportional to the spatial integral of the vector product $\boldsymbol{B} \times \boldsymbol{E}$, where $\boldsymbol{B}$ is the Coulomb magnetic field of a monopole and $\boldsymbol{E}$ is the Coulomb electric field of an electron placed in its vicinity, is an integer multiple of $\hbar/2$. The same quantization condition follows from the requirement that the wave function of an electron interacting with the monopole is uniquely defined everywhere except on the *Dirac string* (which is projected on the north pole of the sphere corresponding to $z = \infty$), where the field (5.86) is singular.

A more accurate mathematical description of the monopole field is due to Wu and Yang [68]. The field may be represented as a connection of a topologically nontrivial line bundle over the sphere. To this end, we cover S^2 with two charts with the topology of disk such that the fields $A^{(1)}$ and $A^{(2)}$ in these charts are not singular and are related in the region where the charts overlap by a gauge transformation $A_\mu^{(1)} = A_\mu^{(2)} + \partial_\mu \chi$, as in (5.74).

One can choose the whole sphere except one point (the north pole) as chart 1, where the field is given by Eq. (5.86), and a small neighborhood of the pole as chart 2 with $A_\mu^{(2)} \approx 0$. A necessary requirement is that the element $\omega = e^{i\chi} \in U(1)$ is uniquely defined at a small circle surrounding the pole. Quantization of q follows in this language from that.

This construction can be generalized for higher n. The gauge field on the chart $\mathcal{C}_0$ can be chosen in the form analogous to Eq. (5.86):

$$A_j = -\frac{iqz^{*j}}{2(1 + z^*z)}, \qquad A_k^* = \frac{iqz^k}{2(1 + z^*z)}. \qquad (5.87)$$

The metric and the gauge field on a chart $\mathcal{C}_{l \neq 0}$ (where the points with $w^l = 0$ are excluded) are also given by the expressions (5.82), (5.87), with $u^j = (w^0/w^l, \ldots)$ being substituted for z^j.

Let us make an important remark. The field (5.87) is the "physical" gauge field (5.63) that enters the Dirac operator. As we have seen, the "mathematical" gauge field $\tilde{A}_M$ that twists the Dolbeault complex does not coincide with A_M, but is shifted as in Eq. (5.68). For $\mathbb{CP}^n$, we derive

$$
\begin{aligned}
\tilde{A}_j &= -i\left(q - \frac{n+1}{2}\right)\frac{z^{*j}}{2(1+z^*z)}, \\
\tilde{A}_k^* &= i\left(q - \frac{n+1}{2}\right)\frac{z^k}{(1+z^*z)}.
\end{aligned}
\tag{5.88}
$$

And it is $\tilde{A}_M$ rather than A_M, which, to assure the fiber bundle to be benign, should be related by uniquely defined gauge transformations when we go from one chart to another. This is possible under the condition that

$$
s = q - \frac{n+1}{2} \qquad \text{is integer.} \tag{5.89}
$$

For $\mathbb{CP}^1$ and for all $\mathbb{CP}^n$ with odd n, this means that also q is integer. If n is even, q must be half-integer. The value $q = 0$ is then excluded — as was mentioned above, the manifolds $\mathbb{CP}^{2k}$ deprived of extra gauge field do not admit a spin structure.

5.2.3 *The index*

Let us calculate the Witten index of this model. For the Fubini-Study metric (5.82) and $SU(n+1)$ - symmetric gauge field (5.87) one can find the wave functions of the vacuum states explicitly [69].[10]

The vacuum states are annihilated by both supercharges in Eq. (5.64): $\hat{Q}$ and $\hat{\bar{Q}}$. This is possible only for the states in the sectors with minimal or maximal fermion numbers: $F = 0$ or $F = n$.

Consider the sector $F = 0$. The condition $\hat{\bar{Q}}\Psi_0 = 0$ is satisfied automatically, while the condition $\hat{Q}\Psi_0 = 0$ gives the equations

$$
\frac{\partial}{\partial z^j}\Psi_0 = -\frac{sz^{*j}}{2(1+z^*z)}\Psi_0 \tag{5.90}
$$

with an integer s defined in (5.89).

We are looking for solutions of these equations normalized with the covariant measure

$$
\int \prod_{k=1}^{n} dz^{*k}dz^k \, \det h \, |\Psi_0|^2 = \int \frac{\prod_k dz^{*k}dz^k}{(1+z^*z)^{n+1}}\,|\Psi_0|^2 = 1. \tag{5.91}
$$

---

[10]In this case, one can actually find the whole spectrum [70], but if the metric and gauge fields are deformed, the wave functions and spectrum are deformed too, whereas the number of vacuum states, the index, is unvariant under deformations.

They exist only for non-negative s, i.e. for $q \geq (n+1)/2$. A general solution reads

$$\Phi_0 = (1 + z^*z)^{-s/2} P(z^*), \tag{5.92}$$

where $P(z^*)$ is a polynomial of z^{*k} of degree not higher than s. The dimension of vector space of such polynomials, i.e. the number of linearly independent solutions of (5.90), is given by a binomial coefficient

$$\#^{\mathrm{vac}}_{F=0} = \binom{n+s}{n} = \binom{q + \frac{n-1}{2}}{n}. \tag{5.93}$$

Consider now the sector $F = n$. In this case, the wave functions are automatically annihilated by $\hat{Q}$, and the condition $\hat{\bar{Q}}\Psi_0 = 0$ gives the equations

$$\frac{\partial}{\partial z^j}\Psi_0 = \frac{(s+n+1)z^{*j}}{2(1+z^*z)}\Psi_0. \tag{5.94}$$

The solutions exist for $s + n + 1 \leq 0$, i.e. for $q \leq -(n+1)/2$. They have the form

$$\Psi_0 = (1 + z^*z)^{(s+n+1)/2} P(z) \prod_{k=1}^{n} \psi^k, \tag{5.95}$$

where $P(z)$ is a polynomial of degree not higher than $|s+n+1|$. There are

$$\#^{\mathrm{vac}}_{F=n} = \binom{-s-1}{n} = \binom{-q + \frac{n-1}{2}}{n}. \tag{5.96}$$

different states.

All the states in the sector $F = 0$ are bosonic. The states in the sector $F = n$ are also bosonic if n is even and fermionic if n is odd. We derive the universal formula for the index:

$$\boxed{I_W = \binom{|q| + \frac{n-1}{2}}{n} [\mathrm{sgn}(q)]^n.} \tag{5.97}$$

It is nonzero and the supersymmetric states are present at $|q| \geq (n+1)/2$. If $|q| < (n+1)/2$, $I_W = 0$ and supersymmetry is spontaneously broken. It is remarkable that basically *the same* formula describes the Witten index for 3-dimensional supersymmetric Yang-Mills-Chern-Simons theory, which we will discuss in Chapter 10 [see Eq. (10.5)].

We solved the problem for $\mathbb{CP}^n$, but how can we do it for more complicated Kähler manifolds where explicit calculations may not be feasible? To this end, we should apply the general methods outlined in the preceding chapter and calculate the functional integral (4.43) for the supersymmetric partition function. Let us first do so in the approximation where the functional integral reduces to the ordinary phase space integral (4.47) with the Hamiltonian (5.62).

Calculating this integral for $\mathbb{C}P^n$, we obtain

$$
\begin{aligned}
I_W &= \int \prod_{k=1}^{n} \frac{dP_k^* dP_k dz^{*k} dz^k}{(2\pi)^2} \prod_{a=1}^{n} d\psi^a d\bar{\psi}^a \, \exp\{-\beta H_W\} \\
&= \left(\frac{q}{2\pi}\right)^n \int \frac{1}{(1+z^*z)^{n+1}} \prod_k dz^{*k} dz^k = \frac{q^n}{n!} .
\end{aligned}
\tag{5.98}
$$

This result looks pretty strange. It does not coincide with the result (5.97) derived above! Only for $n = 1$ both formulas give the same result $I_W = q$. But for $n \geq 2$, the integral (5.98) gives a fractional value for the index, which makes no sense. That means that our calculation was wrong. At some point, we should have made an error.

We have, indeed, but it was not a trivial error. The point is that, no matter how small β is, we are not allowed *in this particular case* to assume that the variables $z^k(\tau)$, $\psi^a(\tau)$, etc. in the functional integral (4.43) do not depend on τ. Instead, we should

- Impose the periodic boundary conditions

$$
z^k(\beta) = z^k(0), \quad \psi^a(\beta) = \psi^a(0), \quad \bar{\psi}^a(\beta) = \bar{\psi}^a(0),
\tag{5.99}
$$

- Expand all the variables in Fourier series,

$$
z^k(\tau) = z_0^k + \sum_{m \neq 0} z_m^k e^{2\pi i m \tau/\beta}, \qquad \text{etc.}
\tag{5.100}
$$

- Substitute this into (4.43).
- Do the integral over the nonzero Fourier modes in the Gaussian approximation. It turns out that the Gaussian approximation is sufficient here — one need not take into account the quartic terms and terms of still higher power. Using the physical slang, it is not sufficient in this case to calculate the functional integral in the *tree approximation*, but the *one-loop approximation* is sufficient and gives the correct answer.

Note that all these complications arise due to nonzero *curvature* of the manifold. In the flat case, the Cecotti-Girardello approximation *works* and we can evaluate the index by the ordinary phase space integral (4.47).

The integration over momenta and fermion variables gives in this case the integral

$$
I_W^{\text{flat}} = \frac{1}{(4\pi)^{D/2}(D/2)!} \, \varepsilon_{A_1 \dots A_D} \int d^D x \, F_{A_1 A_2} \cdots F_{A_{D-1} A_D} .
\tag{5.101}
$$

Introducing the 2-form

$$\mathcal{F} = \frac{1}{2} F_{MN} dx^M \wedge dx^N, \tag{5.102}$$

where $F_{MN} = \partial_M A_N - \partial_N A_M$ with the gauge field (5.63) is the field density, we may represent the result (5.101) in the form

$$I_W = \int e^{\mathcal{F}/2\pi}, \tag{5.103}$$

where[11]

$$e^{\mathcal{F}/2\pi} = 1 + \frac{\mathcal{F}}{2\pi} + \frac{\mathcal{F} \wedge \mathcal{F}}{8\pi^2} + \cdots$$

For $\mathbb{R}^D$, only the term of order $D/2$ in the expansion of the exponential works. When $D = 2$, we reproduce the result (4.61).

The actual calculation of the path integral (4.43) on a curved manifold with the Euclidean Lagrangian[12]

$$L_E = \frac{1}{2} g_{MN}(x) \dot{x}^M \dot{x}^N + \frac{1}{2} \Psi^A (\dot{\Psi}^A + \omega_{AB,M} \dot{x}^M \Psi^B)$$

$$+ i A_M \dot{x}^M - \frac{i}{2} F_{MN} \Psi^M \Psi^N \tag{5.104}$$

is not so simple. This was first done in the papers [48, 49] devoted to the supersymmetric proof of the Atiyah-Singer theorem, but the details of calculation were not given. More details (but not all the details) were displayed in Refs. [71]. For an accurate pedagogical calculation, we address the reader to Chapter 14 of the book [4], where we also explained why one need not go beyond the one-loop approximation. Here we only present the *result* of the calculation and discuss it.

The Witten index for the SQM model (5.48) on a Kähler manifold is given by the integral

$$\boxed{I = \int e^{\mathcal{F}/2\pi} \det{}^{1/2} \left[\frac{\frac{\mathcal{R}}{4\pi}}{\sin \frac{\mathcal{R}}{4\pi}} \right],} \tag{5.105}$$

where $\mathcal{R}$ is the curvature 2-form

$$\mathcal{R}^{AB} = \frac{1}{2} R^{AB}{}_{MN} dx^M \wedge dx^N. \tag{5.106}$$

---

[11] The terms in this expansion are known to mathematicians as *Chern classes* (the first Chern class, the second Chern class, etc.) associated with the line bundle A.

[12] Here $x^{M=1,\ldots,D}$ and $\Psi^{M=1,\ldots,D}$ are real ordinary and Grassmann coordinates. The complex coordinates are related to them according to $z^1 = (x^1 + ix^2)/\sqrt{2}, \psi^1 = (\Psi^1 + i\Psi^2)/\sqrt{2}$, etc. Also $\Psi^A = e^A_M \Psi^M$.

Note that for a pure (untwisted) Dolbeault complex with $W = (1/4)\ln\det h$, $\mathcal{F} = d\mathcal{A}$ is different from zero! Bearing in mind (5.63) and using the representation (5.27) of $I_N{}^P$ in the complex basis, one can represent (5.105) in the form

$$I = \int e^{\frac{\tilde{\mathcal{F}}}{2\pi}}$$

$$\times \ \exp\left[\frac{1}{16\pi}I_N{}^P\partial_N\partial_P(\ln\det g)dx^M \wedge dx^N\right]\det{}^{1/2}\left[\frac{\frac{\mathcal{R}}{4\pi}}{\sin\frac{\mathcal{R}}{4\pi}}\right], \ (5.107)$$

where $\tilde{\mathcal{F}}$ is the "mathematical" gauge field twisting the Dolbeault complex.

The determinant factor can be spelled out as

$$\det{}^{1/2}\left[\frac{\frac{\mathcal{R}}{4\pi}}{\sin\frac{\mathcal{R}}{4\pi}}\right] = 1 + \frac{1}{192\pi^2}\int \mathrm{Tr}\{\mathcal{R}\wedge\mathcal{R}\} + \dots \qquad (5.108)$$

For the actual calculation, one should pick up the volume D-form in the expansion of the integrand.

We see that for 2-dimensional manifolds the determinant curvature factor does not affect the result of the index, which still coincides with the magnetic flux. But for $D \geq 4$ it does. If $D = 4$, the index represents a sum of two terms,

$$I_{\mathbb{CP}^2} = \frac{1}{8\pi^2}\int \mathcal{F}\wedge\mathcal{F} + \frac{1}{192\pi^2}\int \mathrm{Tr}\{\mathcal{R}\wedge\mathcal{R}\}. \qquad (5.109)$$

Take $\mathbb{CP}^2$ with the field (5.87) as an example. The first term is equal to $q^2/2$ — the value given by the "naive" evaluation (5.98) for the index. But the second term is nonzero being equal to $-1/8$. We derive

$$I_{\mathbb{CP}^2} = \frac{q^2}{2} - \frac{1}{8}, \qquad (5.110)$$

which coincides with (5.97) and is integer for a half-integer q.

5.2.3.1 *Dirac index*

Let us now discuss the Dirac complex and its index, defined as

$$I_D = n_R^{(0)} - n_L^{(0)}, \qquad (5.111)$$

where $n_{R,L}^{(0)}$ are, correspondingly, the numbers of the right-handed and left-handed zero modes of $\slashed{\nabla}$. Bearing in mind the SQM-Dolbeault and Dolbeault-Dirac isomorphisms, the index (5.111) should seemingly coincide for Kähler manifolds with the Witten index of the SQM model (5.48) and be given by the formula (5.105).

This is almost true, but there is a subtlety associated with a slight difference in conventions. The Witten index is defined as the difference $n_B^{(0)} - n_F^{(0)}$ between the numbers of bosonic and fermion zero modes. And we mentioned [see the comment after Eq. (4.61)] that in *two* dimensions the states with *negative* chirality are *bosonic* in supersymmetric interpretation, while the positive chirality states are fermionic. Thus, for 2-dimensional systems, the Witten index and the Dirac index have opposite signs. Using the standard convention (5.75) for γ^{D+1}, Definition 5.9, and also the standard conventions for the associated SQM model, one can show that, in a general case, the Dirac index is given by almost the same formula (5.105) as the Witten index, but $\mathcal{F}$ should be replaced by $-\mathcal{F}$. It follows that $I_W = -I_D$ for $D = 4k + 2$, while for $D = 4k$ the two indices coincide.

The Dirac complex also exists for non-Abelian gauge fields. In the flat case, its index is given by the expression

$$I_D = \int \text{Tr} \left\{ e^{-\mathcal{F}/2\pi} \right\}, \qquad (5.112)$$

where $\mathcal{F}$ is now an element of the corresponding Lie algebra. The most known nontrivial examples are $su(2)$ instanton [72] and multiinstanton [73] fields in $\mathbb{R}^4$. The index is given then by the *Pontryagin number*

$$I_D = \frac{1}{32\pi^2} \varepsilon_{\mu\nu\rho\sigma} \int \text{Tr} \{ F_{\mu\nu} F_{\rho\sigma} \}. \qquad (5.113)$$

In particular, the index is equal to unity for the instanton field, signaling the presence of a single right-handed zero mode [74].

If the manifold is curved, the index is given by a similar formula, but one has to insert in the integrand the "gravitational factor" depending on the curvature. It has the same form as in Eq. (5.105).

An important remark is now in order. To make the integral (5.113) finite, the fields should rapidly decrease at large Euclidean distances. Then the spectrum of the Dirac operator is continuous and the index refers to the number of *normalized* zero modes of $\slashed{\nabla}$ that exist "on the continuous background" (a similar system will be discussed in Chapter 9). But one can get rid of the continuous spectrum in this case by considering the Dirac operator on 4-sphere rather than $\mathbb{R}^4$. The gravitational factor $\sim \int \text{Tr} \{ \mathcal{R} \wedge \mathcal{R} \}$ is zero for S^4 and does not contribute in the index.

5.2.3.2 *Non-Kähler manifolds*

Up to now we discussed only Kähler manifolds. But the Dolbeault complex is also well defined for a generic complex manifold and its index can also be evaluated in that case. This was first done by Hirzebruch [75] (and the result is called *Hirzebruch-Riemann-Roch theorem*) in a pure geometric framework.

But one can also calculate the index using supersymmetric methods. A direct calculation of the supersymmetric path integral (4.43) for the index is not feasible in this case due to the presence of the 4-fermion term $(\partial_k \partial_{\bar{q}} h_{m\bar{n}}) \psi^k \psi^m \bar{\psi}^{\bar{q}} \bar{\psi}^{\bar{n}}$ in the Lagrangian (5.49). This term brings about potential singularities in the path integral in the limit $\beta \to 0$, which are difficult to handle.

In Ref. [76], non-Kähler manifolds of a special class, the so-called *strong HKT manifolds* were considered, where, in spite of the fact that the Kähler form $\omega = h_{j\bar{k}} dz^j \wedge d\bar{z}^{\bar{k}}$ is not closed, the identity $\partial \bar{\partial} \omega = 0$ still holds so that the 4-fermion term vanishes. In this case, the Euclidean Lagrangian still has the form (5.104), but it includes now not the ordinary spin connections, but the so-called *Bismut* spin connections [76, 77],

$$\omega^{(B)}_{BC,M} = \omega_{BC,M} - \frac{1}{2} C_{MKP} e^K_B e^P_C, \qquad (5.114)$$

where C_{MKP} is the completely antisymmetric torsion tensor (5.80) chosen so that the components $\omega^{(B)}_{ab,\bar{m}}$ and $\omega^{(B)}_{\bar{a}\bar{b},m}$ (which are nonzero in the non-Kähler case for the ordinary spin connections and were a reason for some complications discussed on p. 101) vanish. The corresponding affine connections,

$$G^{(B)}_{P,MN} = \Gamma_{P,MN} + \frac{1}{2} C_{PMN}, \qquad (5.115)$$

also involve torsions.

For Kähler manifolds, the ordinary Levi-Civita covariant derivatives of the complex structure tensor I_{MN} vanish [see Eq.(5.25)]. They do not in a generic case, but the covariant derivatives of I_{MN} involving $G^{(B)}_{P,MN}$ rather than the ordinary Christoffel symbols $\Gamma_{P,MN}$ do vanish. The complex version of the Bismut Riemann tensor has the same structure (5.54) as the ordinary Riemann tensor on a Kähler manifold, having only the components $R^{(B)}_{m\bar{n}k\bar{l}}$ and those derived from them by symmetry.

For strong HKT manifolds, the functional integral (4.43) with the modified L_E can be calculated in the same way as in the Kähler case with the only difference that the ordinary Riemann curvature form in Eq. (5.105) is replaced by the Bismut one; this confirms [78] the result derived in Ref. [76] by a refined mathematical reasoning.

Having proven this, one can deform the torsions away capitalizing on the fact that the value of the index (always integer) cannot be changed under a smooth deformation. We arrive at the result (5.105) with the ordinary Riemann curvature!

For a generic non-Kähler manifold, the functional integral for the Witten index of the Dolbeault SQM model is difficult to calculate due to the presence of the 4-fermion term in L_E, but, fortunately, there is an alternative: we can make use of the the Dolbeault-Dirac isomorphism established above. We showed that, in a generic non-Kähler case, the sum $i(\hat{Q} + \hat{\bar{Q}})$ maps to the torsionful Dirac operator (5.79). Again, we can unwind the torsions away so that the index does not change. The index of the ordinary Dirac operator can be calculated for *any* manifold, not necessarily complex, as a functional integral with the Euclidean Lagrangian (5.104) expressed via the real variables $x^M(\tau)$ and $\Psi^M(\tau)$. The result of the calculation of this integral[13] is still given by (5.105). This formula describes the Dolbeault index for Kähler and also for non-Kähler complex manifolds [75, 76, 78], which can be represented in the form (5.107).

Consider two non-Kähler examples.

(1) The manifold S^4 is not complex.[14] As is well known [80], the eigenvalues of a "gravitational" Dirac operator on a sphere are positive definite, so that the index is zero. And the same follows from the formula (5.105). The integrand is proportional in this case to $\mathcal{R}_{MN} \wedge \mathcal{R}^{MN}$. For the sphere, $\mathcal{R}^{MN} \propto dx^M \wedge dx^N$ and the integrand vanishes.

(2) The *Hopf manifold* is a complex manifold of complex dimension $d = 2$ with the conformally flat metric[15]

$$ds^2 = \frac{d\bar{z}_j dz_j}{\bar{z}_k z_k}, \qquad (5.116)$$

where the complex coordinates $z_{j=1,2}$ dwell in the region $1 \leq |z_j| \leq 2$ and the points z_j and $2z_j$ are identified. The metric (5.116) is consistent with this identification.

This is a non-Kähler manifold of strong HKT variety, $\bar{\partial}\partial\omega = 0$. Let $\mathcal{F} = 0$. One can show [78] that the corresponding SQM Hamiltonian has a zero mode in the sector $F = 0$, $\Psi(\bar{z}, z; \psi) = 1$, and a zero mode in the sector $F = 1$, $\Psi(\bar{z}, z; \psi) = \bar{z}_j \psi_j / (\bar{z}_l z_l)$, so that the index vanishes. This is confirmed by the calculation of the integral (5.107).

---

[13]We address again the reader to Sect. 14.4 of the book [4].

[14]One can consider S^4 with one point removed, in which case the Dolbeault complex can still be defined [79]. But this complex has some specific features, and it does not map to the Dirac complex.

[15]We need not distinguish here between covariant and contravariant, holomorphic and antiholomorphic indices.

Indeed, the symmetry dictates

$$
\begin{aligned}
\mathcal{R}_{MN} &= A(x)\, dx_M \wedge dx_N \\
&+ B(x)\,(x_M dx_N - x_N dx_M) \wedge (x_Q dx_Q) \quad (5.117)
\end{aligned}
$$

with some irrelevant for us scalar functions $A(x)$ and $B(x)$. The integrand $\propto \mathcal{R}_{MN} \wedge \mathcal{R}^{MN}$ vanishes.

5.3   Hirzebruch signature

The Hirzebruch signature is a topological characteristic of a manifold of real dimension $D = 4k$, which is relative to the Euler characteristic. Like the topological invariants considered above, it can be interpreted as the Witten index of a certain SQM system.

Consider the de Rham complex on a $4k$-dimensional manifold. Consider the Hodge duality operator acting on a p-form α:

$$
\star \alpha = \frac{1}{(D-p)!}\, E^{M_1 \ldots M_p}{}_{M_{p+1} \ldots M_D}\, \alpha_{M_1 \ldots M_p}\, dx^{M_{p+1}} \wedge \cdots \wedge dx^{M_D}, (5.118)
$$

where

$$
E_{M_1 \ldots M_D} = \sqrt{g}\, \varepsilon_{M_1 \ldots M_D} \tag{5.119}
$$

is a covariant antisymmetric tensor. The square of the star operator is sometimes $+1$ and sometimes -1, which is not convenient for our purposes. But a remarkable fact is that, if $D = 4k$, one can define the *signature operator*, the composition[16]

$$
\hat{\tau} = (-1)^{p(p+1)/2} \circ \star \tag{5.121}
$$

which always squares to 1 and anticommutes with the operator $\hat{R} = d + d^\dagger$ (hence it commutes with the Laplace-de Rham operator $\triangle = -dd^\dagger - d^\dagger d$). It follows that the eigenvalues of τ are either $+1$ or -1. Thus, the whole set of p-forms can be divided in this case into two subsets: the subset S_+ of the forms with positive signature for which $\hat{\tau}\alpha = \alpha$ and the subset S_- of the forms with negative signature. The property $\{\hat{R}, \hat{\tau}\} = 0$ implies that the action of the operator $\hat{R}$ on a positive-signature state gives a state with negative signature and vice versa.

---

[16] This actually means

$$
\hat{\tau}\alpha_p = (-1)^{(4k-p)(4k-p+1)/2}\, \star \alpha_p = (-1)^{p(p-1)/2}\, \star \alpha_p. \tag{5.120}
$$

In other words, $\hat{R}$ plays, in this setting, the same role that the Dirac operator plays for the Dirac complex, while $\hat{\tau}$ plays the role of γ^{D+1}. In the full analogy with $\not\nabla (1 \pm \gamma^{D+1})$, the operators $\hat{R}_\pm = \hat{R}(1 \pm \hat{\tau})$ are nilpotent, Hermitian conjugate to one another and play the role of the supercharges.

Note that these Hirzebruch supercharges do not coincide at all with the de Rham supercharges d and $d^\dagger$. However, their anticommutator gives *the same* Hamiltonian (coinciding up to a sign with the Laplace-de Rham operator) as in the de Rham complex. Note also that the extended $\mathcal{N} = 4$ supersymmetry does not hold here: the anticommutators $\{\hat{R}(1 \pm \tau), d\}$ and $\{\hat{R}(1 \pm \tau), d^\dagger\}$ do not vanish.

Definition 5.10. The *Hirzebruch signature* of a $4k$–dimensional manifold is defined as

$$\boxed{sign = b_+ - b_-,}$$

(5.122)

where $b_\pm$ are the numbers of closed but not exact forms, belonging, correspondingly, to S_+ and S_-.

Remarks

- The terminology is a little awkward here, but we hope that the reader will not confuse two different notions: the signature of a form (the eigenvalue of $\hat{\tau}$) and the Hirzebruch signature of a manifold.
- Do not mix up the Hirzebruch signature with the Euler characteristic (5.32), which also can be presented in the form (5.122), but the numbers b_+ and b_- would have in that case a completely different meaning being the sums of the Betti numbers b_p with even and odd p.
- In fact, only the forms of degree $p = D/2 = 2k$ contribute to (5.122). Indeed, the duality operator transforms p-forms to $(D - p)$-forms. Hence, for any form $\alpha_{p \neq D/2}$, the form $\alpha_p + \hat{\tau}\alpha_p$ belongs to S_+, while the form $\alpha_p - \hat{\tau}\alpha_p$ belongs to S_-. The first form contributes to b_+, while the second one to b_-. The difference (5.122) is not sensitive for that.
- Also the space of forms of degree $D/2$ with *nonzero* eigenvalues of the de Laplace-de Rham operator $-\hat{R}^2$ can be split into doublets involving a form belonging to S_+ and a form belonging to S_-. They go to one another by the action of $\hat{R}_\pm$ (and do not affect the index).

On the supersymmetric side, we are dealing with the system described by the Hamiltonian (5.21).

Theorem 5.6. *The operator (5.121) has in this language a nice represen-*
tation

$$\hat{\tau} = \prod_{M=1}^{4k} \left(\frac{\partial}{\partial \psi^M} - \psi^M \right). \tag{5.123}$$

Proof. The operator (5.123) can be represented as a product of k blocks:

$$\hat{\tau} = \prod_{M=1}^{4} \left(\frac{\partial}{\partial \psi^M} - \psi^M \right) \times \prod_{M=5}^{8} \left(\frac{\partial}{\partial \psi^M} - \psi^M \right) \times \cdots . \tag{5.124}$$

Each block commutes with "alien" fermion variables (for example, the first
block commutes with $\psi^{M=5,\ldots,4k}$). We leave it for the reader to check that
the action of, for example, the first block on the variables $\psi^{M=1,2,3,4}$ and
their different products is the same as is prescribed by (5.120). □

The Hirzebruch signature may be expressed as

$$sign = \text{Tr}\{\hat{\tau} e^{-\beta \hat{R}^2}\}, \tag{5.125}$$

where Tr is the operator Hilbert space trace. This is isomorphic to the
supersymmetric partition function $\tilde{Z} = \text{Tr}\{(-1)^{\hat{F}} e^{-\beta \hat{H}}\}$ [cf. Eq. (4.51)]
and admits a path integral representation. The corresponding path integral
can be calculated. We will not do so here and only quote the result.

Theorem 5.7. *The Hirzebruch signature is given by the integral*

$$sign = \int \det^{-1/2} \left[\frac{\tan \frac{\mathcal{R}}{2\pi}}{\frac{\mathcal{R}}{2\pi}} \right], \tag{5.126}$$

where $\mathcal{R}$ is the curvature 2-form (5.106).

Proof. It consists in a not so simple calculation of the appropriate path
integral. The reader can find it in the original papers [48, 49]. □

For 4-dimensional manifolds,

$$sign = -\frac{1}{24\pi^2} \int \text{Tr}\{\mathcal{R} \wedge \mathcal{R}\}. \tag{5.127}$$

The simplest nontrivial example is $\mathbb{CP}^2$. This manifold is 4-dimensional.
The nonzero Betti numbers are $b_0 = b_2 = b_4 = 1$. The closed but not
exact 2-form α_2 happens to have the eigenvalue $+1$ with respect to $\hat{\tau}$ and
the Hirzebruch signature is equal to 1. The integral (5.127) also gives this
value.

Chapter 6

Vacuum structure of 4-dimensional supersymmetric theories I

In Chapters 2,4 and 5, we discussed the vacuum structure and Witten indices of different SQM models. But, as we mentioned above, the main physical (possibly, phenomelogical) interest of such studies lies in 4-dimensional supersymmetric field theories — our world is 4-dimensional and it is important to know whether supersymmetry is broken spontaneously or not in this or that model that may describe it. We will mainly discuss supersymmetric gauge theories with the Lagrangians written in Sect. 3.3, which attracted an intense attention of the theorists. But before doing that, let us consider the simplest nontrivial supersymmetric field model — the Wess-Zumino model.

6.1 Wess-Zumino model

The component Lagrangian of the renormalizable Wess-Zumino model is given by the expression (3.19) with the superpotential

$$\mathcal{W}(\phi) = \frac{m}{2}\phi^2 + \frac{\lambda}{6}\phi^3. \qquad (6.1)$$

The spectrum of the model includes interacting massive bosons and fermions. A gap separates these states from the vacuum. Under such conditions, we can safely deform the model putting it in a spatial box of size $L \ll m^{-1}$ and leave in the Hamiltonian only the zero spatial Fourier modes of the fields. We arrive at the SQM Hamiltonian

$$H = p^*p + \left(m\phi^* + \frac{\lambda(\phi^*)^2}{2}\right)\left(m\phi + \frac{\lambda\phi^2}{2}\right)$$
$$+ \frac{1}{2}(m + \lambda\phi)\psi\psi + \frac{1}{2}(m + \lambda\phi^*)\bar{\psi}\bar{\psi} \qquad (6.2)$$

[for $m = 0, \lambda = -2$, it coincides with (2.68)]. The index of the SQM model thus obtained coincides with the index of the original field-theory model.

The potential

$$V(\phi^*, \phi) = \left| m\phi + \frac{\lambda}{2}\phi^2 \right|^2$$

has two minima: at $\phi = 0$ and at $\phi = -2m/\lambda$. If λ is small (but different from zero to ensure the existence of the second minimum), the barrier between the minima is high and two classical vacua go over to two quantum vacua. They have the bosonic nature, and hence the Witten index is equal to 2. And the same result one obtains by calculating the functional integral in the CG approximation (4.47). We derive[1]

$$
\begin{aligned}
I_W &= \frac{1}{4\pi^2} \int dp^* dp \, d\phi^* d\phi \prod_{\alpha=1,2} d\psi_\alpha d\bar{\psi}^\alpha \, e^{-\beta H} \\
&= -i\frac{\beta}{2\pi} \int d\phi^* d\phi \, \mathcal{W}''(\phi^*) \mathcal{W}''(\phi) \, e^{-\beta \mathcal{W}'(\phi^*)\mathcal{W}'(\phi)} .
\end{aligned}
\tag{6.3}
$$

We introduce now the variable $w = \sqrt{\beta}\,\mathcal{W}'(\phi)$ and reduce the integral to

$$\boxed{I_W = -i\mathcal{N}\frac{1}{2\pi} \int dw^* dw \, e^{-w^* w} = \mathcal{N},}
\tag{6.4}$$

where $\mathcal{N}$ is the degree of the mapping $\phi \to \sqrt{\beta}\,\mathcal{W}'(\phi)$ of the complex plane onto itself. For the superpotential (6.1), $\mathcal{N} = 2$.

6.1.1   *O'Raifeartaigh model*

Consider a model including *three* chiral superfields $\Phi_{1,2,3}$ with the superpotential [81]

$$\mathcal{W} = \kappa\Phi_1 + m\Phi_2\Phi_3 + \frac{g}{2}\Phi_1\Phi_3^2,
\tag{6.5}$$

where κ, m, g are some nonzero constants, which we assume to be real. The potential part of the Hamiltonian is

$$V(\phi_j, \phi_j^*) = \left| \frac{\partial \mathcal{W}}{\partial \phi_j} \right|^2 .
\tag{6.6}$$

One can now observe that the zero vacuum energy conditions

$$
\begin{aligned}
\frac{\partial \mathcal{W}}{\partial \phi_1} &= \kappa + \frac{g}{2}\phi_3^2 = 0, \\
\frac{\partial \mathcal{W}}{\partial \phi_2} &= m\phi_3 = 0, \\
\frac{\partial \mathcal{W}}{\partial \phi_3} &= m\phi_2 + g\phi_1\phi_3 = 0
\end{aligned}
\tag{6.7}
$$

---

[1]Here $\phi = (x + iy)/\sqrt{2}$, $p = (p_x - ip_y)/\sqrt{2}$ and hence $d\phi^* d\phi = idxdy$ and $dp^* dp = -idp_x dp_y$.

cannot be satisfied simultaneously. Thus, the vacuum has a nonzero energy: supersymmetry is spontaneously broken.

The Witten index of such a model should vanish, and it does. After momentum and fermion integrations, the CG integral acquires the form

$$I \propto \int \prod_j d\phi_j^* d\phi_j \left| \det \left(\frac{\partial^2 \mathcal{W}}{\partial \phi_j \partial \phi_k} \right) \right|^2 \exp\{-\beta V(\phi_j, \phi_j^*)\} . \qquad (6.8)$$

In our case,

$$\det \left(\frac{\partial^2 \mathcal{W}}{\partial \phi_j \partial \phi_k} \right) = \det \begin{pmatrix} 0, & 0, & g\phi_3 \\ 0, & 0, & m \\ g\phi_3, & m, & g\phi_1 \end{pmatrix} = 0 . \qquad (6.9)$$

Note that the simple WZ model with the superpotential $\mathcal{W}(\Phi) = \kappa\Phi$ would also have a nonzero vacuum energy signaling supersymmetry breaking. But that model is *too* simple and dull, being just a free theory with a complex massless scalar and fermion. The overall shift of energy does not affect dynamics in that case.

On the other hand, O'Raifeartaigh model involves nontrivial interactions. The interaction constant g is dimensionless and the theory is renormalizable. At the tree level, the potential does not depend on ϕ_1, bringing about a massless scalar particle in the spectrum, but this degeneracy is lifted after taking into account loop corrections [82]. Due to supersymmetry breaking, the mass spectra of the bosons and fermions do not coincide. In particular, the spectrum of the theory includes an exactly massless fermion *goldstino* mode. The existence of such mode follows from a generalization of the Goldstone theorem. Spontaneous breaking of an ordinary global symmetry entails the existence of a massless boson, and spontaneous breaking of supersymmetry entails the existence of a massless fermion [83].

We now go over to the discussion of *gauge* supersymmetric theories. The common problem of all such theories is the presence of massless modes which make the spectrum of the Hamiltonian continuous. And we know that the index is not so well defined in such cases. For example, if one tries to calculate the functional integral for the index in the model (3.40) using the CG method, one obtains a fractional result [57] [2] $I_W = 1/4$ [cf. Eq. (4.57) for an attempt to evaluate the index in superconformal quantum mechanics]. To cope with this problem, one has to regularize the theory in the infrared, and a universal method to do so is putting the theory in

---

[2] The first calculation of this integral was performed in Ref. [84]a, but I missed there the overall factor 1/4. See Chapter 9 for more discussion.

a spatial box of finite size[3] L. In the next two sections, I will present the results of Witten's analysis [47].

6.2   Supersymmetric electrodynamics

The physical fields in this theory are gauge potentials [their dynamical components are $A_{j=1,2,3}(\boldsymbol{x}, t)$], photino λ_α, matter scalars s, t and matter fermions ψ, χ. The component Lagrangian was written in Eq. (3.34). Note that this Lagrangian is invariant under a discrete (charge conjugation) symmetry:

$$\text{C}: \qquad A \to -A, \ \lambda \to -\lambda, \ \psi \leftrightarrow \xi, \ s \leftrightarrow t. \qquad (6.10)$$

So, we put the system in a finite spatial box of size L imposing on all the fields the periodic boundary conditions:

$$A_j(x + L, y, z) = A_j(x, y + L, z) = A_j(x, y, z + L) = A_j(x, y, z),$$
$$s(x + L, y, z) = s(x, y + L, z) = s(x, y, z + L) = s(x, y, z), \qquad (6.11)$$

etc. and expand them into the Fourier series. The characteristic excitation energy of all nonzero Fourier harmonics is then of order $1/L$. For the zero Fourier modes of the matter fields, it is of order of their mass m, which we assume to be large enough (at the level $\sim 1/L$ or larger). We assume also the coupling constant e^2 to be small. Under this conditions, the low-energy dynamics of the theory is determined by the zero Fourier modes of the gauge field and its fermion superpartner.

Note that, at finite L, the constant potentials $A_j^{(0)}$ cannot be completely gauged away, as they can in the infinite volume. The admissible gauge transformations,

$$A_j \to A_j + \partial_j \chi(\boldsymbol{x}), \quad (s, \psi) \to (s, \psi)e^{-ie\chi(\boldsymbol{x})}, \quad (t, \xi) \to (t, \xi)e^{ie\chi(\boldsymbol{x})}, (6.12)$$

should respect the periodicity conditions (6.11). This implies[4] $A_j^{(0)}$ live on the dual torus, $A_j^{(0)} \in [0, \frac{2\pi}{eL})$, and the different potentials within this domain are not gauge-equivalent. As we will shortly see, the characteristic excitation energy for this mode is

$$E_{\text{char}} \sim e^2/L, \qquad (6.13)$$

which is small compared to $1/L$ and m.

---

[3]In this chapter, we will assume that $L \neq 0$. The systems with $L = 0$ (gauge supersymmetric quantum mechanics), where the continuous spectrum reappears, will be treated in Chap. 9.

[4]Implies in a certain sense — see the discussion in Sect. 6.2.1.

The low-energy dynamics of the system is then described by the effective Lagrangian including only three bosonic variables $c_j = A_j^{(0)}$ and two Grassmann variables $\eta_\alpha = \lambda_\alpha^{(0)}$. In the physical slang (which goes back to the classic paper [85] devoted to the spectrum of the hydrogen molecule — the positions of the electrons were treated there as *fast* variables and the position of the protons as *slow* ones), the variables c_j and η_α are slow and all other variables are fast to be "integrated out".

The effective slow Lagrangian has a very simple form:[5]

$$L^{\text{eff}} = L^3 \left(\frac{\dot{c}_j^2}{2} + i\eta\dot{\bar{\eta}} \right) . \tag{6.14}$$

The corresponding quantum effective Hamiltonian,

$$\boxed{\hat{H}^{\text{eff}} = \frac{\hat{P}_j^2}{2L^3} = -\frac{1}{2L^3}\triangle_c ,} \tag{6.15}$$

describes free motion over the dual torus of length $a = 2\pi/(eL)$. The characteristic momentum for the low-energy states is therefore $P_{\text{char}} \sim eL$. Bearing this in mind, we reproduce the estimate (6.13).

The Hamiltonian (6.15) is supersymmetric: it satisfies the algebra (2.65) with the Hermitially conjugated supercharges

$$\hat{Q}_\alpha \propto (\sigma_j)_\alpha{}^\beta \eta_\beta \hat{P}_j, \qquad \hat{\bar{Q}}^\alpha \propto (\sigma_j)_\beta{}^\alpha \bar{\eta}^\beta \hat{P}_j . \tag{6.16}$$

It acts on the wave functions $\Psi(c_j, \eta_\alpha)$. The full spectrum includes four sectors:

$$\Psi_0(\boldsymbol{c}, \eta_\alpha) = f_0(\boldsymbol{c}), \qquad \Psi_{1,2}(\boldsymbol{c}, \eta_\alpha) = \eta_{1,2} f_{1,2}(\boldsymbol{c}),$$
$$\Psi_3(\boldsymbol{c}, \eta_\alpha) = \eta_1 \eta_2 f_3(\boldsymbol{c}) . \tag{6.17}$$

We impose on all $f_k(\boldsymbol{c})$ periodic boundary conditions:[6]

$$f_k(\boldsymbol{c} + \boldsymbol{a}n) = f_k(\boldsymbol{c}) \tag{6.18}$$

with integer $\boldsymbol{n}$. Then

$$f_k(\boldsymbol{c}) = \sum_{\boldsymbol{m}} b_{k\boldsymbol{m}} e^{ieLc\cdot\boldsymbol{m}} \tag{6.19}$$

with integer $\boldsymbol{m}$. The spectrum splits into degenerate quartets with energies

$$E_{\boldsymbol{m}} = \frac{e^2}{2L}\boldsymbol{m}^2 . \tag{6.20}$$

---

[5] Just delete in (3.34) all the terms including the matter fields, neglect nonzero modes of A_j and λ_α and multiply the result by the spatial volume $V = L^3$ (the L in the RHS is not the Lagrangian, of course).

[6] More general boundary conditions will be considered in Sect. 6.2.1.

It involves four zero-energy states:

$$\Psi_0^{(0)} = 1, \qquad \Psi_1^{(0)} = \eta_1, \qquad \Psi_2^{(0)} = \eta_2, \qquad \Phi_3^{(0)} = \eta_1\eta_2. \qquad (6.21)$$

The states $\Psi_{0,3}^{(0)}$ are bosonic, while the states $\Psi_{1,2}^{(0)}$ are fermionic. It follows that the Witten index vanishes:

$$I_W = 2 - 2 = 0. \qquad (6.22)$$

Generically, this would not allow us to conclude anything about the vacuum structure of the original undeformed Hamiltonian. The supersymmetric quartet of lowest states could in principle move up from zero, however they do not, and there is a *special reason* for that.[7]

As was mentioned, the Lagrangian (3.34) has the discrete symmetry (6.10). This allows us to consider alongside with the supersymmetric partition function $\mathrm{Tr}\{(-1)^{\hat{F}} e^{-\beta\hat{H}}\}$ an object

$$I_C = \mathrm{Tr}\left\{\hat{C}(-1)^{\hat{F}} e^{-\beta\hat{H}}\right\} \qquad (6.23)$$

where the states with negative C-parity enter the sum with an extra minus. For the effective Hamiltonian (6.15), the symmetry (6.10) boils down to

$$c \to -c, \qquad \eta_\alpha \to -\eta_\alpha. \qquad (6.24)$$

The states $\Psi_{0,3}^{(0)}$ are even under this transformation, whereas the states $\Psi_{1,2}^{(0)}$ are odd. As a result, the contribution of the vacuum states in I_C is $2 + 2 = 4 \neq 0$.

And the excited states of the Hamiltonian do not contribute there. The reason is simple: the supercharges (6.16) commute with $\hat{C}$ and hence all the states in a quartet including the states that are related by the action of the supercharges have the same C-parity and do not contribute in (6.23).

We can now proceed with the same reasoning as for the ordinary Witten index. Given that only the vacuum states contribute in I_C, its value, being integer, cannot depend on smooth deformations, and we may say that $I_C = 4$ as well for the original undeformed system. And this implies that the latter has four supersymmetric zero-energy vacuum states and supersymmetry is not broken.

The following important remark is, however, in order. The validity of this calculation for the original field theory may look not so obvious because the result (6.15) for the effective Hamiltonian is valid only in the

---

[7]The reader might recall at this juncture a remark on p. 66.

leading Born-Oppenheimer approximation. There are perturbative corrections which become large in the regions close to the corner of the box where
the dimensionless parameter

$$\kappa = \frac{1}{e|L\boldsymbol{c}|^3} \tag{6.25}$$

ceases to be small. These corrections have been evaluated [86] in the first
order in κ both for SQED and for the pure SYM theory to be discussed
in the next section. They do not bring about an effective potential, but
instead modify the kinetic term in (6.15) to endow it with a nontrivial
metric. The full supersymmetric effective Hamiltonian acquires the form

$$\hat{H}^{\text{eff}} = -\frac{1}{2} f \overrightarrow{\partial_k^2} f + i\varepsilon_{kpl} \hat{\bar{\eta}}\sigma_l \eta f(\partial_p f)\partial_k + \frac{1}{6} f(\partial_k^2 f) (\hat{\bar{\eta}}\sigma_l \eta)^2, \tag{6.26}$$

where we have set $L = 1$, the derivatives in the first term act on whatever
they find on the right, and $f(\boldsymbol{c}) = 1 - \kappa/4 + o(\kappa)$. One can observe now
that, in the main body of the dual torus, the Hamiltonian (6.26) is *close* to
(6.15) and the admissible vacuum wave functions,

$$\Psi_0 = f^{-1}, \qquad \Psi_3 = f^{-1}\eta_1\eta_2, \tag{6.27}$$

are close to the functions $\Psi_{0,3}$ in (6.21). It is true that, when one takes into
account only one-loop corrections, the functions (6.27) become singular
at the small sphere $|\boldsymbol{c}| = 1/[L(4e)^{1/3}]$ around the corner of the torus,[8]
but the BO approximation breaks down in this region. One may *assume*
that the corner region does not lead to a trouble and does not change
the vacuum counting (one can say that the BO corrections *deform* the
effective Hamiltonian, and the Witten index must stay invariant under such
deformation), though a rigorous *proof* of this assertion is absent.

6.2.1 *Universes*

There is a subtlety which we have ignored up to now in our analysis. On
p. 118 we wrote that the relevant domain of $A_j^{(0)}$ is

$$A_j^{(0)} \in \left[0, \frac{2\pi}{eL}\right]. \tag{6.28}$$

---

[8]Note that the effective Hamiltonian for the non-Abelian pure supersymmetric Yang-
Mills theory with the $SU(2)$ group to be discussed in the next section also has the form
(6.26), but in the next-to-leading BO order, the function f is

$$f(\boldsymbol{c}) = 1 + \frac{3}{4g|L\boldsymbol{c}|^3}$$

and the corresponding wave functions are *not* singular at the corner.

In a certain sense, it is correct. Indeed, the field configurations A_j and $A_j + 2\pi n_j/(eL)$ are related by admissible gauge transformations that respect the periodicity condition (6.11). If we require the wave function Ψ to be invariant under gauge transformations, the statement above is correct without reservations. However, strictly speaking, the Gauss law constraints, which should be imposed on wave functions, only imply that the latter are invariant under *infinitesimal* gauge transformations and hence under topologically trivial finite gauge transformations that can be reduced to trivial ones by a set of continuous deformations. And a gauge transformation of the type

$$\hat{\omega}(n_j) : \ A_j \to A_j + \frac{2\pi}{eL} n_j,$$

$$(s, \psi) \to e^{-2\pi i n \cdot x/L}(s, \psi), \qquad (t, \xi) \to e^{2\pi i n \cdot x/L}(t, \xi)\,, (6.29)$$

which can bring any $U(1)$ gauge field to a field whose zero Fourier mode lies in the dual torus (6.28), is not topologically trivial, but represents an uncontractible loop.

In such a case, we may impose generalized constraints:

$$\hat{\omega}(n_j)\Psi \ = \ e^{i\vartheta \cdot n}\Psi, \qquad \vartheta_j \in [0, 2\pi)\,. \tag{6.30}$$

The wave functions in a universe[9] with given ϑ_j do not "talk" to the functions in some other universe in a sense that the result of the action of a local physically relevant operator on a function from a given universe belongs to the same universe.

It follows from (6.29) and (6.30) that the effective wave functions $\Psi_{\vartheta}^{\text{eff}}(c_j)$ in the universe ϑ satisfy the constraints

$$\boxed{\Psi_{\vartheta}^{\text{eff}}\left(c_j + \frac{2\pi}{eL} n_j\right) \ = \ e^{i\vartheta \cdot n}\,\Psi_{\vartheta}^{\text{eff}}(c_j)\,.} \tag{6.31}$$

Then any such function has the form

$$\Psi_{\vartheta}^{\text{eff}}(c_j) \ = \ e^{ieL\vartheta \cdot c/(2\pi)}\,\Psi_{0}^{\text{eff}}(c_j)\,, \tag{6.32}$$

where $\Psi_{0}^{\text{eff}}(c_j)$ is periodic on the dual torus. The ground states in the spectrum of the effective Hamiltonian (6.15) have the energy

$$E_0(\vartheta) \ = \ \frac{e^2\vartheta^2}{8\pi^2 L}\,. \tag{6.33}$$

---

[9]A more traditional term is "superselection sector", but recently the name "universe" has gone into use (see e.g. Ref. [87]) and we are following the crowd.

Thus, the true supersymmetric vacua dwell in the universe with $\vartheta = 0$. Their wave functions are invariant under all gauge transformations — topologically trivial or otherwise. The results $I_W = 0$ and $I_C = \text{Tr}\{\hat{C}e^{-\beta\hat{H}}\} = 4$ for the index remain intact.

Looking at (6.33), a reader may think that the system has continuous spectrum, may recall that we mentioned on several occasions that the notion of index is ill-defined in such systems, and get confused.

However, the spectrum becomes continuous only if the spectra of all the universes are brought together in one heap. One should not do that. A correct interpretation of Eqs. (6.31) – (6.33) is the following:

When we put SQED in a finite box and impose periodic boundary conditions, infinitely many quantum systems corresponding to infinitely many universes can be defined. In all the universes with nonzero ϑ, supersymmetry is spontaneously broken (indeed, in that case, the ground states are not annihilated by the action of the supercharges). In the universe $\vartheta = 0$, supersymmetry is not broken and the spectrum includes two bosonic and two fermionic zero-energy states.

6.2.2   *Fayet-Illiopoulos model*

Let us modify the model (3.32) by adding to the Lagrangian the extra term [88],

$$\mathcal{L}_{FI} = \frac{\xi}{4} \int d^4\theta \, V . \tag{6.34}$$

The modified theory is still gauge-invariant because the shift of V in the supergauge transformation (3.24) gives zero after integration over $d^4\theta$. The Fayet-Illiopolous theory is also supersymmetric, but, as we will shortly see, supersymmetry is *spontaneously broken* in this case.

The analysis is similar to what we did for the O'Raifeartaigh model in the previous section. Excluding the auxiliary fields F, F^*, G, G^*, and D, we arrive at the following expression for the classical potential:

$$V(s, s^*, t, t^*) = m^2(s^*s + t^*t) + \frac{[\xi + e(s^*s - t^*t)]^2}{2} . \tag{6.35}$$

If $\xi \neq 0$, this potential is positive definite, the vacuum energy is not zero, and supersymmetry is broken. Assume for definiteness that ξ is positive. There are two distinct physical possibilities.

- $m^2 > e\xi$. In this case, the minimum of the potential is achieved at $s = t = 0$. Gauge symmetry is not broken by Higgs effect, and the physical spectrum includes the massless photon and massless photino playing the role of goldstino.

- $m^2 < e\xi$. The minimum occurs at

$$\langle s \rangle = 0, \quad \langle |t| \rangle^2 = \frac{e\xi - m^2}{e^2}.$$

Gauge symmetry is broken and photon acquires mass. The fermion mass matrix can be extracted from (3.34). Its diagonalization gives the massless goldstino mode,

$$G_\alpha = \lambda_\alpha - \frac{ie\sqrt{2}\langle t^* \rangle}{m} \psi_\alpha, \qquad (6.36)$$

and a Dirac fermion carrying the mass

$$M = \sqrt{m^2 + 2e^2 \langle |t^2| \rangle} = \sqrt{2e\xi - m^2}. \qquad (6.37)$$

Thus, gauge symmetry may be broken or not, but supersymmetry always *is* as soon as ξ is not zero.

This is in line with the Witten index analysis. As we have seen, in the undeformed theory, Witten index is zero, but supersymmetry is still not broken because of the nonzero value of the generalized index (6.23). But the latter does not have much meaning in the deformed theory, because the FI term *breaks* the charge conjugation symmetry (6.10). We are only left with the ordinary Witten index, it is zero, and there is no wonder that supersymmetry breaks down.

6.3   Supersymmetric Yang-Mills theory: unitary gauge groups

We start our discussion of non-Abelian gauge theories with supersymmetric YM theories not including matter superfields and described by the Lagrangian (3.36), (3.40). We also assume in this section that the gauge group $\mathcal{G}$ is $SU(N)$. More complicated systems based on other classical Lie groups will be considered later.

To regularize the system in the infrared and make the spectrum of the Hamiltonian discrete, we proceed as before and put the system in a finite spatial box. One can then impose periodic boundary conditions as in (6.11), but in the non-Abelian case, one can also impose the so-called *twisted* boundary conditions to be discussed in the second part of the section. We say right away that the two ways of handling the system bring about the same physical answer: $I_W \neq 0$ and supersymmetry is *not* spontaneously broken.

6.3.1 *Periodic boundary conditions*

We take the Lagrangian (3.40), eliminate the auxiliary field D^a using the equation of motion $D^a = 0$, impose the boundary conditions

$$A_j^a(x+L, y, z) = A_j^a(x, y+L, z) = A_j^a(x, y, z+L) = A_j^a(x, y, z),$$
$$\lambda_\alpha^a(x+L, y, z) = \lambda_\alpha^a(x, y+L, z) = \lambda_\alpha^a(x, y, z+L) = \lambda_\alpha^a(x, y, z), \quad (6.38)$$

and expand $A_j^a(\boldsymbol{x})$ and $\lambda_\alpha^a(\boldsymbol{x})$ in the Fourier series.

In the Abelian case, the vacuum dynamics was associated with the constant gauge field modes, because the configurations $A_j(x, y, z) = c_j$ (and only them) corresponded to the zero field strength and zero vacuum energy. One could interpret the motion over the dual torus discussed in the previous section as the motion on the *moduli space* of classical vacua.

What are classical vacuum field configurations in the non-Abelian case? Locally, any configuration $\hat{A}_j(\boldsymbol{x})$ with zero field strength[10]

$$\hat{F}_{jk} = \partial_k \hat{A}_j - \partial_j \hat{A}_k - ig[\hat{A}_j, \hat{A}_k]$$

(a *flat connection* in the mathematical language) can be represented as

$$\hat{A}_j = -\frac{i}{g} U^{-1} \partial_j U, \quad (6.39)$$

where $U(\boldsymbol{x})$ is an element of the gauge group $\mathcal{G}$. We may pick out a particular point in our cube, say the vertex $(0, 0, 0)$, and define a set of holonomies (Wilson loops along nontrivial cycles of the torus)

$$\Omega_1 = P \exp\left\{ ig \int_0^L \hat{A}_1(x, 0, 0) dx \right\},$$

$$\Omega_2 = P \exp\left\{ ig \int_0^L \hat{A}_2(0, y, 0) dy \right\}, \quad (6.40)$$

$$\Omega_3 = P \exp\left\{ ig \int_0^L \hat{A}_3(0, 0, z) dz \right\},$$

where $P \exp\{\cdots\}$ is a path-ordered exponential — a product of an infinite number of factors:

$$\Omega_1 = \lim_{n \to \infty} \prod_{l=0}^{n-1} \left[\mathbb{1} + ig \frac{L}{n} \hat{A}_1\left(\frac{lL}{n}, 0, 0 \right) \right] \quad (6.41)$$

and similarly for Ω_2 and Ω_3.

---

[10]Comparing this with (1.20), do not forget our conventions: $\partial_\mu = (\partial_0, \partial_j)$, but $A_\mu = (A_0, -A_j)$.

Let us prove a theorem [89].

Theorem 6.1. *For a periodic flat connection on the 3-torus, the holonomies (6.40) commute. Inversely, for any set of commuting elements $\Omega_{1,2,3}$ of a connected, simply connected group $\mathcal{G}$, a periodic flat connection exists such that Ω_j are its holonomies.*

Proof. The proof of the direct theorem is simple. Infinitesimally,

$$\mathbb{1} + \frac{i}{g} A_j(\boldsymbol{x})dx^j = \mathbb{1} + U^{-1}(\boldsymbol{x}) \, \partial_j U(\boldsymbol{x}) \, dx^j$$

and hence $U(x+dx, y, z) = U(\boldsymbol{x}) \left[\mathbb{1} + \frac{i}{g} A_j(\boldsymbol{x})dx^j\right]$, etc. Multiplying over all the factors in Eq. (6.41), we deduce that the function $U(x, y, z)$ satisfies the boundary conditions

$$\begin{aligned}
U(x+L, y, z) &= U(x, y, z)\Omega_1 \,, \\
U(x, y+L, z) &= U(x, y, z)\Omega_2 \,, \\
U(x, y, z+L) &= U(x, y, z)\Omega_3
\end{aligned} \tag{6.42}$$

with constant commuting holonomies Ω_j (commutativity of Ω_j is necessary for the matrix U to be uniquely defined).

To prove the second part of the theorem, we have to construct the matrix $U(x, y, z)$ satisfying the boundary conditions (6.42) for an arbitrary commuting triple $(\Omega_1, \Omega_2, \Omega_3)$. We do so in several steps.

- First, we define

$$\begin{aligned}
U(x, 0, 0) &= \exp\left\{iT_1 \frac{x}{L}\right\} \\
U(0, y, 0) &= \exp\left\{iT_2 \frac{y}{L}\right\} \\
U(0, 0, z) &= \exp\left\{iT_3 \frac{z}{L}\right\}
\end{aligned} \tag{6.43}$$

where $\Omega_j = \exp\{iT_j\}$. (The choice of T_j once Ω_j are given is not unique, but it is irrelevant. Take *some* set of the logarithms of holonomies Ω_i.) Having done this, we can extend the construction over all other edges of the 3-cube so that the boundary conditions (6.42) are fulfilled. For example, we define

$$U(L, y, 0) = \exp\left\{iT_2 \frac{y}{L}\right\}\Omega_1, \quad U(x, L, 0) = \exp\left\{iT_1 \frac{x}{L}\right\}\Omega_2$$

[then $U(L, L, 0) = \Omega_2\Omega_1 = \Omega_1\Omega_2$], etc.

- With $U(x, y, z)$ defined on the edges of the cube in hand, we can continue U also to the *faces* of the cube due to the fact that, according to our assumption, $\pi_1(\mathcal{G}) = 0$ i.e. any loop in the group is contractible. Let us do that first for 3 faces adjacent to the vertex $(0,0,0)$.
- With $U(x, y, 0)$, $U(x, 0, z)$, and $U(0, y, z)$ in hand, we can find $U(x, y, z)$ on the other 3 faces of the cube:

$$U(x, y, L) = U(x, y, 0)\Omega_3, \qquad U(x, L, z) = U(x, 0, z)\Omega_2,$$
$$U(L, y, z) = U(0, y, z)\Omega_1. \tag{6.44}$$

- With $U(x, y, z)$ defined on the surface of the cube, we can continue it into the interior using the fact that $\pi_2(\mathcal{G}) = 0$ for all simple Lie groups.

By construction, $U(x, y, z)$ satisfies the boundary conditions (6.42) and hence $\hat{A}_j(x, y, z)$ in Eq. (6.39) is periodic.   □

The skeleton construction above is rather common in homotopy theory and can be found also in physical literature (see e.g. Ref. [90]).

Now, for the unitary and also for the symplectic groups (but not for higher ortogonal and exceptional groups — see the next section), three commuting group elements always belong to a *maximal torus* such that we can represent $\Omega_j = e^{iT_j}$ with commuting T_j that belong to a Cartan subalgebra. The function $U(x, y, z)$ satisfying the boundary conditions (6.42) may then be chosen as

$$U(x, y, z) = \exp\left\{iT_1\frac{x}{L}\right\} \exp\left\{iT_2\frac{y}{L}\right\} \exp\left\{iT_3\frac{z}{L}\right\} \tag{6.45}$$

According to (6.39), this gives constant gauge potentials $\hat{A}_j^{(0)} = T_j/(gL)$.

By conjugation we bring the Cartan subalgebra $\mathfrak{C}$, where T_j belong, to a convenient form $\mathfrak{C}_0$. For $SU(N)$, $\mathfrak{C}_0$ is spanned by $r = N - 1$ diagonal matrices[11] of order N with zero sum of their elements: $t^3, t^8, \ldots, t^{N^2-1}$, the standard normalization $\text{Tr}\{t^a t^b\} = \delta^{ab}/2$ being chosen. The moduli space of classical vacua is parameterized by $N - 1$ three-vectors:

$$\boldsymbol{c}^1 = \boldsymbol{A}^{3(0)}, \qquad \boldsymbol{c}^2 = \boldsymbol{A}^{8(0)}, \ldots, \tag{6.46}$$

In a similar way as it was the case in the Abelian theory, the range of c_j^a is finite. Indeed, two different sets of c_j^a that correspond to the same holonomies Ω_j can be obtained from one another by a gauge transformation that respects the boundary conditions (6.42). There are periodic transformations similar to (6.29) that shift c_j^a by certain constants. Besides, there

---

[11] Physicists usually do not distinguish between a Lie algebra and its envelopping algebra. Alas, your author also has this bad habit.

are *Weyl reflections* acting on the three copies of the Cartan subalgebra. These transformations may bring each spatial component of c_j^a to a finite region called *Weyl alcove*.

Let us explain how it works in the simplest $SU(2)$ example. The slow variables are $c_j = A_j^{3(0)}$ and their superpartners $\eta_\alpha = \lambda_\alpha^{3(0)}$. The wave function must be invariant under the periodic gauge transformations,

$$t^3 A_j^{3(0)} \to \omega t^3 A_j^{3(0)} \omega^{-1} - \frac{i}{g}(\partial_j \omega)\omega^{-1}, \quad t^3 \lambda^{3(0)} \to \omega t^3 \lambda^{3(0)} \omega^{-1}, \quad (6.47)$$

with

$$\omega(\boldsymbol{x}) = \exp\left\{\frac{4\pi i x_j n_j t^3}{L}\right\}, \qquad n_j \in \mathbb{Z}. \quad (6.48)$$

In contrast to the Abelian gauge transformations (6.29), the transformation (6.47), (6.48) is topologically trivial — the function (6.48) represents a map $S^1 \to SU(2)$ and $\pi_1[SU(2)] = 0$.

The Weyl reflection in this case is simply

$$c_j \to -c_j, \qquad \eta_\alpha \to -\eta_\alpha. \quad (6.49)$$

It amounts to a constant gauge transformation with

$$\omega = e^{i\pi t^1} = 2it^1 \quad \text{or} \quad \omega = e^{i\pi t^2} = 2it^2. \quad (6.50)$$

The wave function must be invariant under any of these topologically trivial transformations. It is hence completely defined by its values in the region

$$0 \le c_j \le \frac{2\pi}{gL}, \quad (6.51)$$

representing the product of three Weyl $SU(2)$ alcoves.

For $SU(N)$, the wave functions are invariant under the shifts of c_j^a similar to (6.47), (6.48) with a change

$$\omega(\boldsymbol{x}) = \exp\left\{\frac{4\pi i x_j n_j t^*}{L}\right\}, \quad (6.52)$$

where t^* are the elements of the Cartan subalgebra corresponding to the coroots of $su(N)$ and normalized in the same way as t^3. The wave functions are also invariant under Weyl reflections ($N!$ of them) permuting the entries of the diagonal matrix representing an element of $\mathfrak{C}_0$. The Weyl alcove for $SU(3)$ is depicted (in an orthonormal basis) in Fig. 6.1.

If the coupling constant g is small, $c^{a=1,\dots,r}$ represent slow variables, which determine together with their superpartners η_α^a the vacuum dynamics

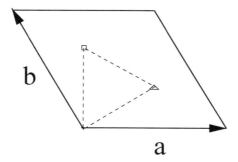

Fig. 6.1: The dashed triangle represents the Weyl alcove for $SU(3)$. Its area is 6 times smaller than the area of the solid rhombus formed by the elements of $\mathfrak{C}_0$, $a = 2\pi/(gL)\,\text{diag}(1, -1, 0)$ and $b = 2\pi/(gL)\,\text{diag}(0, 1, -1)$, proportional to the simple coroots. The vertices of the alcove $\square$ and $\triangle$ are proportional to *fundamental coweights*: $\square = 2\pi/(3gL)\text{diag}(1, 1, -2)$, $\triangle = 2\pi/(3gL)\,\text{diag}(2, -1, -1)$.

of our system. In the lowest BO order and in the orthogonal basis, the effective Hamiltonian has a very simple form:

$$\hat{H}^{\text{eff}} = -\frac{1}{2L^3} \sum_{a=1}^{r} \triangle_{c^a}, \tag{6.53}$$

where $r = N - 1$ is the rank of $SU(N)$. The zero-energy eigenfunctions of (6.53) do not depend on c^a and represent polynomials of η_α^a whose order does not exceed $2r$. However, not all such polynomials are admissible. The wave functions should be invariant under Weyl reflections. Imposing this requirement, we are left with only N different vacuum functions:

$$\Psi(c^a, \eta^a) = 1, \eta^a \eta^a, \ldots, (\eta^a \eta^a)^{N-1}. \tag{6.54}$$

We derive

$$I_W = N. \tag{6.55}$$

For the SYM theory based on a symplectic gauge group $Sp(2r)$, a similar analysis gives the value of the index $I_W = r + 1$.

6.3.1.1 *Instantons*

We continue the discussion of $SU(N)$.

In Theorem 6.1, we proved the *existence* of a periodic flat connection for any set of holonomies (6.40). But, obviously, there are infinitely many such connections related to one another by a gauge transformations

$$U(x, y, z) \;\to\; \tilde{U}(x, y, z) U(x, y, z)$$

with periodic $\tilde{U}(x, y, z)$. Due to trivial $\pi_1[SU(N)]$ and $\pi_2[SU(N)]$, we can continuously deform $\tilde{U}(x, y, z)$ so that $\tilde{U} = \mathbb{1}$ on the whole surface of the cube. But $\pi_3[SU(N)] = \mathcal{Z}$ is nontrivial. This means that topologically nontrivial $\tilde{U}(x, y, z)$ exist that cannot be continuously deformed to $\tilde{U} = \mathbb{1}$ in the interior of the cube.

As a result, admissible $U(x, y, z)$ and the corresponding periodic $\hat{A}_j(x, y, z)$ are split into distinct topological classes characterized by a degree q of mapping $S^3 \to SU(N)$. This degree may be expressed as the *Chern-Simons* topological invariant,

$$q \;=\; \frac{g^2}{8\pi^2} \int_{\text{cube interior}} \mathrm{Tr}\left\{ A \wedge dA - \frac{2ig}{3} A \wedge A \wedge A \right\}, \qquad (6.56)$$

where $A = \hat{A}_j \, dx^j$.

Let $\hat{\omega}$ be a topologically nontrivial gauge transformation relating the trivial classical vacuum $\hat{A}_j = 0$ to a nontrivial one with $q = 1$. Then, similar to what we had in the Abelian theory, the whole Hilbert space is split into the *universes*. All wave functions in a universe characterized by the *vacuum angle* $\vartheta \in [0, 2\pi)$ satisfy the property

$$\hat{\omega}\Psi_\vartheta(A, \lambda) \;=\; e^{i\vartheta}\Psi_\vartheta(A, \lambda). \qquad (6.57)$$

There are two essential differences between the Abelian and non-Abelian universes.

(1) The Abelian universes are defined only at finite volume — they are related to nontrivial cycles wound on the torus. But non-Abelian universes are there also in the infinite volume, for the YM theory living on $\mathbb{R}^3$ with the spatial infinity treated as one point (topologically, it is S^3). An example of a topologically nontrivial vacuum carrying Chern-Simons charge $q = 1$ in the $SU(2)$ theory on $\mathbb{R}^3$ is $gA = -i(d\omega)\omega^{-1}$ with

$$\omega \;=\; \exp\left\{ i\pi f(r)\sigma_j x_j \right\}, \qquad (6.58)$$

where σ_j are the Pauli matrices, $f(0) = 1$ and $\lim_{r \to \infty} rf(r) = 1$.

The tunneling trajectories between the vacua of different charge q are [92] the *instantons* — topologically nontrivial solutions of the Euclidean

Yang-Mills equations of motion [72]. In a certain gauge, an instanton interpolating between the trivial vacuum and the vacuum (6.58) with unit Chern-Simons charge has the form

$$A_\mu^a = \frac{1}{g} \frac{2\eta_{\mu\nu}^a x_\nu}{x^2 + \rho^2},$$ (6.59)

where ρ (the size of the instanton) is an arbitrary real number and $\eta_{\mu\nu}^a$ are the so-called 't Hooft symbols:

$$\eta_{00}^a = 0, \qquad \eta_{jk}^a = \varepsilon_{ajk}, \qquad \eta_{j0}^a = -\eta_{0j}^a = \delta_{aj}.$$ (6.60)

The configuration (6.59) carries a unit Pontryagin number (5.113), which can also be represented as

$$q^{\text{Pontr}} = \frac{g^2}{8\pi^2} \int_{\mathbb{R}^4} \text{Tr}\{F \wedge F\} = 1$$ (6.61)

with $F = dA - igA \wedge A$.

(2) In the Abelian case, the supersymmetric vacua dwell in the universe $\vartheta = 0$. In other universes, only the states with positive energies are present. Similarly, in the pure Yang-Mills theory or in QCD with massive quarks, the energies of the states in the universe with nonzero ϑ exceed by a fixed amount $\Delta E \sim c(1 - \cos\vartheta)$ the energies of the analogous states in the universe $\vartheta = 0$. But in any theory with *massless* fermions including the SYM theory, c turns to zero due to an exact fermion zero mode in the instanton background [74]. As a result, the spectra in all the universes coincide. Supersymmetry is not broken in any of them. In a finite box with periodic boundary conditions, each such spectrum involves N vacuum states.

6.3.2 *Twisted boundary conditions*

To define a precise quantum problem for a field theory placed on the torus, we need to specify the boundary conditions. In the analysis above, we imposed the periodic boundary conditions (6.38). This gave us the moduli space (6.46) of classical vacua. The existence of N quantum vacuum states was derived by studying the spectrum of the effective Hamiltonian that described the moduli space dynamics. This analysis was quite sufficient to arrive at the essential conclusion that the supersymmetry is not spontaneously broken in the original infinite-volume SYM field theory, but it is interesting to explore what happens if another way to regularize the theory in the infrared, another kind of boundary conditions are chosen.

For simplicity, we set $L = 1$ and consider the following boundary conditions [91]:

$$\hat{A}_j(x+1,y,z) = P\hat{A}_j(x,y,z)P^{-1},$$
$$\hat{A}_j(x,y+1,z) = Q\hat{A}_j(x,y,z)Q^{-1},$$
$$\hat{A}_j(x,y,z+1) = \hat{A}_j(x,y,z), \qquad (6.62)$$

where P, Q are constant $SU(N)$ matrices forming the so-called *Heisenberg pair* — they satisfy the condition

$$QP = \epsilon PQ, \qquad (6.63)$$

with $\epsilon \mathbb{1}$ representing an element of the group center. For example, we can choose

$$P = \begin{pmatrix} 1 & 0 & 0 & 0 & \dots \\ 0 & \epsilon & 0 & 0 & \dots \\ 0 & 0 & \epsilon^2 & 0 & \dots \\ \vdots & \vdots & & \vdots & \vdots \end{pmatrix}, \qquad Q = \begin{pmatrix} 0 & 1 & 0 & 0 & \dots \\ 0 & 0 & 1 & 0 & \dots \\ 0 & 0 & 0 & 1 & \dots \\ \vdots & \vdots & \vdots & & \vdots & \vdots \\ 1 & 0 & 0 & 0 & \dots \end{pmatrix}. \qquad (6.64)$$

with $\epsilon = e^{2i\pi/N}$.

The boundary conditions for $U(\boldsymbol{x})$ that we should impose in order to respect the twisted boundary conditions (6.62) for the flat gauge potentials (6.39) have the form

$$U(x+1,y,z) = PU(x,y,z)P^{-1}\epsilon^{k_x},$$
$$U(x,y+1,z) = QU(x,y,z)Q^{-1}\epsilon^{k_y},$$
$$U(x,y,z+1) = U(x,y,z)\epsilon^{k_z} \qquad (6.65)$$

with integer $\boldsymbol{k}$. An essential difference with (6.42) is that the extra factors $\epsilon^{k_{x,y,z}}$ that appear in the RHS of Eq.(6.65) are not arbitrary commuting $SU(N)$ matrices, as was the case in (6.42), but belong to the *center* of the group. As a result, we do not have a continuous moduli space of classical vacua anymore. The classical vacua now represent *isolated points* (up to topologically trivial gauge transformations) in the field configuration space.

Let $\boldsymbol{k} = (1,0,0)$. In this case, the matrix U satisfying (6.65) can be chosen to be constant: $U = Q$. If $\boldsymbol{k} = (0,1,0)$, we may choose $U = P^{-1}$. Such gauge transformations leave the classical vacuum $\hat{A}_j = 0$ intact and do not bring about anything new. But the transformations (6.65) with $\boldsymbol{k} = (0,0,1)$ give us something interesting.

An explicit solution for $U(x, y, z)$ to the equations

$$
\begin{aligned}
\hat{U}(x+1, y, z) &= P U(x, y, z) P^{-1}, \\
U(x, y+1, z) &= Q U(x, y, z) Q^{-1}, \\
U(x, y, z+1) &= \epsilon U(x, y, z)
\end{aligned}
\tag{6.66}
$$

was found in Ref. [93]. It reads

$$
U = e^{2\pi i z \hat{T}(x,y)}, \tag{6.67}
$$

where $\hat{T}(x, y)$ is the following Hermitian $su(N)$ matrix:

$$
\boxed{T_j^k(x, y) = \frac{1}{N} \delta_j^k - \psi_j(x, y) \psi^{\dagger k}(x, y),} \tag{6.68}
$$

with ψ_j being a complex unitary vector of the following form:

$$
\psi_j(x, y) = \frac{\exp\left\{-\pi \left(\frac{y+j-1}{N}\right)^2 + 2\pi i x \left(\frac{y+j-1}{N}\right)\right\} \theta\left(x + i\frac{y+j-1}{N}\right)}{\sqrt{\sum_{k=1}^{N} \exp\left\{-2\pi \left(\frac{y+k-1}{N}\right)^2\right\} \left|\theta\left(x + i\frac{y+k-1}{N}\right)\right|^2}}. \tag{6.69}
$$

Here $\theta(w)$ is a Jacobi θ function [see Eq. (B.2) and Ref. [94] for an extensive review]:

$$
\theta(w) = \sum_{n=-\infty}^{\infty} \exp\{-\pi n^2 + 2\pi i n w\}. \tag{6.70}
$$

Indeed, the vector (6.69) satisfies the boundary conditions

$$
\begin{cases}
\psi(x+1, y) = e^{2\pi i y/N} P \psi(x, y) \\
\psi(x, y+1) = Q \psi(x, y)
\end{cases}
\tag{6.71}
$$

with P and Q chosen as in Eq. (6.64). It follows that the matrix $\hat{T}$ satisfies the conditions

$$
\begin{cases}
\hat{T}(x+1, y) = P \hat{T}(x, y) P^{-1} \\
\hat{T}(x, y+1) = Q \hat{T}(x, y) Q^{-1}
\end{cases},
\tag{6.72}
$$

and the same holds for $\exp\{2\pi i z \hat{T}\}$. The appearance of Jacobi functions is natural in the problem with toroidal geometry.

Now note that the matrix (6.68) can be brought by a conjugation to the form

$$
\hat{T} \xrightarrow{\text{conj}} \hat{T}_0 = \frac{1}{N} \text{diag}(1, \ldots, 1, 1 - N).
$$

This is achieved by rotating $\psi_j \to \psi_j^{(0)} = \delta_{jN}$. The matrix $\hat{T}_0$ is a *fundamental coweight* of $su(N)$, so that

$$\exp\{2\pi i \hat{T}_0\} = \epsilon \mathbb{1},$$

as in Eq. (A.10), and the same for $\exp\{2\pi i \hat{T}\}$. It follows that the matrix (6.67) satisfies all the conditions (6.66).

The flat connection (6.39) based on (6.67) has a nonzero Chern-Simons charge (6.56). An explicit calculation [93] gives the value

$$q = \frac{1}{N}. \tag{6.73}$$

Thus, this classical vacuum plays on the torus the same role as the topologically nontrivial classical vacua on S^3 with a unit Chern-Simons number, which we discussed at the end of the previous subsection. Tunneling Euclidean trajectories discovered by 't Hooft [91], which relate the trivial vacuum $\hat{A}_j = 0$ to nontrivial ones, are called *torons*. In contrast to ordinary instantons, their Pontryagin number is an integer multiple of $1/N$,

$$q^{\text{Pontr}} = \frac{g^2}{8\pi^2} \int_{T^4} \text{Tr}\{F \wedge F\} = \frac{k}{N}, \tag{6.74}$$

and may be fractional.

The toron dynamics can be treated in a similar way as we treated the instanton dynamics a couple of pages earlier. Let $\hat{\tilde{\omega}}$ be an operator of a topologically nontrivial gauge transformation that shifts the Chern-Simons number by $1/N$. Then the spectrum is split into different universes. The wave functions in each such universe, represented schematically as

$$|\vartheta\rangle = \sum_q e^{iq\vartheta} |q\rangle \tag{6.75}$$

with

$$q = \ldots, -\frac{1}{N}, 0, \frac{1}{N}, \ldots$$

satisfy the property

$$\boxed{\hat{\tilde{\omega}}\Psi_\vartheta = e^{i\vartheta/N}\Psi_\vartheta.} \tag{6.76}$$

If only integer values of q were admissible (as was the case for the problem with periodic boundary conditions), the parameter ϑ would belong to the interval $[0, 2\pi)$. But if the fractional q are also present, as it is the case for twisted boundary conditions, $\vartheta \in [0, 2\pi N)$. There is a single quantum vacuum state in each universe.

We have to warn the reader: the interpretation suggested above is not the standard one. Following [47], people usually only write (6.57) with the gauge transformation shifting the Chern-Simons number by 1. Then there are less universes, $\vartheta \in [0, 2\pi)$ rather than $\vartheta \in [0, 2\pi N)$, but in each universe, there are N distinct vacuum states. And this conforms with the counting $I_W = N$ derived above for periodic boundary conditions.

There is a certain reason to do so. Putting theory in a finite box is a way to regularize it in the infrared, but what interests us more is the dynamics of SYM field theory in infinite volume. In infinite volume, we know classical vacuum field configurations (6.58) with an integer Chern-Simons number and the corresponding Euclidean instanton solutions (6.59) with integer Pontryagin number. But we are not aware of similar nice configurations with fractional Chern-Simons or Pontryagin number.[12]

But let us separate two issues: *(i)* the dynamics of the theory on a finite torus with twisted boundary conditions and *(ii)* the dynamics of the theory in infinite volume.

As far as the twisted torus is concerned, I do not see any specifics of the fractional q configurations compared to the configurations with integer q. The universe interpretation with $\vartheta \in [0, 2\pi N)$ seems, in this case, to be the most natural.

What really happens in the infinite volume limit is a difficult and not so clear[13] question. First of all, the torons appear only if twisted boundary conditions are imposed. With periodic conditions, there is no trace of torons and the Chern-Simons number is always an integer. And we believe that two ways of regularization should give the same dynamics in the infinite volume limit.

Furthermore, we know that in SYM theory the *gluino condensate* $\langle \lambda_\alpha^a \lambda^{a\alpha} \rangle$ is formed [96]. It follows from the fact that the correlator

$$C = \langle \lambda^{a\alpha} \lambda_\alpha^a (x_1) \ldots \lambda^{a\alpha} \lambda_\alpha^a (x_N) \rangle \qquad (6.77)$$

evaluated in the instanton background is a real nonzero constant not depending on x_k.[14] As this is true both at small separations $x_j - x_k$, where the instanton calculation is reliable, and also at large separations, it im-

---

[12]Well, there are some configurations with fractional topological charge, the so-called *merons* [95], but they are not so nice, being singular and having an infinite action. Their role in Yang-Mills dynamics is not clear.

[13]At least, not so clear to your author.

[14]The fact that it does not depend on x_k follows from certain supersymmetric Ward identities, and the fact that this constant is nonzero follows from the presence of $2N$ gluino zero modes on the instanton background.

plies by cluster decomposition that $\langle \lambda_\alpha^a \lambda^{a\alpha} \rangle$ is a nonzero complex constant whose phase may acquire N complex values, $\phi_k = 2\pi k/N$.

But does it mean, as people usually say, spontaneous breaking of discrete chiral symmetry Z_N (after $U(1)$ chiral symmetry of the Lagrangian (3.40) is broken explicitly by instanton effects)? Or do the condensates with different phases ϕ_k live in different universes?..

6.3.3 *Domain walls*

Physically, spontaneous breaking of a discrete symmetry means the coexistence of different phases separated by domain walls in one physical space. Unfortunately, SYM theory does not describe nature, and we cannot stage an experiment where these domain walls would or would not be observed. We have to find theoretical arguments pro or contra their existence.

One can recall in this regard a very confusing story about would-be spontaneous breaking of center symmetry in hot Yang-Mills theory. Many people claimed that such breaking associated with different phases of the Polyakov loop,

$$P(\boldsymbol{x}) \;=\; \frac{1}{N_c}\mathrm{Tr}\exp\{ig\beta\hat{A}_0(\boldsymbol{x})\}, \qquad\qquad \beta = 1/T, \qquad (6.78)$$

really occurs. Even the domain walls separating the regions with different phases of $P(\boldsymbol{x})$ were ostensibly found [97].

But this is not correct [98]. Physical spontaneous breaking of Z_N does not occur in YM theory at high temperature. The Polyakov loop [in contrast to the correlator $\langle P(\boldsymbol{x})P^*(0)\rangle$] is not a physically observable quantity. It plays the same role in hot YM theory as the dual plaquette variables η_j in the Ising model.[15]

And how can one confirm or disprove the existence of domain walls between the regions with different phases of the gluino condensate in SYM theory?

The only instrument that I am aware of is the Veneziano-Yankielowicz effective Lagrangian [100], which is formulated in terms of the composite colorless chiral superfield

$$S \;=\; -2\mathrm{Tr}\{\hat{W}^\alpha \hat{W}_\alpha\}. \qquad\qquad (6.79)$$

---

[15]We have deviated from the main subject of our book, but I cannot resist to comment here that the deconfinement phase transition, if it takes place in hot YM theory (which, I think, it does) is associated not with this nonexistent spontaneous breaking, but probably with percolation of color fluxes [99].

The lowest component of S is the bifermion operator $\lambda^{\alpha a}\lambda^a_\alpha$. Its vacuum expectation value is the gluino condensate discussed above.

The VY Lagrangian has the form

$$\mathcal{L}^{VY} = \alpha \int d^4\theta\,(\overline{S}S)^{1/3} + \beta \left(\int d^2\theta\,S\ln\frac{S}{\Lambda^3} + \text{h.c.} \right), \qquad (6.80)$$

where α is an arbitrary numerical coefficient, the coefficient β has some particular, irrelevant for our purposes numerical value, and Λ is a dimensionful parameter of quantum SYM theory — the scale where its effective coupling becomes strong.[16]

The SYM Lagrangian (3.36) is invariant under the scale transformations and also under the chiral transformations,

$$\hat{W}_\alpha \to e^{i\gamma}\hat{W}_\alpha, \qquad\qquad \theta_\alpha \to e^{i\gamma}\theta_\alpha. \qquad (6.81)$$

In supersymmetric theory, these two transformations have the same nature and the corresponding currents enter the same supermultiplet. These symmetries are, however, anomalous — they are broken by quantum effects. The VY effective Lagrangian takes account of this anomaly. The first (kinetic) term in Eq. (6.80) is invariant under (6.81), but the second (potential) term is not. Its variation is proportional to $\int d^2\theta\,S - \int d^2\bar\theta\,\overline{S}$, and this is exactly what the anomalous Ward identity for chiral rotations gives.

The Lagrangian (6.80) looks similar to the Lagrangian (3.14), (3.17) of the Wess-Zumino model (it is especially clear if one expresses it in terms of the superfield $\Phi = S^{1/3}$ of canonical dimension 1). The Witten index for the WZ model was evaluated at the beginning of this chapter. For the polynomial superpotential $\mathcal{W}(\Phi)$, we derived that the index is equal to the degree of the polynomial $\mathcal{W}'(\Phi)$.

In our case,

$$\mathcal{W}'(\Phi) \sim \Phi^2\ln\frac{\Phi}{\tilde\Lambda} \qquad (\tilde\Lambda = \Lambda e^{-1/3})$$

is not a polynomial, but includes an extra logarithmic factor. It has an unphysical cut to be handled somehow, but its presence does not change

---

[16] I mostly tried to write our book without assuming that the reader is an expert in quantum field theory. In principle, the information provided in Chap.1 and Chap.3 of the book is sufficient to understanding everything else. But these few paragraphs explaining the meaning of the VY effective Lagrangian are addressed to a reader familiar with the notions of *asymptotic freedom*, *dimensional transmutation* and *quantum anomalies*. The same largely applies to the next chapter. Nonexperts (if they are still with us) may skip the explanations and take Eq. (6.80) at face value.

the estimate for the index. The classical vacua are

$$\phi = 0, \qquad \text{and} \qquad \phi = \tilde{\Lambda}. \tag{6.82}$$

and the index is equal to 2.

This does not conform with the evaluations above. The main pecularity is the presence of the *chirally symmetric* vacuum [101] $\langle \lambda^{a\alpha} \lambda^a_\alpha \rangle = \langle \phi \rangle^3 = 0$ on top of the chirally asymmetric one with $\langle \lambda^{a\alpha} \lambda^a_\alpha \rangle = \tilde{\Lambda}^3$.

Do not worry, this is not a paradox. The matter is that the VY Lagrangian is *not* derived following a correct Born-Oppenheimer (people also call it *Wilsonian*) procedure, where slow variables on which the effective Lagrangian depends are carefully separated from the fast ones to be integrated out.

The VY Lagrangian describes *some* features of the low-energy dynamics of the SYM theory, but unfortunately does not describe some other features. In particular, it does not describe well the vacuum structure of the theory. Besides the existence of the unphysical chirally symmetric vacuum, the VY Lagrangian as written in Eq. (6.80) has another unpleasant feature: unlike the original SYM theory, it does not exhibit the discrete chiral Z_N symmetry. In particular, it has only one asymmetric vacuum at $\phi = \tilde{\Lambda}$ and not N vacua that the original theory [in the standard description with $\vartheta \in [0, 2\pi)$] must have.

In Ref. [101], an amendment of the Lagrangian (6.80) was suggested. It includes the standard instanton parameter $\vartheta_{\text{inst}} \in [0, 2\pi)$ and has the discrete Z_N chiral symmetry. The effective potential includes N sectors. There is a chirally asymmetric vacuum in each sector *and* the extra chirally symmetric minimum in the origin (common for all sectors) — there is no way to get rid of it in this approach.

The Kovner-Shifman effective potential for $SU(3)$ in the universe $\vartheta_{\text{inst}} = 0$ is represented in Fig. 6.2. The dashed lines on the plane $s = \phi^3$ delimit 3 sectors. In the right sector, the potential is given by the VY expression,

$$V_0(s, s^*) = (ss^*)^{2/3} \ln \frac{s}{\tilde{\Lambda}^3} \ln \frac{s^*}{\tilde{\Lambda}^3}. \tag{6.83}$$

In the upper sector, the potential is

$$V_1(s, s^*) = V_0(e^{-2i\pi/3} s, \ e^{2i\pi/3} s^*). \tag{6.84}$$

In the lower sector:

$$V_2(s, s^*) = V_0(e^{2i\pi/3} s, \ e^{-2i\pi/3} s^*). \tag{6.85}$$

On the border $\arg(s) = \pi/3$, the potentials V_0 and V_1 coincide. At $\arg(s) = -\pi/3$, $V_0 = V_2$. And at $\arg(s) = \pi$, $V_1 = V_2$. But the derivatives of the potential are not continuous.[17]

The potential has 3 chirally asymmetric minima at $s = \tilde{\Lambda}^3$ and $s = \tilde{\Lambda}^3 e^{\pm 2i\pi/3}$, which are marked by boxes, and the chirally symmetric minimum at $s = 0$, which is marked by a blob.

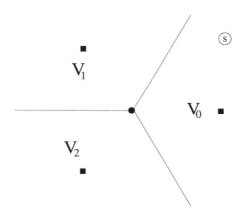

Fig. 6.2: Kovner-Shifman effective potential. $N = 3$ and $\vartheta_{\text{inst}} = 0$.

The question that we originally posed was whether domain walls between the different chirally asymmetric vacua exist. And the analysis in the framework of VY-KS effective potential gives a negative answer: there are domain walls connecting a box to the blob, but there are no walls between different boxes [102]. This can be considered as an argument in favor of the universe picture rather than that involving spontaneous chiral symmetry breaking. But the presence of the unphysical chirally symmetric vacuum in the VY approach makes the latter not so convincing. We will revisit this issue in the next chapter where we will bring matter fields in the consideration.

And for more detailed discussion of this walls vs. torons controversy, we address the reader to Sect. 7 of Ref. [102]. The question still remains open. But whatever is true: is discrete chiral Z_N symmetry in SYM theory

---

[17]We will see in the next chapter that the effective VY potential may be derived from a more general TVY effective potential, including as well matter fields, in the limit when the mass of the latter becomes large. And the VY potential thus derived *has* the patched form as in Fig. 6.2.

with $SU(N)$ gauge group spontaneously broken or are configurations with fractional topological charge also relevant in infinite volume, one thing is clear: supersymmetry is *not* broken there and an exact supersymmetric vacuum (or several vacua) exist.

6.4   Other groups

In the original paper [47], the reasoning of Sect. 6.3.1, which allowed one to evaluate the Witten index on the torus with periodic boundary conditions for SYM theory based on an $SU(N)$ group, was extended to all other simple gauge groups. The effective wave function on a small torus had the same form as in Eq. (6.54) with the only difference that the highest power of $\eta^a \eta^a$ was equal to the rank r of the group. This led to the result

$$I_W = r + 1. \tag{6.86}$$

However, as was mentioned above, there is another way to evaluate the index. It is based on the standard scenario with spontaneous breaking of discrete chiral symmetry in the infinite volume and the vacuum angle restricted to the interval $\vartheta_{\text{inst}} \in [0, 2\pi)$. For a generic group, the instanton admits $2h^\vee$ gluino zero modes, where $h^\vee$ is the *dual Coxeter number*, which is defined as the sum

$$h^\vee = \sum_{\text{nodes}} a_\alpha^\vee \tag{6.87}$$

of the modified Dynkin labels

$$a_\alpha^\vee = a_\alpha |\boldsymbol{\alpha}|^2 . \tag{6.88}$$

in the nodes of the extended Dynkin diagram (the length of the long roots being normalized to 1). It coincides with the adjoint Casimir eigenvalue $c_V = T^a T^a$ when a proper normalization for the adjoint generators T^a is chosen [see (A.20)].

As was the case for $SU(N)$, supersymmetric Ward identities dictate that the Euclidean correlator

$$C(x_1, \ldots, x_{h^\vee}) = \langle \lambda^a \lambda^a (x_1) \ldots \lambda^a \lambda^a (x_{h^\vee}) \rangle \tag{6.89}$$

does not depend on the coordinates. Evaluating it in the instanton background, we obtain a nonzero real constant. And this implies that $\langle \lambda^{a\alpha} \lambda_\alpha^a \rangle$ is a complex constant whose phase may acquire $h^\vee$ complex values, so that

$$\boxed{I_W = h^\vee .} \tag{6.90}$$

Table 6.1: Vacuum counting for exceptional groups

group $\mathcal{G}$	G_2	F_4	E_6	E_7	E_8
$r + 1$	3	5	7	8	9
$h^\vee$	4	9	12	18	30
mismatch	1	4	5	10	21

For the unitary and also for the symplectic groups, the estimates (6.86) and (6.90) coincide.

But for higher orthogonal groups $SO(N \geq 7)$ and all the exceptional groups, they do not. For $SO(N \geq 7)$, $h^\vee = N - 2 > r + 1$. Also for the exceptional groups $G_2, F_4, E_{6,7,8}$, the index (4.1) is larger than Witten's original estimate (see Table 6.1).

The origin of this troublesome mismatch remained unclear until 1997 when Witten realized [103] that one of the assumptions on which the estimate (6.86) was based — that *a triple of commuting holonomies Ω_j can always be brought by conjugation to the maximal torus* — is correct for the unitary and symplectic groups, but is *wrong* for higher orthogonal and exceptional groups.

To understand why, note first that for an orthogonal group, which is not simply connected, even a *pair* of commuting group elements cannot always be conjugated to a maximal torus. Consider for example $SO(3)$. The elements

$$g_1 = \mathrm{diag}(-1, -1, 1) \qquad \text{and} \qquad g_2 = \mathrm{diag}(-1, 1, -1) \qquad (6.91)$$

obviously commute, but their "logarithms" proportional to the generators of the correponding rotations T^3 and T^2 do not.

However, for a simply connected group, a pair of commuting elements Ω_1, Ω_2 can always be conjugated to the maximal torus. This follows from the so-called Bott theorem: *a centralizer*[18] $\mathcal{G}_\Omega$ *of any element Ω in a simply connected Lie group $\mathcal{G}$ is connected.* Put Ω_1 on the torus of $\mathcal{G}$ and put Ω_2 on the torus of $\mathcal{G}_{\Omega_1}$. Due to the fact that $\mathcal{G}_{\Omega_1}$ is connected, these two tori coincide.

---

[18] A reader-physicist who might be not so shrewd in group theory wisdom (I was not until I started to work on this problem) can consult Appendix A for some basic facts and terminology.

On the other hand, the centralizer of g_1 in $SO(3)$ (which is not simply connected) consists of two disconnected components:

$$\begin{pmatrix} \cos\phi & \sin\phi & 0 \\ -\sin\phi & \cos\phi & 0 \\ 0 & 0 & 1 \end{pmatrix} \quad \text{and} \quad \begin{pmatrix} \cos\phi & \sin\phi & 0 \\ \sin\phi & -\cos\phi & 0 \\ 0 & 0 & -1 \end{pmatrix}. \tag{6.92}$$

One can note in this regard that the preimages of the commuting in $SO(3)$ elements g_1 and g_2 in the universal simply connected covering $SU(2)$ are σ^3 and σ^2, and they do not commute, representing a Heisenberg pair as in Eq. (6.63).

But exceptional commuting *triples* that cannot be conjugated to the maximal torus may exist also for simply connected groups. To give the simplest example of such a triple, consider the group $Spin(7)$. An element of $Spin(7)$ can be represented as

$$\Omega \in Spin(7) = \exp\{\omega_{\mu\nu}\gamma_\mu\gamma_\nu\}, \tag{6.93}$$

with skew-symmetric $\omega_{\mu\nu}$. Here γ_μ are the gamma matrices 8×8 satisfying the Clifford algebra $\gamma_\mu\gamma_\nu + \gamma_\nu\gamma_\mu = 2\delta_{\mu\nu}$, $\mu, \nu = 1, \ldots, 7$.

Consider the triple [89, 103]

$$\begin{aligned} \Omega_1 &= \exp\left\{\frac{\pi}{2}(\gamma_1\gamma_2 + \gamma_3\gamma_4)\right\} = \gamma_1\gamma_2\gamma_3\gamma_4, \\ \Omega_2 &= \exp\left\{\frac{\pi}{2}(\gamma_1\gamma_2 + \gamma_5\gamma_6)\right\} = \gamma_1\gamma_2\gamma_5\gamma_6, \\ \Omega_3 &= \exp\left\{\frac{\pi}{2}(\gamma_1\gamma_3 + \gamma_5\gamma_7)\right\} = \gamma_1\gamma_3\gamma_5\gamma_7. \end{aligned} \tag{6.94}$$

The corresponding triple in $SO(7)$ is

$$\begin{aligned} \tilde{\Omega}_1 &= \operatorname{diag}(-1, -1, -1, -1, 1, 1, 1), \\ \tilde{\Omega}_2 &= \operatorname{diag}(-1, -1, 1, 1, -1, -1, 1), \\ \tilde{\Omega}_3 &= \operatorname{diag}(-1, 1, -1, 1, -1, 1, -1). \end{aligned} \tag{6.95}$$

One can easily check that all Ω_j commute, while not all the exponents in the LHS of Eq.(6.94) do. The representation (6.93) is not unique, but one can prove the following simple theorem.

Theorem 6.2. *A representation* $\Omega_j = e^{iS_j}$ *with* $S_j \in spin(7)$ *and* $[S_j, S_k] = 0$ *does not exist.*

Proof. Suppose, such a triple (S_1, S_2, S_3) exists. The property $[S_j, \Omega_k] = 0$ should then also hold. Let $S = i\omega_{[\mu\nu]}\gamma_\mu\gamma_\nu$. The requirement $[S, \Omega_j] = 0$

gives

$$\omega_{15}T_{2345} + \omega_{16}T_{2346} + \omega_{17}T_{2347} - \omega_{25}T_{1345} - \omega_{26}T_{1346} - \omega_{27}T_{1347}$$
$$+ \ \omega_{35}T_{1245} + \omega_{36}T_{1246} + \omega_{37}T_{1247} - \omega_{45}T_{1235} - \omega_{46}T_{1236} - \omega_{47}T_{1237} = 0,$$
$$\omega_{13}T_{2563} + \omega_{14}T_{2564} + \omega_{17}T_{2567} - \omega_{23}T_{1563} - \omega_{24}T_{1564} - \omega_{27}T_{1567}$$
$$+ \ \omega_{53}T_{1263} + \omega_{54}T_{1264} + \omega_{57}T_{1267} - \omega_{63}T_{1253} - \omega_{64}T_{1254} - \omega_{67}T_{1257} = 0,$$
$$\omega_{12}T_{1352} + \omega_{14}T_{1354} + \omega_{16}T_{1356} - \omega_{32}T_{1572} - \omega_{34}T_{1574} - \omega_{36}T_{1576}$$
$$+ \ \omega_{52}T_{1372} + \omega_{54}T_{1374} + \omega_{56}T_{1376} - \omega_{72}T_{1352} - \omega_{74}T_{1354} - \omega_{76}T_{1356} = 0,$$

$$(6.96)$$

where $T_{2345} = \gamma_2\gamma_3\gamma_4\gamma_5$ etc. It follows that all $\omega_{\mu\nu}$ vanish. $\qquad\square$

The underlying reason of nontriviality of the triple (6.94) is the fact that the centralizer of the element Ω_1 (or Ω_2 or Ω_3) in $Spin(7)$ is not simply connected. And it is not just not simply connected, but not simply connected in not a simple way, if you will.

I mean the following: For a unitary or symplectic group a centralizer of any element represents a product of a semi-simple simply connected group and some number of $U(1)$ factors. For example, the centralizer of an element of $SU(3)$ can be $SU(3)$, $SU(2) \times U(1)$ or $[U(1)]^2$. But the centralizer of Ω_1 in $Spin(7)$ has a more complicated structure. Ω_1 commutes with three generators $\gamma_5\gamma_6$, $\gamma_5\gamma_7$ and $\gamma_6\gamma_7$ of the subgroup $Spin(3) \subset Spin(7)$ and also with six generators $\gamma_1\gamma_2$, etc. of the subgroup $Spin(4) \subset Spin(7)$. It may seem that the centralizer is $Spin(3) \times Spin(4)$, but it is not exactly so because the product $Spin(3) \times Spin(4)$ is not a subgroup of $Spin(7)$.

Indeed, both $Spin(4)$ and $Spin(3)$ have non–trivial centers. The center of $Spin(3) \equiv SU(2)$ is Z_2 with a nontrivial element corresponding to rotation by 2π in, say, (56) plane, $z = \exp\{\pi\gamma_5\gamma_6\}$. The center of $Spin(4) \equiv SU(2) \times SU(2)$ is $Z_2 \times Z_2$. It has a diagonal element z representing the product of non–trivial center elements in each $SU(2)$ factor:

$$z = \exp\left\{\frac{\pi}{2}(\gamma_1\gamma_2 - \gamma_3\gamma_4)\right\} \exp\left\{\frac{\pi}{2}(\gamma_1\gamma_2 + \gamma_3\gamma_4)\right\} = \exp\{\pi\gamma_1\gamma_2\}. \ (6.97)$$

Now note that $\exp\{\pi\gamma_1\gamma_2\} = \exp\{\pi\gamma_5\gamma_6\} = -\mathbb{1}$ in the full $Spin(7)$ group. Hence the true centralizer of Ω_1 is $[Spin(4) \times Spin(3)]/Z_2 \equiv [SU(2)]^3/Z_2$ where the factorization is done over the diagonal center element $(-1, -1, -1)$ in $[SU(2)]^3$. The matrices Ω_2 and Ω_3 correspond to the elements $(i\sigma^2, i\sigma^2, i\sigma^2)$ and $(i\sigma^3, i\sigma^3, i\sigma^3)$ in $[SU(2)]^3$, i.e. they form a Heisenberg pair in each $SU(2)$ factor. They still commute in the centralizer and in $Spin(7)$ exactly because the factorization over the common center should be done. The elements Ω_2, Ω_3 cannot be conjugated to the maximal torus in the centralizer and hence the whole triple (6.94) cannot be conjugated to the maximal torus in $Spin(7)$.

Let us call the group elements whose centralizers involve a not simply connected semi-simple factor *exceptional*. Up to a conjugation, there is only one exceptional element in Spin(7). To understand an underlying reason for existence of such elements, some group theoretic study is required, in particular — a study of the *Dynkin diagrams* of $Spin(7)$ and then of other orthogonal and exceptional groups.

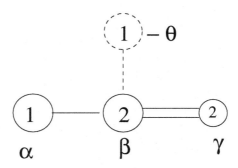

Fig. 6.3: Dynkin diagram for $Spin(7)$ with the highest root and the Dynkin labels.

The extended Dynkin diagram for $Spin(7)$ is depicted in Fig. 6.3. α, β, and γ are the simple roots. In a conveniently chosen orthonormal basis of the Cartan subalgebra, $e_1 = i\gamma_1\gamma_2/2$, $e_2 = i\gamma_3\gamma_4/2$, $e_3 = i\gamma_5\gamma_6/2$, the roots are[19] $\alpha = (1, -1, 0)$, $\beta = (0, 1, -1)$, $\gamma = (0, 0, 1)$. The highest root is $\theta = \alpha + 2\beta + 2\gamma = (1, 1, 0)$. The root vectors are

$$
\begin{aligned}
E_{\pm\alpha} &= \frac{1}{4}[\gamma_2\gamma_3 + \gamma_4\gamma_1 \pm i(\gamma_1\gamma_3 + \gamma_2\gamma_4)], \\
E_{\pm\beta} &= \frac{1}{4}[\gamma_4\gamma_5 + \gamma_6\gamma_3 \pm i(\gamma_3\gamma_5 + \gamma_4\gamma_6)], \\
E_{\pm\gamma} &= \frac{1}{2}(\gamma_6 \pm i\gamma_5)\gamma_7 .
\end{aligned}
\tag{6.98}
$$

---

[19]It is convenient in this case to choose the normalization where the length of the long roots is $\sqrt{2}$, while the length of the short root is 1.

The coroots are

$$\alpha^\vee = [E_\alpha, E_{-\alpha}] = \frac{i}{2}(\gamma_3\gamma_4 + \gamma_2\gamma_1),$$

$$\beta^\vee = [E_\beta, E_{-\beta}] = \frac{i}{2}(\gamma_5\gamma_6 + \gamma_4\gamma_3),$$

$$\gamma^\vee = [E_\gamma, E_{-\gamma}] = i\gamma_6\gamma_5. \tag{6.99}$$

Note now that the Dynkin label for the root β is equal to 2. And the existence of the exceptional group element Ω_1 with a nontrivial centralizer is related to *this* fact. Actually, Ω_1 can be represented as

$$\Omega_1 = e^{i\pi\omega_\beta}, \tag{6.100}$$

where $\omega_\beta = \theta^\vee = \frac{i}{2}(\gamma_1\gamma_2 + \gamma_3\gamma_4)$ is the *fundamental coweight* of the root β (i.e. $[\omega_\beta, E_\alpha] = [\omega_\beta, E_\gamma] = 0$ and $[\omega_\beta, E_\beta] = E_\beta$) .

The centralizer of Ω_1 can be found by inspecting the extended Dynkin diagram. Just delete the node β and look at what is left. One sees three disconnected circles giving $[SU(2)]^3$, but, as was explained above, this should be factorized over the diagonal Z_2 center element. Note that it is only the node β which plays a special role giving rise to an exceptional element. Similarly constructed elements $e^{i\pi\omega_\alpha}$ and $e^{i\pi\omega_\gamma}$ are not exceptional. The former is just an element of the center, and its centralizer is the whole $Spin(7)$, while the centralizer of the latter is $Spin(6) \equiv SU(4)$, which is simply connected.

We mentioned that the underlying reason for the existence of exceptional elements is the presence of the roots whose Dynkin label is greater than one. The root β has this property, but the same is true for root γ and one may wonder why there is no associated exceptional element. We will not prove it here, instead directing the reader to Ref. [104], but it turns out that the exceptional elements are only generated by the roots α_j where not just a_j, but also the modified Dynkin labels (6.88) are greater than 1. The root γ is short, $a_\gamma^\vee = 1$ and there is no associated exceptional element.

That is why there are no exceptional triples for the groups $SU(N)$ and $Sp(2r)$. The former has only unit Dynkin labels, while the latter has Dynkin labels $a_j = 2$, but they all sit on the short nodes so that $a_j^\vee = 1$ everywhere. In all the nodes of the Dynkin diagrams for $SU(N)$ and $Sp(2r)$, $a_\alpha^\vee = 1$, and $h^\vee$ given by the sum (6.87) coincides with the number of nodes, which is $r + 1$.

For the groups $Spin(7), Spin(8)$ and G_2 where only one root has $a_j^\vee > 1$, the exceptional element and the corresponding exceptional triple are isolated. This gives one single extra vacuum state. But for the groups

$Spin(N \geq 9)$, F_4 and $E_{6,7,8}$, there are many such roots. This brings about extra *moduli spaces* of exceptional triples and of classical vacua. For higher exceptional groups, they have a rather complicated structure. The spectra of effective Hamiltonians on these moduli spaces lead to appearance of extra quantum vacua so that the final counting is $I_W = h^\vee$ in all the cases.

All the details of this counting are spelled out in the original papers [104, 105]. But to give the reader an idea of how it is done, we will briefly explain here the essentials.

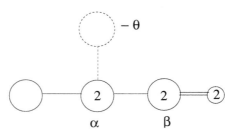

Fig. 6.4: Extended Dynkin diagram for $Spin(9)$.

- *Orthogonal groups.* The Dynkin diagrams for $Spin(9)$ and $Spin(10)$ include two nodes (let's call them α and β) with $a_\alpha^\vee = a_\beta^\vee = 2$. This brings about a 1-parametric moduli space of exceptional elements[20] $\Omega = \exp\{2\pi i(s_\alpha\omega_\alpha + s_\beta\omega_\beta)\}$ with $s_\alpha \geq 0$, $s_\beta \geq 0$ such that $2(s_\alpha + s_\beta) = 1$. That gives a 3-parametric moduli space of exceptional triples and hence of classical vacua. The effective Hamiltonian in this moduli space includes 3 bosonic variables c_j and their fermion superpartners η_α.

  The problem is basically the same as for the ordinary moduli space for $SU(2)$ where we had two vacuum states with wave functions $\Psi = 1$ and $\Psi = \eta^\alpha\eta_\alpha$. In the case under consideration, we get two *extra* vacuum states. Adding to this $r + 1$ vacuum states associated with the maximal torus, we obtain 7 quantum vacua for $Spin(9)$ and 8 quantum vacua for $Spin(10)$.

  For $Spin(11)$ and $Spin(12)$, there are 3 nodes with $a^\vee = 2$. This gives a 2-parametric moduli space of exceptional elements and 6-parametric moduli space of classical vacua, similar to the ordinary moduli space for $SU(3)$. We obtain 3 extra quantum vacua and all together $(r + 1) + 3$ vacua, which gives $I_W = 9$ for $Spin(11)$ and $I_W = 10$ for $Spin(12)$.

---

[20] *2-exceptional* in the terminology of Ref. [104].

This reasoning is easily generalized for $Spin(N)$ with higher N, and we derive

$$I_W[Spin(N)] = N - 2 = h^\vee. \tag{6.101}$$

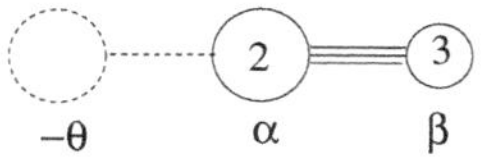

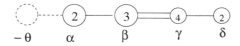

Fig. 6.5: Extended Dynkin diagram for G_2.

- G_2. Its Dynkin diagram is shown in Fig. 6.5. It includes a long node α with the label $a_\alpha^\vee = a_\alpha = 2$ and the short node of length $1/\sqrt{3}$ with the label $a_\beta^\vee = a_\beta/3 = 1$. There is an isolated exceptional element $\Omega = \exp\{i\pi\omega_\alpha\}$ and the associated exceptional triple.[21] This gives an extra vacuum state. All together:

$$I_W[G_2] = (r + 1) + 1 = 4 = h^\vee(G_2). \tag{6.102}$$

- F_4. The Dynkin diagram is shown in Fig. 6.6.

Fig. 6.6: Extended Dynkin diagram for F_4.

The node β carries the Dynkin label $a_\beta = a_\beta^\vee = 3$. Up to the conjugation, there is only one associated 3-exceptional element $\Omega_1 = \exp\{2\pi i\omega_\beta/3\}$. Its centralizer is $[SU(3) \times SU(3)]/Z_3$. Two other components of the triple represent the Heisenberg pairs (6.63) in each $SU(3)$ factor of the centralizer: $\Omega_2 = (P, P)$ and $\Omega_3 = (Q, Q)$. Now, Z_3 has two nontrivial elements and hence there are two different Heisenberg pairs: the pair (6.64) and its permutation. Correspondingly, there are two isolated exceptional triples and two isolated vacua.

The nodes α and γ carry the labels $a_\alpha^\vee = a_\gamma^\vee = 2$. As was the case for $Spin(9)$ and $Spin(10)$, this brings about 2 extra quantum states. We have all together 4 extra vacua and the total count is

$$I_W[F_4] = (r + 1) + 2_\beta + 2_{\alpha\gamma} = 9 = h^\vee[F_4]. \tag{6.103}$$

- E_6. The Dynkin diagram is shown in Fig. 6.7.

---

[21] The latter can in fact be lifted to $Spin(7)$ [as is explained in Appendix A, $G_2 \subset Spin(7)$], giving (6.94).

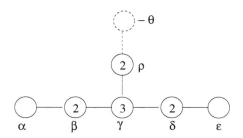

Fig. 6.7: Extended Dynkin diagram for E_6.

There are 2 extra isolated states associated with the node γ and 3 quantum states coming from the moduli space associated with the nodes β, δ, ρ. The total count is

$$I_W[E_6] = (r+1) + 2_\gamma + 3_{\beta\delta\rho} = 12 = h^\vee[E_6]. \qquad (6.104)$$

- E_7. The Dynkin diagram is shown in Fig. 6.8.

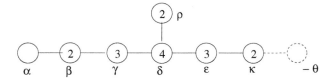

Fig. 6.8: Extended Dynkin diagram for E_7.

The nodes $\beta, \kappa, \rho, \delta$ bring about a 3-parametric moduli space of 2-exceptional elements. This gives 4 extra states.

Two nodes γ, ϵ with $a_\gamma^\vee = a_\epsilon^\vee = 3$ give a 1-parametric moduli space of 3-exceptional elements having the form $\Omega_1 = \exp\{2\pi i(s_\gamma\omega_\gamma + s_\epsilon\omega_\epsilon)\}$ with $3(s_\gamma + s_\epsilon) = 1$. A centralizer of each such element is $[SU(3)]^3/Z_3 \times U(1)$. This gives two distinct 3-parametric moduli spaces of the Heisenberg pairs in this centralizer and hence of commuting triples. Each such moduli space brings about 2 quantum vacua, and the total contribution to the vacuum counting coming from these nodes is $2 \times 2 = 4$.

Besides, there are two isolated exceptional triples associated with the root δ. They are formed by a 4-exceptional element $\Omega_1 = \exp\{\pi i\omega_\delta/2\}$ with the centralizer $[SU(4) \times SU(4) \times SU(2)]/Z_4$ and two different Heisenberg pairs (6.63) in the $SU(4)$ factors with the center elements $\pm i\mathbb{1}$ (the Heisenberg pair with the center element $-\mathbb{1}$ being already counted in the $\beta\kappa\rho\delta$ contribution). This gives two isolated vacua. The total count is

$$I_W[E_7] = (r+1) + 4_{\gamma\epsilon} + 4_{\beta\kappa\rho\delta} + 2_\delta = 18 = h^\vee[E_7]. \quad (6.105)$$

• E_8. The Dynkin diagram is shown in Fig. 6.9.

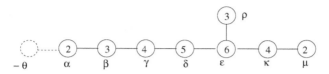

Fig. 6.9: Extended Dynkin diagram for E_8.

There is a 4-parametric moduli space of 2-exceptional elements spanned on the nodes $\alpha, \gamma, \epsilon, \kappa, \mu$. This gives 5 states.

Besides, the nodes γ, κ give a 1-parametric moduli space of 4-exceptional elements $\Omega_1 = \exp\{2\pi i(s_\gamma \omega_\gamma + s_\kappa \omega_\kappa)\}$ with $4(s_\gamma + s_\kappa) = 1$. Their centralizer is $[SU(4) \times SU(4) \times SU(2)]/Z_4 \times U(1)$. Two different Heisenberg pairs in $SU(4)$ with the center elements $\pm i\mathbb{1}$ give two distinct 3-parametric moduli spaces of exceptional triples and 4 quantum states.

There is a 2-parametric moduli space of 3-exceptional elements: $\Omega_1 = \exp\{2\pi i(s_\beta \omega_\beta + s_\rho \omega_\rho + s_\epsilon \omega_\epsilon)\}$ with $3(s_\beta + s_\rho + s_\epsilon) = 1$. Their centralizer is $[SU(3)]^3/Z_3 \times [U(1)]^2$. Two Heisenberg pairs for $SU(3)$ give two 6-parametric moduli spaces of the exceptional triples. This gives $2 \times 3 = 6$ quantum states.

The node δ gives an isolated 5-exceptional element $\Omega = \exp\{2\pi i \omega_\delta/5\}$ with the centralizer $[SU(5) \times SU(5)]/Z_5$, four different Heisenberg pairs, and 4 quantum states.

The node ϵ gives an isolated 6-exceptional element $\Omega = \exp\{\pi i s_\delta \omega_\delta/3\}$ with the centralizer $[SU(6) \times SU(3) \times SU(2)]/Z_6$. The elements $e^{\pi i/3}\mathbb{1}$ and $e^{5\pi i/3}\mathbb{1}$ of Z_6 give two isolated Heisenberg pairs, two isolated triples and two vacua. The total count is

$$I_W[E_8] = (r+1) + 5_{\alpha\gamma\epsilon\kappa\mu} + 6_{\beta\rho\epsilon} + 4_{\gamma\kappa} + 4_\delta + 2_\epsilon = 30 = h^\vee[E_8].$$
$$(6.106)$$

Up to now, we only discussed in this section the dynamics of the theories with periodic boundary conditions. However, as soon as the center of the group is nontrivial, one can also impose twisted boundary conditions. For the unitary groups, the center is Z_N and, as we discussed in Sect. 6.3.2, if one does not allow Euclidean configurations with a fractional topological charge, the twisted conditions lead to N distinct vacuum states in a universe with a given vacuum angle $\vartheta_{\text{inst}} \in [0, 2\pi)$. For other classical groups, the center is smaller so that the classical vacua are not isolated, but form moduli spaces, and the analysis is more complicated. But it can be done [106] along the lines of [104,105]. The total number of quantum vacua in a given instanton universe is always equal to $h^{\vee}$.

Chapter 7

Vacuum structure of $4D$ supersymmetric theories II: supersymmetric QCD

In the previous chapter we discussed the Wess-Zumino model, supersymmetric QED, and SYM theories without matter. Now we will turn to the analysis of the vacuum dynamics of non-Abelian $4D$ supersymmetric gauge theories with matter and concentrate in this chapter on the models with left-right symmetric fermion content — the supersymmetric versions of QCD — the model (3.42) and its generalizations including several matter multiplets and/or based on the other gauge groups.

If we include in the $SU(N)$ theory, besides the vector multiplet $\hat{V}(x^\mu, \theta^\alpha, \bar{\theta}_{\dot\alpha})$, an equal number of chiral multiplets $S_j^f(x_\mu^L, \theta^\alpha)$ and $T^{fj}(x_\mu^L, \theta^\alpha)$ belonging to the fundamental and antifundamental representations of the gauge group, $f = 1, \ldots, N_f$, we can endow all the fields with the mass by including the term

$$\mathcal{L}_\mathcal{M} = \frac{\mathcal{M}_{fg}}{2} \left(\int T^{fj} S_j^g \, d^2\theta + \text{c.c.} \right) \tag{7.1}$$

in the Lagrangian. The complex mass matrix $\mathcal{M}_{fg}$ can be arbitrary, but to avoid unnecessary complications, we will assume $\mathcal{M}_{fg} = m\delta_{fg}$. If the mass m is large compared to Λ_{SYM}, one seems to be able to get rid of the matter fields altogether by integrating them out so that the Witten index for such a theory coincides with the index for the SYM theory [47].

This is indeed so for the model (3.42), but we will see later that the presence of matter may affect the index in the theories involving trilinear couplings between matter multiplets. Such is the theory based on G_2 gauge group and including three fundamental matter 7-plets with a nonzero cubic superpotential to be considered at the end of the chapter.

But to begin with, we will describe in detail the vacuum dynamics of the ordinary supersymmetric QCD based on the $SU(N)$ gauge group. As far as the mass is nonzero, the Witten index there is the same as in the

SYM theory, $I_W = N$, but at low masses the vacuum dynamics is quite different from the dynamics of pure SYM theories and represents a separate interest.

7.1   Unitary groups

First of all, in a theory with fundamental fermions, fractional topological charges (6.61) are not admissible — neither in the infinite volume nor on a finite torus: the Euclidean path integral including fundamental fermion determinants is not defined in the gauge background with fractional Pontryagin number. Thus, there are only the ordinary instanton universes with $\vartheta \in [0, 2\pi)$.

Take the simplest theory (3.42) based on the $SU(2)$ gauge group and including a single pair of the chiral matter multiplets. Assume at first that the matter field are *massless*. As we shall see, the dynamics in this case is quite peculiar.

The component Lagrangian was written in Eq.(3.44). When $m = 0$, it reads

$$
\begin{aligned}
\mathcal{L} =\ & -\frac{1}{4} F^a_{\mu\nu} F^{a\mu\nu} + i\lambda^a \sigma^\mu (\partial_\mu \bar{\lambda}^a + g f^{abc} A^b_\mu \bar{\lambda}^c) - s^* (\partial_\mu - i g \hat{A}_\mu)^2 s \\
& -t(\partial_\mu + i g \hat{A}_\mu)^2 t^* + i\bar{\psi}\bar{\sigma}^\mu (\partial_\mu - i g \hat{A}_\mu)\psi + i\xi\sigma^\mu(\partial_\mu + i g \hat{A}_\mu)\bar{\xi} \\
& + ig\sqrt{2}[s^*(\hat{\lambda}\psi) - (\bar{\psi}\hat{\bar{\lambda}})s - (\xi\hat{\lambda})t^* + t(\hat{\bar{\lambda}}\bar{\xi})] - \frac{g^2}{2}(s^* t^a s - t t^a t^*)^2 \ . \quad (7.2)
\end{aligned}
$$

It does not involve a dimensional parameter and is invariant (at the tree level) under scale transformations. The corresponding conserved (at the classical level) current is $\Theta_{\mu\nu} x^\nu$, where $\Theta_{\mu\nu}$ is the energy-momentum tensor.

But the theory is also supersymmetric and the dilatational current has superpartners. One of the superpartners is the supercurrent and another is an axial current, the so-called R current [107] — the generator of the following transformations:

$$
\lambda \rightarrow e^{i\alpha}\lambda, \qquad (s, t) \rightarrow e^{2i\alpha/3}(s, t), \qquad (\psi, \xi) \rightarrow e^{-i\alpha/3}(\psi, \xi), \quad (7.3)
$$

while A_μ does not transform. The invariance of (7.2) under (7.3) is manifest. The transformations (7.3) correspond to the following transformations of the superfields supplemented by a transformation of the odd superspace coordinate:

$$
W \rightarrow e^{i\alpha} W, \qquad (S, T) \rightarrow e^{2i\alpha/3}(S, T), \qquad \theta \rightarrow e^{i\alpha}\theta \ . \quad (7.4)
$$

We already met a similar transformation in the previous chapter when we discussed pure SYM theory [see Eq. (6.81)]. As we mentioned there, the symmetry (6.81) is *anomalous*, being broken by quantum effects. And the same concerns the symmetry (7.4).

One also may observe, however, that the SQCD Lagrangian (3.42) is invariant under two extra symmetries when the superfield $\hat{W}$ and the coordinate θ are left intact, but the matter superfields are multiplied by phase factors:

$$S \rightarrow e^{i\beta}S, \qquad T \rightarrow e^{i\gamma}T. \tag{7.5}$$

The generator of the transformation (7.5) with $\beta = -\gamma$ is the *vector* $U(1)$ current. Such a transformation is a true symmetry of the theory even when the mass term is switched on. And the generator of the transformation (7.5) with $\beta = \gamma$ is the *axial* $U(1)$ current. This symmetry is broken by the mass term. And not only by the mass term. As was the case for the R symmetry (7.4), the symmetry $(S,T) \rightarrow e^{i\beta}(S,T)$ is anomalous — it is broken by quantum effects.

A remarkable fact, however, is that there exist a certain combination of two anomalous chiral symmetries that is *not* anomalous and represents a *true* symmetry of the massless theory (7.2). It reads [108][1]

$$W \rightarrow e^{i\alpha}W, \qquad (S,T) \rightarrow e^{-i\alpha}(S,T), \qquad \theta \rightarrow e^{i\alpha}\theta. \tag{7.7}$$

The physics here is rather similar to what happens in ordinary QCD.[2] The tree Lagrangian of QCD with one massless quark $q(x)$ admits two conserved currents: the vector current $\bar{q}\gamma_\mu q$ and the axial current $\bar{q}\gamma_\mu\gamma^5 q$.

---

[1]To understand that the symmetry (7.7) is not anomalous, consider the corresponding transformations of the fermion components of W, S, T:

$$\lambda \rightarrow e^{i\alpha}\lambda, \qquad (\psi, \xi) \rightarrow e^{-2i\alpha}(\psi, \xi).$$

The corresponding axial current may be expressed in the form

$$J_\mu^5 = 2\bar{\Psi}\gamma_\mu\gamma^5\Psi - \overline{\Lambda}\gamma_\mu\gamma^5\Lambda, \tag{7.6}$$

where Ψ is the fundamental Dirac spinor (1.52) made of ψ and $\bar{\xi}$ and $\Lambda = \begin{pmatrix} \bar{\lambda} \\ \lambda \end{pmatrix}$ is an adjoint Majorana spinor.

The anomalous divergence of the first term in (7.6) has a factor $1/2$ — the Dynkin index of the fundamental representation [see (A.19)], and the anomalous divergence of the second term has the factor 2 [the Dynkin index of the adjoint representation in $SU(2)$] and in addition the factor $1/2$ following from the Majorana nature of Λ. The cancellation is manifest.

[2]While the first six chapters of the book were almost self-sufficient and did not require of the reader a preliminary acquaintance with quantum field theory, this chapter is mainly addressed to experts in QFT. Not an expert who would wish to continue reading may first consult e.g. Chapters 12 and 14 of the book [8].

The vector current is not anomalous, but the axial current is. In addition, the corresponding symmetry is broken explicitly if the quark has mass.

If one considers QCD with N_f massless quark flavours, the symmetry of the tree Lagrangian is $U_L(N_f) \times U_R(N_f)$ describing independent unitary rotations of the left-handed and right-handed quarks. The flavor-singlet axial symmetry is anomalous, but other symmetries are not. Spontaneous breaking of $SU_L(N_f) \times SU_R(N_f) \times U_V(1)$ down to $U_V(N_f)$ associated with the formation of quark condensate leads by Goldstone theorem to appearance of $N_f^2 - 1$ massless particles — pseudoscalar mesons.[3]

In massless supersymmetric QCD with one flavor (one pair of chiral multiplets), the situation is similar, the only essential difference is the presence of an extra massless Weyl fermion — the gluino. As a result, at the tree level, we have 2 conserved axial currents, one of them being anomalous while the other is not. In a massless $SU(N)$ theory with N_f pairs of multiplets, one has $N_f^2 - 1$ flavor-nonsinglet conserved axial currents, as in the ordinary QCD, and one extra flavor-singlet nonanomalous current corresponding to the symmetry

$$W \to e^{i\alpha}W, \quad (S^f, T^f) \to e^{-i\alpha(N-N_f)/N_f}(S^f, T^f), \quad \theta \to e^{i\alpha}\theta. \quad (7.8)$$

Let us go back to the simplest $SU(2)$ theory with one flavor and write a structure

$$\propto \int \frac{d^2\theta}{T^j S_j}, \quad (7.9)$$

which is *invariant* under the transformations (7.7) (we are restoring the color index, $j = 1, 2$). This structure has a quite precise and important physical meaning: it enters the effective Born-Oppenheimer Lagrangian of massless SQCD in the region where the values of the scalar fields $|s|, |t|$ are much larger than the parameter Λ such that the effective coupling constant $g^2(\Lambda)$ is of order 1, while $g^2(|s| \sim |t|)$ is small.[4]

The structure (7.9) is allowed by symmetry, but it does not *a priori* mean that it is actually generated. In fact, it *is*. The mechanism is purely nonperturbative associated with the instanton solutions (6.59) of the Euclidean Yang-Mills field equations. We will not describe it here, rather referring the reader to the original papers [108, 109] and the reviews [110, 111].

---

[3]We hasten to add, however, that the physical u, d and s quarks are light, but not massless, the axial symmetry is not exact, and pseudoscalar mesons carry mass, though this mass is comparatively small.

[4]As was the case for pure SYM theory, the theory (3.42) is asymptotically free and the effective coupling falls at large energies and grows at small energies.

The *Affleck-Dine-Seiberg effective theory* represents a Wess-Zumino model including the chiral supermultiplets S_j and T^j. The Lagrangian reads

$$\mathcal{L}_{\text{ADS}} = \frac{1}{4} \int d^4\theta (\overline{S}^j S_j + T^j \overline{T}_j) + \frac{\Lambda^5}{2} \left(\int \frac{d^2\theta}{T^j S_j} + \text{c.c.} \right). \qquad (7.10)$$

The factor $\sim \Lambda^5$ appears by dimensional reasons,[5] The first term in (7.10) comes from the first line of the original SQCD Lagrangian (3.42), where we omitted the supergauge fields, which play the role of fast variables in the region $|s| \sim |t| \gg \Lambda$ and are irrelevant for the low-energy dynamics. The second term includes the superpotential generated by nonperturbative effects:

$$\mathcal{W}^{\text{np}}(S,T) = \frac{\Lambda^5}{T^j S_j}. \qquad (7.11)$$

The mass term is not included yet.

To find the vacuum states in this theory, one has to solve the equations

$$\frac{\partial \mathcal{W}^{\text{np}}}{\partial s_j} = \frac{\partial \mathcal{W}^{\text{np}}}{\partial t^j} = 0. \qquad (7.12)$$

And such solutions are *absent*! The energy density,

$$\epsilon(s,t) = \left| \frac{\partial \mathcal{W}^{\text{np}}}{\partial s_j} \right|^2 + \left| \frac{\partial \mathcal{W}^{\text{np}}}{\partial t^j} \right|^2 = \Lambda^{10} \frac{s^{j*}s_j + t^j t_j^*}{|t^k s_k|^4}, \qquad (7.13)$$

is positive definite, but it tends to zero when s or t go to infinity. We face the situation of *run-away vacuum*. The ground state in the quantum Hamiltonian is *absent* (so that supersymmetry is spontaneously broken), but there is a continuous spectrum including the states with arbitrarily low energies. Such is a rather peculiar (as promised) low-energy dynamics of massless $SU(2)$ SQCD with one flavor.

But what happens if we switch on the mass? In this case, we have to add to the ADS superpotential (7.11) the term $mT^j S_j$, assuming $m \ll \Lambda$. In this case, the equations (7.12) (with $\mathcal{W}^{\text{np}} \to \mathcal{W}^{\text{tot}}$) have finite solutions with

$$(t^j s_j)^2 = \frac{\Lambda^5}{m} \overset{\text{def}}{=} v^4. \qquad (7.14)$$

As indicated in Eq.(7.2), there is also a quartic scalar contribution in the potential energy.[6] In vacuum, it should also vanish giving

$$s^* t^a s - t t^a t^* = 0. \qquad (7.15)$$

---

[5] A precise numerical coefficient in the second term depends on the precise definition of Λ. We have inserted $1/2$ bearing in mind the convention in Eq.(3.17).

[6] It is often called "D-term", because the quartic scalar contribution in (7.2) takes its origin from the term $\sim D^a D^a$ in (3.40).

We have altogether $2 + 3 = 5$ real conditions for $4 \times 2 = 8$ real components of s_j and t^j. Modulo a 3-parametric gauge freedom, this equation system has two distinct solutions. They can be chosen in the form

$$
\mathbf{I}: \; s_j \;=\; \begin{pmatrix} v \\ 0 \end{pmatrix}, \qquad\qquad t^j \;=\; (v, 0);
$$

$$
\mathbf{II}: \; s_j \;=\; \begin{pmatrix} iv \\ 0 \end{pmatrix}, \qquad\qquad t^j \;=\; (iv, 0). \tag{7.16}
$$

The existence of two distinct vacua conforms with the result $I_W = 2$ for pure SYM theory with $SU(2)$ gauge group.

Nonzero vacuum expectation values of the scalar fields break $SU(2)$ gauge symmetry by Higgs mechanism giving mass to gauge superfields. When the matter mass m decreases, the absolute value v of the scalar vevs increases. In the limit $m \to 0$, the vacua run away to infinity.

For a small finite m, the characteristic mass of the massive vector superfields is of order

$$
M_V \sim g(v)v \sim v \sim \left(\frac{\Lambda^5}{m} \right)^{1/4}
$$

up to calculable logarithmic corrections associated with the scale dependence of the coupling constant. This mass is much larger than the characteristic mass m of scalar excitations, which endows the ADS Lagrangian the true BO status and justifies the analysis above.

As was the case for pure SYM theory, the vacua are characterized by a nonzero gluino condensate. Once the scalar vevs (7.16) are known, the values of the condensate can be calculated using the anomalous *Konishi relation* [112] (see also [113]):

$$
\boxed{\{\hat{\bar{Q}}_{\dot\alpha}, \bar\psi^{\dot\alpha j} s_j\} \;\propto\; -mt^j s_j + \frac{g^2}{32\pi^2} \lambda^a \lambda^a.} \tag{7.17}
$$

Here the first term is the "naive" classical result: the supercharge $\hat{\bar{Q}}_{\dot\alpha}$ makes the auxiliary field F^{*j} out of $\bar\psi^j$, and $F^{*j} = -mt^j$ by an equation of motion. The second term is a quantum anomaly arising from a certain triangle graph. It is akin to chiral anomaly in ordinary QCD.

Take the vev of both sides of (7.17). The average of the left-hand side over a supersymmetric vacuum vanishes, and so should the right-hand side. We derive

$$
g^2 \langle \lambda^a \lambda^a \rangle \;=\; \pm 32\pi^2 \sqrt{m\Lambda^5}. \tag{7.18}
$$

This analysis can be generalized to higher N. We dwell on the theory involving $N_f = N - 1$ pairs of chiral multiplets having the same small mass m. In this case, the scalar vevs break gauge symmetry completely. For $N > 2$, the instantons generate the following effective superpotential:

$$\mathcal{W}^{\mathrm{np}} = \frac{\Lambda^{2N+1}}{\det(T^{jf} S_j^g)}, \tag{7.19}$$

where $f, g = 1, \ldots, N_f = N - 1$ are the flavor indices. The integral $\int d^2\theta \, \mathcal{W}^{\mathrm{np}}$ is invariant under the symmetry (7.8). Add to (7.19) the superpotential $m T^{jf} S_j^f$. The vacuum states appear as the solutions to the equation system

$$\frac{\partial}{\partial s_j^f} \left[\mathcal{W}^{\mathrm{np}}(s, t) + m \, t^{jf} s_j^f \right] = \frac{\partial}{\partial t^{jf}} \left[\mathcal{W}^{\mathrm{np}}(s, t) + m \, t^{jf} s_j^f \right] = 0. \tag{7.20}$$

supplemented with the "D-flatness condition" (7.15). The equations (7.20) imply that

$$t^{jf} s_j^g = e^{2\pi i k/N} \left(\frac{\Lambda^{2N+1}}{m} \right)^{1/N} \delta^{fg} \stackrel{\text{def}}{=} v^2 e^{2\pi i k/N} \delta^{fg}, \tag{7.21}$$

$k = 0, \ldots, N - 1$. Taking in account also (7.15), we obtain (modulo a gauge transformation) N distinct solutions:

$$s_1 = \begin{pmatrix} v e^{i\pi k/N} \\ 0 \\ \cdots \\ 0 \end{pmatrix}, \quad \ldots, \quad s_{N-1} = \begin{pmatrix} 0 \\ \cdots \\ v e^{i\pi k/N} \\ 0 \end{pmatrix},$$

$$t^1 = (v e^{i\pi k/N}, 0, \ldots, 0), \quad \ldots, \quad t^{N-1} = (0, \ldots, v e^{i\pi k/N}, 0). \tag{7.22}$$

Again, this conforms with the vacuum counting $I_W = N$ for pure SYM theory. As was mentioned, the scalar vevs (7.22) break the gauge $SU(N)$ symmetry completely. The Konishi relation implies the formation of the gluino condensate:

$$\boxed{\langle \lambda^a \lambda^a \rangle \sim e^{2\pi i k/N} \left(\Lambda^{2N+1} m^{N-1} \right)^{1/N}.} \tag{7.23}$$

If $N_f < N - 1$, $SU(N)$ gauge symmetry is broken down not completely, but to $SU(N - N_f)$. Still, for small masses, the low-energy dynamics is determined by scalar fields, and the analysis exhibits N vacuum states also in this case [108]. Take for example the $SU(3)$ theory with one pair of chiral multiplets. The full effective superpotential is[7]

$$\mathcal{W}(S, T) = \frac{\Lambda^4}{\sqrt{T^j S_j}} + m T^j S_j. \tag{7.24}$$

---

[7]Then $\int d^2\theta \, \mathcal{W}^{\mathrm{np}}(S, T)$ is invariant under the transformation (7.8).

There are three vacua at

$$v^2 = t^j s_j = e^{2\pi i k/3} \left(\frac{\Lambda^4}{2m}\right)^{2/3}. \tag{7.25}$$

The corresponding gluino condensates are

$$\langle \lambda^a \lambda^a \rangle \sim e^{2\pi i k/3} (m\Lambda^8)^{1/3}. \tag{7.26}$$

In this case, the spectrum involves three different physical scales:

- The low scale $\sim m$ of the scalar excitations and their superpartners.
- The mass $M_V \sim v \sim (\Lambda^4/m)^{1/3}$ of the massive vector multiplets.
- An intermediate scale Λ_2 — the characteristic dimensional parameter for the remaining SQCD theory with the unbroken $SU(2)$ group.

Both the original $SU(3)$ theory and the $SU(2)$ theory describing the dynamics at energies below v are asymptotically free, and the effective coupling drops with energy. But the speeds by which it drops at $E \lesssim v$ and $E \gtrsim v$ are different. Generically, for SQCD with N colors and $N_f < 3N$ flavors, one can write at the one-loop level

$$g^2(\mu) = \frac{2\pi}{(3N - N_f) \ln \frac{\mu}{\Lambda_N}}. \tag{7.27}$$

In the case under consideration, we have

$$g^2(\mu) \propto \frac{2\pi}{8 \ln \frac{\mu}{\Lambda_3}} \tag{7.28}$$

for $\mu \gtrsim v$ and

$$g^2(\mu) \propto \frac{2\pi}{5 \ln \frac{\mu}{\Lambda_2}} \tag{7.29}$$

for $\mu \lesssim v$.

The parameter Λ_2 can be found from the condition that the estimates (7.28) and (7.29) coincide at $\mu \sim v$. We obtain

$$\Lambda_2 \sim \Lambda_3^{8/5} v^{-3/5} \tag{7.30}$$

or, bearing in mind (7.25),

$$\Lambda_2 \sim \Lambda_3^{8/5} \left(\frac{\Lambda_3^4}{m}\right)^{-1/5} = m^{1/5} \Lambda_3^{4/5}. \tag{7.31}$$

For a generic $SU(N)$ theory with N_f pairs of fundamental matter multiplets, where the symmetry is broken down to $SU(N - N_f)$, we derive

$$\Lambda_{N-N_f} \sim \Lambda_N \left(\frac{m}{\Lambda_N}\right)^{\frac{3N_f(N-N_f)}{2N(3N-4N_f)}}. \tag{7.32}$$

This estimate[8] is valid as long as $3N > 4N_f$. Otherwise, the $SU(N - N_f)$ theory has too many matter fields and is not asymptotically free.

If $N_f \geq N$, the effective nonperturbative potential is not generated and the physics is the same as in pure SYM $SU(N)$ theory. This again gives $I_W = N$.

7.1.1   *Domain walls*

The existence of distinct classical vacua suggests the existence of domain walls that interpolate between them. And in the limit $m \ll \Lambda$ when the ADS Lagrangian describes the low-energy dynamics of SQCD, these walls indeed exist. One can observe furthermore that they are *BPS saturated*.

The abbreviation BPS stands for Bogomolny, Prasad, and Sommerfeld who found [114] that, in the limit when the scalar self-interaction is switched off, the mass of the 't Hooft-Polyakov monopole can be found by solving a simple first-order differential equation. The surface energy and the profile of the domain walls interpolating between the different vacuum states of the ADS effective potential can be found by a similar method.

In the theories of interest, the superpotentials depend on several superfields S_j, T^j. But to understand what happens, consider first the Wess-Zumino model (3.19) including only one superfield Φ with the superpotential

$$\mathcal{W}(\Phi) = \mu^2 \Phi - \frac{\alpha \Phi^3}{3}. \tag{7.33}$$

If $\alpha > 0$ (we assume for simplicity that α and μ are real), the model includes two vacua at $\phi_{\text{vac}} = \pm \mu/\sqrt{\alpha}$. Consider a static wall configuration interpolating between two minima:[9] $\phi(z = -\infty) = -\mu/\sqrt{\alpha}$ and $\phi(z = \infty) = \mu/\sqrt{\alpha}$. Its surface energy density is

$$\epsilon = \int_{-\infty}^{\infty} dz \left(|\partial_z \phi|^2 + |\mathcal{W}'(\phi)|^2 \right). \tag{7.34}$$

---

[8]We disagree at this point with the estimate quoted in Ref. [108],

$$\Lambda_{N-N_f} \sim \Lambda_N \left(\frac{m}{\Lambda_N} \right)^{N_f/3N}.$$

The latter estimate could be reproduced under the assumption that the matter does not affect the evolution of the coupling below v.

[9]Note that this configuration does not depend on two transverse coordinates and the problem reduces to evaluating the mass of the corresponding *soliton* in a 2D model [115, 116]. We will discuss it further, when we talk about the CFIV index at the very end of the last chapter.

We can rewrite it as

$$\epsilon = \int_{-\infty}^{\infty} dz \, [\partial_z \phi - e^{-i\delta} \mathcal{W}'(\phi^*)][\partial_z \phi^* - e^{i\delta} \mathcal{W}'(\phi)]$$

$$+ \int_{-\infty}^{\infty} dz \, [e^{i\delta} \partial_z \phi \, \mathcal{W}'(\phi) + e^{-i\delta} \partial_z \phi^* \mathcal{W}'(\phi^*)]$$

$$= \int_{-\infty}^{\infty} dz |\partial_z \phi - e^{-i\delta} \mathcal{W}'(\phi^*)|^2 + 2\text{Re}\{e^{i\delta}[\mathcal{W}(\infty) - \mathcal{W}(-\infty)]\} \quad (7.35)$$

with a phase δ to be shortly determined. The solution to the classical equations of motion represents a minimum of the energy functional. We arrive at the first-order equation

$$\partial_z \phi - e^{-i\delta} \mathcal{W}'(\phi^*) = 0. \tag{7.36}$$

In the case considered, we may set $\delta = 0$. The BPS wall configuration satisfying the equation $\partial_z \phi = \mathcal{W}'(\phi^*)$ is real:

$$\boxed{\phi(z) = \frac{\mu}{\sqrt{\alpha}} \tanh(\mu\sqrt{\alpha}z).} \tag{7.37}$$

The surface energy is

$$\epsilon = 2[\mathcal{W}(\infty) - \mathcal{W}(-\infty)] = \frac{8\mu^3}{3\sqrt{\alpha}}. \tag{7.38}$$

Suppose now that α is negative. Then the vacua occur at purely imaginary values: $\phi_{\text{vac}} = \pm i\mu/\sqrt{|\alpha|}$. The wall solution is also purely imaginary, it satisfies the equation (7.36) with $\delta = \pi/2$ or $\delta = -\pi/2$, depending on how z is chosen. In other words, the parameter δ depends on the positions of the vacua on the complex plane, between which the wall interpolates [117].

This method can be easily generalized to a WZ model including several superfields. It is not difficult to prove that the wall surface energy is always equal to or exceeds the *BPS bound*,[10]

$$\boxed{\epsilon_{\text{BPS}} = 2|\mathcal{W}(\infty) - \mathcal{W}(-\infty)|.} \tag{7.40}$$

---

[10]The existence of this bound is in fact associated with a modification of the standard supersymmetry algebra (3.2). Nonzero central charges appear [118] ! They read

$$\{\hat{Q}_\alpha, \hat{Q}_\beta\} = 2i(\sigma)_\alpha{}^\gamma \varepsilon_{\gamma\beta} \int d^3x \, \nabla \mathcal{W}[\phi^*(\boldsymbol{x})] \tag{7.39}$$

[see Eq. (11.72) and discussion thereof].

For the SU(2) ADS model with the superpotential

$$\mathcal{W}(S,T) = \frac{\Lambda^5}{T^j S_j} + mT^j S_j, \tag{7.41}$$

we find [102] the following profile of the BPS domain wall interpolating between the vacua (7.16):

$$t^j s_j(z) = v^2 \frac{(e^{4mz} + i)^2}{e^{8mz} + 1}. \tag{7.42}$$

The absolute value of the moduli $t^j s_j$ is constant, being equal to $v^2 = \sqrt{\Lambda^5/m}$, but its phase changes from π at $z = -\infty$ to 0 at $z = \infty$. The surface energy of this wall is

$$\epsilon = 2|\mathcal{W}(\infty) - \mathcal{W}(-\infty)| = 4\left(\frac{\Lambda^5}{v^2} + mv^2\right) = 8\sqrt{m\Lambda^5}. \tag{7.43}$$

Using (7.18), we can also express it via the gluino condensate in the original theory:

$$\epsilon = \frac{g^2}{8\pi^2} |\langle \lambda^a \lambda^a \rangle_{z=\infty} - \langle \lambda^a \lambda^a \rangle_{z=-\infty}|. \tag{7.44}$$

For higher N, there are N complex vacua (7.21). There are domain walls interpolating between any pair of them and all these walls are BPS. The profile of the wall relating the adjacent vacua with $\langle t^f s^g \rangle = v^2 \delta^{fg}$ and $\langle t^f s^g \rangle = v^2 e^{2\pi i/N} \delta^{fg}$ was found numerically in Ref. [119]. In contrast to the case $N = 2$ where $|ts|$ did not depend on z, at higher N, the wall exhibits a small bump in the middle (see Fig. 7.1).

What happens if we increase the mass? If $m \sim \Lambda$, the ADS Lagrangian does not describe the low-energy dynamics of the theory anymore and the latter depends on the scalar superfields and gauge superfields on equal footing. In pure SYM theory, we tried to extract information about the vacua and the walls by studying the effective *non-Wilsonian* Veneziano-Yankielowicz Lagrangian (6.80). The results were surprising: besides the chirally asymmetric vacua, a chirally symmetric vacuum (probably, an artifact of the VY description) was observed, and the only kind of walls which we saw there were the walls relating a chirally asymmetric vacuum to the chirally symmetric one. The walls between different asymmetric vacua did not exist.

In the theory with matter, we can write a generalization of (6.80), the *Taylor-Veneziano-Yankielowicz* effective Lagrangian [120]. This Lagrangian depends on the composite gauge superfield (6.79), whose cubic

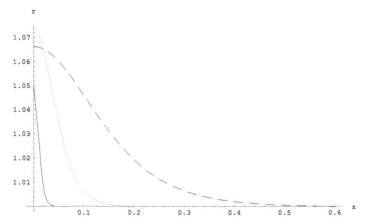

Fig. 7.1: The plots of $r(z) = |ts|(z)/|ts|(\pm\infty)$ for the BPS walls in the ADS theory with $N_f = N - 1$ for N = 3 (dashed line), N = 5 (dotted line) and N = 10 (solid line). In each case, only a half of the wall (from its middle to the right) is displayed.

root Φ has canonical dimension of mass, and on the matter moduli $\mathcal{M}^{fg} = T^{jf}S_j^g$, $f, g = 1, \ldots, N_f$. If the mass matrix of the matter fields has the form $m_{fg} = m\mathbb{1}$, the matrix of matter moduli can also be chosen in the diagonal form, $\mathcal{M}^{fg} = X^2 \delta^{fg}$. In this case and if $N_f = N - 1$, the TVY Lagrangian reads

$$
\begin{aligned}
\mathcal{L}_{\text{TVY}} &= \alpha \int d^4\theta\, \overline{\Phi}\Phi + \beta \int d^4\theta\, \overline{X}X \\
&\quad + \left[\int d^2\theta \left(\gamma \Phi^3 \ln \frac{\Phi^3 X^{2(N-1)}}{\Lambda^{2N+1}} + \frac{m}{2}X^2 \right) + \text{c.c.} \right] \quad (7.45)
\end{aligned}
$$

with some α, β, γ.

When the mass is much smaller than Λ and the characteristic values of the lowest component of Φ, one can integrate out the gauge fields and arrive at the ADS Lagrangian. When $m \gg \Lambda$, one can integrate out the matter fields and arrive at the VY effective Lagrangian. The intermediate values of m can be explored numerically. The results are rather amusing [119, 121].

• For all values of mass, there exist a "real" domain wall interpolating between a chirally asymmetric vacuum and the chirally symmetric one, where ϕ and χ (the lowest components of Φ and X) vanish and which the TVY Lagrangian enjoys in the same way as the VY Lagrangian discussed in the previous chapter did.

- For $m \ll \Lambda$, the complex ADS domain wall is reproduced, but on top of that, one observes *another* BPS wall, in the middle of which the values of ϕ and χ are rather close to zero: the wall passes close to the chirally symmetric vacuum and dwells for a while in its vicinity. When $N = 2$, it simply splits in two real walls in the limit of $m \to 0$. For $N > 2$, it stays at some distance from the chirally symmetric point even in this limit.
- When the mass increases, the second BPS wall retreats from the chirally symmetric point and approaches the first wall. At some critical value of the mass $m^* \sim \Lambda$, the two walls merge, a bifurcation occurs, and at still larger values of mass, the BPS solutions disappear.
- In a mass interval $m^* < m \le m^{**}$, the wall solution to the second-order field equations still exists, but it is not BPS anymore.
- For $m > m^{**}$, complex domain walls are absent, as they were absent in the VY case.

The last fact is especially strange. If we believe in the picture of spontaneous chiral symmetry breaking in supersymmetric QCD (and, as we mentioned, in the presence of the matter, there are no torons and no universe interpretation), the walls between different phases *must* exist there. Maybe they do. But there is no trace of them in the TVY approach for the theory $N_f = N - 1$ with equal masses of all matter fields.

Note, however, that the situation is different when the number of flavors N_f is less than $N/2$. In this case, the TVY walls between different chiral phases are *tenacious* — they stay there even in the limit of $m \to \infty$ [122][11] But this limit is not smooth. For large masses, the tenacious TVY walls acquire "hard cores" near the cuts of the Kovner-Shifman potential (solid lines in Fig. 6.2). In these narrow cores, the field ϕ does not change much, but the field χ does, so that the cores carry a significal fraction of the total wall energy.

Well, the TVY dynamics is very rich and interesting. I am afraid, however, that it does not recount well what happens (would physically happen if SQCD described nature) in supersymmetric QCD...

7.2   Orthogonal groups

The constructions above can be generalized to all other simple Lie groups [123–126].

---

[11]This may happen also when $N_f = N - 1$ if the masses of the matter fields are different, $m_{fg} \ne m\delta_{fg}$.

Consider e.g. a theory based on an orthogonal group[12] $\mathcal{G} = SO(N \geq 5)$ and involving a light matter chiral multiplet S_a in the defining vector representation. Instantons and other nonpertubative gauge field configurations generate an effective Lagrangian depending on the gauge-invariant composite matter superfield $S_a S_a$. (The vector representation is real and one does not need a pair of mutiplets to cook up a gauge invariant.) To determine its form, we use the same logic as for the unitary groups. For all such theories, there exists a certain combination of the R current and the matter axial current that is free from anomaly. In the case considered, the corresponding exact symmetry of the action is

$$W \rightarrow e^{i\alpha}W, \qquad S \rightarrow e^{-i\alpha(N-3)}S, \qquad \theta \rightarrow e^{i\alpha}\theta. \qquad (7.46)$$

The effective potential that is invariant under this symmetry reads

$$\mathcal{L}^{\mathrm{np}} = \int d^2\theta \, \frac{\Lambda^{(3N-7)/(N-3)}}{(S_a S_a)^{1/(N-3)}} + \text{c.c.} \qquad (7.47)$$

Add here the mass term,

$$\mathcal{L}_m \sim m \int d^2\theta \, S_a S_a \,,$$

and solve the equation $\partial \mathcal{W}/\partial S_a = 0$. We obtain $N-2$ distinct solutions:

$$\boxed{\langle s_a s_a \rangle_{\mathrm{vac}} \sim \frac{\Lambda^{(3N-7)/(N-2)}}{m^{(N-3)/(N-2)}} e^{2\pi i k/(N-2)}, \qquad k = 0, \ldots, N-3.} \qquad (7.48)$$

Now, $N-2$ coincides with the dual Coxeter number $h^\vee$ for $SO(N \geq 5)$ and hence the Witten index for our theory coincides with the result (6.101) for the pure SYM theory based on $SO(N)$ group.

For the $SO(N)$ theory involving N_f vector multiplets S^f, the anomaly-free symmetry is[13]

$$W \rightarrow e^{i\alpha}W, \qquad S^f \rightarrow e^{-i\alpha(N-2-N_f)/N_f}S^f, \qquad \theta \rightarrow e^{i\alpha}\theta. \qquad (7.49)$$

---

[12]The cases $N = 3, 4$ are special. $SO(3) \simeq SU(2)/Z_2$, $SO(4) \simeq [SU(2) \times SU(2)]/Z_2$, and the formula $h^\vee[SO(N)] = N - 2$ is not valid.

[13]Take Eq.(7.8) and replace there $N = h^\vee[SU(N)]$ by $N - 2 = h^\vee[SO(N)]$. Note that a *single* vector multiplet S_a for $SO(N)$ plays the same role as a *pair* of multiplets S_j and T^j for $SU(N)$. It follows from the fact that the triangle diagrams describing the anomalous nonconservation of the matter axial currents give in these two cases the same contributions.

And this in turn follows from two facts: *(i)* the Dynkin index (A.19) of a vector representation in $SO(N)$ is 2 times larger than the Dynkin index of a fundamental representation in $SU(N)$; *(ii)* the vector representation is real and the anomalous triangle involves an additional symmetry factor $1/2$.

The invariant nonperturbative potential is generated for $N_f < N - 2$ and reads

$$\mathcal{L}^{\mathrm{np}}(N_f) = \int d^2\theta \, \frac{\Lambda^{(3N-N_f-6)/(N-2-N_f)}}{\det(\mathcal{M})^{1/(N-2-N_f)}} + \text{c.c.} \qquad (7.50)$$

with $\mathcal{M}^{fg} = S_a^f S_a^g$.

The Witten index is equal to $N - 2$ also in this case.

7.3 G_2

Consider the SQCD theory based on the G_2 gauge group. We discuss first its simplest version with the Lagrangian including a single matter fundamental supermultiplet $S_{\alpha=1,...,7}$. The Dynkin index of the fundamental representation in G_2 is equal to 1, the anomaly-free symmetry is

$$W \to e^{i\alpha}W, \qquad S \to e^{-3i\alpha}S, \qquad \theta \to e^{i\alpha}\theta, \qquad (7.51)$$

and the requirement of the invariance under (7.51) dictates the following form of the invariant non-perturbative effective potential:

$$\mathcal{L}^{\mathrm{np}} = \int d^2\theta \left(\frac{\Lambda^{11}}{S_\alpha S_\alpha} \right)^{1/3} + \text{c.c.} \qquad (7.52)$$

Adding the mass term $\sim m \int d^2\theta \, S_\alpha S_\alpha$ and solving the equation $\partial \mathcal{W}^{\mathrm{tot}}/\partial S_\alpha = 0$, we find four distinct vacuum states:

$$\langle s_\alpha s_\alpha \rangle_{\mathrm{vac}} \sim \left(\frac{\Lambda^{11}}{m^3} \right)^{1/4} e^{i\pi k/2}. \qquad (7.53)$$

The Witten index is equal to

$$I_W = 4 = h^\vee(G_2), \qquad (7.54)$$

which coincides with (6.102). The same result is obtained in the theory including two or three massive fundamental supermultiplets. In the theories with four or more multiplets, the nonperturbative superpotential is not generated and the index also coincides with the pure SYM result (6.102).

A similar analysis can be made for higher exceptional groups [126]. In the SQCD theories whose Lagrangian represents a sum of the pure SYM term and the kinetic and mass terms for the matter fields, the index is always equal to $h^\vee(\mathcal{G})$.

Note, however, that in the G_2 theory including *three* fundamental su-permultiplets S_α^f, we are allowed to write in the Lagrangian, besides the mass term, also the *Yukawa* term [127]:

$$\mathcal{L}^{\text{Yukawa}} = \frac{h}{6} \int d^2\theta f_{\alpha\beta\gamma} \varepsilon_{fgh} S_\alpha^f S_\beta^g S_\gamma^h \tag{7.55}$$

with $f_{\alpha\beta\gamma}$ defined in (A.18). The coupling constant h carries no dimension and the theory keeps its renormalizability.

The structure

$$B = \frac{1}{6} f_{\alpha\beta\gamma} \varepsilon_{fgh} S_\alpha^f S_\beta^g S_\gamma^h \tag{7.56}$$

appears in this theory as an extra modulus parameter on which the effective nonperturbative superpotential depends besides the familiar bilinear modulus $\mathcal{M}^{fg} = S_\alpha^f S_\alpha^g$. This superpotential reads [128]

$$\mathcal{W}^{\text{np}}(\mathcal{M}, B) = \frac{\Lambda^9}{\det \mathcal{M} - B^2}. \tag{7.57}$$

Indeed, symmetry considerations dictate

$$\mathcal{W}^{\text{np}}(\mathcal{M}, B) = \frac{\Lambda^9}{\det \mathcal{M}} f\left(\frac{B^2}{\det \mathcal{M}}\right), \tag{7.58}$$

and the easiest way to derive that $f(x) = 1/(1 - x)$ is to note that, if one breaks G_2 down to $SU(3)$ by Higgs mechanism, assuming a nonzero vacuum expectation value of one of the scalar fields, $\langle s_\alpha^1 \rangle = v_0 \delta_{\alpha 7}$, the nonperturbative superpotential (7.57) should coincide with the one in the $SU(3)$ theory with two chiral triplets $\tilde{S}_j^{f=1,2}$ and two antitriplets $T^{(f=1,2)j}$ into which the 7-plets S^2 and S^3 split:

$$\mathcal{W}_{G_2}^{\text{np}} \to \frac{1}{v_0^2} \mathcal{W}_{SU(3)}^{\text{np}} \propto \frac{1}{\det(T\tilde{S})} \tag{7.59}$$

[see Eq. (7.19)]. And that is what the structure $1/(\det \mathcal{M} - B^2)$ gives.

The full superpotential of the effective theory is

$$\boxed{\mathcal{W} = \frac{1}{2(\det \mathcal{M} - B^2)} + \frac{m}{2} \text{Tr}\, \mathcal{M} + hB.} \tag{7.60}$$

(From now on, everything will be measured in the units of Λ. The factors $1/2$ are introduced for convenience.)

Our goal is to find the classical vacua. Were the Yukawa term absent, we could use the ansatz like

$$\langle s_\alpha^1 \rangle_{\text{vac}} = v_0 \delta_{\alpha 1}, \qquad \langle s_\alpha^2 \rangle_{\text{vac}} = v_0 \delta_{\alpha 2}, \qquad \langle s_\alpha^3 \rangle_{\text{vac}} = v_0 \delta_{\alpha 3} \tag{7.61}$$

(any other relevant configurations could be obtained from this by a gauge transformation). Then the equation $\partial \mathcal{W}/\partial v_0 = 0$ would reduce to

$$m(v_0^2)^4 \;=\; 1 \tag{7.62}$$

with 4 distinct solutions for the modulus $u = v_0^2$.

If $h \neq 0$, the ansatz (7.61) should be modified. One can choose it as [127]:

$$\langle s_\alpha^1 \rangle \;=\; \begin{pmatrix} 0 \\ 0 \\ 0 \\ 0 \\ 0 \\ 0 \\ v_0 \end{pmatrix}, \quad \langle s_\alpha^2 \rangle \;=\; \frac{1}{\sqrt{2}} \begin{pmatrix} v_1 \\ 0 \\ 0 \\ 0 \\ 0 \\ v_2 \\ 0 \end{pmatrix}, \quad \langle s_\alpha^3 \rangle \;=\; \frac{1}{\sqrt{2}} \begin{pmatrix} 0 \\ v_1 \\ v_2 \\ 0 \\ 0 \\ 0 \\ 0 \end{pmatrix} \tag{7.63}$$

with the constraint $v_0^2 = (v_1^2 + v_2^2)/2 \equiv u$. In this case, the matrix $\mathrm{m}^{fg} = \langle s_\alpha^f s_\alpha^g \rangle$ is still diagonal, $\mathrm{m}^{fg} = u\,\delta^{fg}$, whereas

$$b \;=\; \frac{1}{6} f_{\alpha\beta\gamma} \varepsilon_{fgh} \, \langle s_\alpha^f s_\beta^g s_\gamma^h \rangle \;=\; \frac{1}{2} v_0 (v_1^2 - v_2^2). \tag{7.64}$$

We have to solve the equations

$$\frac{\partial \mathcal{W}(u,b)}{\partial u} \;=\; -\frac{3u^2}{2(u^3 - b^2)^2} + \frac{3m}{2} \;=\; 0\,,$$

$$\frac{\partial \mathcal{W}(u,b)}{\partial b} \;=\; \frac{b}{(u^3 - b^2)^2} + h \;=\; 0\,. \tag{7.65}$$

It follows that $b = -hu^2/m$ while u satisfies the equation

$$mu^4 \left(1 - \frac{h^2 u}{m^2} \right)^2 \;=\; 1\,. \tag{7.66}$$

It has *six* roots bringing about six distinct vacua!

Let us see what happens in the limit of small $|h|$, the smallness being characterized by a dimensionless parameter

$$\kappa = \left| \frac{h^2}{m^{9/4}} \right| \equiv \left| h^2 (\Lambda/m)^{9/4} \right| \ll 1 \tag{7.67}$$

(do not forget that h and m are generically complex). Then 4 of the roots are very close to the known solutions of Eq. (7.62), but on top of this, we see two extra roots at *large* values of u: $u = m^2/h^2 \pm h^2/m^{5/2} + \ldots$. In the limit $h \to 0$, these new vacua run away at infinity of the moduli space and decouple.

The latter statement can be attributed a precise meaning. Different vacua are separated by the domain walls. Their surface energy (which is the measure of the height of the barrier separating different vacua) satisfies a strict lower BPS bound,

$$\epsilon \geq 2|\mathcal{W}_1 - \mathcal{W}_2|, \qquad (7.68)$$

where $\mathcal{W}_{1,2}$ are the values of the superpotential at the vacua 1,2 between which the wall interpolates [see Eq. (7.40)].

When $\kappa = 0$ and only 4 vacua are left, the domain walls between them are BPS–saturated and have the energy density $\epsilon = 2\sqrt{2}|m|^{3/4}$ for the walls connecting the "adjacent" vacua (with, say, $u = m^{-1/4}$ and $u = im^{-1/4}$) and $\epsilon = 4|m|^{3/4}$ for the walls connecting the "opposite" vacua with $u = \pm m^{-1/4}$ or with $u = \pm im^{-1/4}$. In fact, the effective ADS Lagrangian is exactly the same here as for the $SU(4)$ theory with 3 chiral quartets and 3 antiquartets, whose domain walls were studied in Ref. [119].

When κ is nonzero but small so that a couple of new vacua at large values of u appear, the values of the superpotential (7.60) at these vacua are also large $\sim m^3/h^2$. The bound (7.68) dictates that the energy density of a wall connecting an "old" and a "new" vacua is of order $|m^3/h^2| = m^{3/4}/\kappa$ and tends to infinity in the limit $\kappa \to 0$.

Now we can understand why a heuristic argument of Ref. [47], saying that the matter sector decouples for large masses and does not affect the index counting, does not work in this case. In the limit $m \to \infty$, it decouples. But if the mass is very large but finite, extra vacua at very large values of u appear. They are separated from four conventional vacua at small u by a very high barrier and are irrelevant, as far as the conventional low-energy dynamics is concerned. But a traveller who managed to overpass this barrier would find himself in a completely different fascinating world. The presence of this world should be taken into account in evaluating the Witten index.

When we increase κ, the new vacua move in from infinity and, at $\kappa \sim 1$, the values of u_{vac} for the vacua of both types are roughly the same. It is interesting that, at $\kappa = \frac{2}{3\sqrt{3}} \approx .385$, two of the vacua (an "old" and a "new" one) become degenerate. At $\kappa \gg 1$, six vacua find themselves at the vertices of a nearly perfect hexagon in the complex u-plane.[14] The complex

---

[14]The size of this hexagon is of order $|u_{\text{vac}}| \sim |m^3/h^4|^{1/6}$. To assure that $|u_{\text{vac}}|$ are still large (in physical units, $|u_{\text{vac}}| \gg \Lambda^2$) so that we are still in the Higgs phase and the light and heavy degrees of freedom are well separated, we should keep $|h^4/m^3| \ll 1$. But if $|m|$ is small enough, this condition can be fulfilled at no matter how large κ.

roots of Eq. (7.66) in the units of $|m|^{-1/4}$ for 3 illustrative values of κ are displayed in Fig. 7.2.

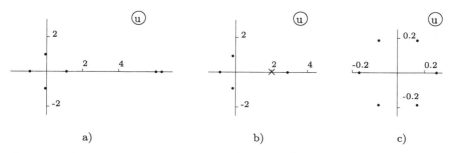

Fig. 7.2: Solutions of the equation $u^4(1 - \kappa u)^2 = 1$ for *a)* $\kappa = .16$, *b)* $\kappa = .385$, and *c)* $\kappa = 100$. The cross marks out the double degenerate root at $\kappa = .385$.

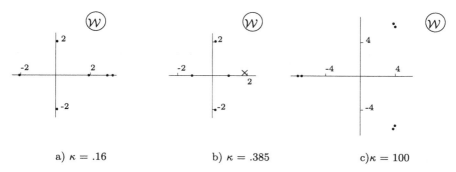

a) $\kappa = .16$ b) $\kappa = .385$ c) $\kappa = 100$

Fig. 7.3: Values of the superpotential $\mathcal{W}_{\text{vac}}$

The values of the superpotential $\mathcal{W}(u_{\text{vac}}, b_{\text{vac}})$ for the same values of κ in the units of $|m|^{3/4}$ are shown in Fig. 7.3. We see that, for large κ, 6 vacua are clustered in 3 pairs (each pair corresponding to the opposite vertices of the hexagon in Fig. 7.2c). The values of the superpotential for two vacua of the same pair are close and hence the energy barrier between them is small.

Indeed, for very small masses and fixed h, one can neglect the mass term in the superpotential, in which case $|u_{\text{vac}}|^3 \ll |b_{\text{vac}}|^2$ and the vacuum values of b are determined from the equation $hb_{\text{vac}}^3 = -1$. In this limit, we see only three vacua, while their splitting inside a pair is the effect of higher order in $1/\kappa$.

Note finally that this phenomenon, the emergence of extra vacuum states when Yukawa couplings are switched on, is not specific for G_2 theory. It is common for all models where such Yukawa couplings are admissible. Probably the simplest example of such a model is the theory based on the gauge group $SU(2)$ and involving a pair of chiral multiplets S_j and $T^j = \varepsilon^{jk}T_k$ in the fundamental doublet representation (in $SU(2)$, the fundamental representation is pseudoreal, being equivalent to the antifundamental one) and an adjoint multiplet Φ_j^k with $\Phi_j^j = 0$. The theory, including the Yukawa term,

$$h \int d^2\theta \, S^j \Phi_j^k T_k \; + \; \text{c.c.}, \tag{7.69}$$

has *three* rather than two vacuum states [129], with the third state going to infinity in the limit $h \to 0$.

7.4 Extended SYM theories

In $\mathcal{N} = 1$ terms, the $\mathcal{N} = 2$ and $\mathcal{N} = 4$ supersymmetric Yang-Mills theories represent the ordinary SYM theories including matter multiplets in the adjoint representation.

The $\mathcal{N} = 2$ theory includes only one such chiral multiplet [see Eq. (3.45)]. Similarly to the $\mathcal{N} = 1$ theory (7.2) with a pair of massless chiral multiplets in the fundamental and antifundamental representations, it involves a scalar vacuum valley. The dynamics along this valley (called *Coulomb branch*) is very rich and interesting — especially at the beginning of the valley when nonperturbative effects come into play [32]. But we will not discuss it here and only note that, similarly to the model (7.2), the $\mathcal{N} = 2$ model may be deformed by giving a mass to the matter fields so that the valley disappears. Hence its Witten index is exactly the same as in the pure $\mathcal{N} = 1$ SYM theory and coincides with the dual Coxeter number $h^\vee$ of the group.

The $\mathcal{N} = 4$ theory includes three such multiplets and the corresponding superfield Lagrangian (3.52) involves a Yukawa coupling term. Then the dynamics is similar to the dynamics of the G_2 model described at length above. When we deform the theory giving a mass M to the scalar fields, we find, on top of the $h^\vee$ vacua coming from the effective Hamiltonian for the zero Fourier modes of the gauge field, also some vacua having nonzero scalar field averages $\langle \phi \rangle \sim M/g$. For the $SU(2)$ model, there is only one such vacuum (up to an overall gauge rotation) [130]:

$$\phi_f^a = \frac{M}{g\sqrt{2}} \delta_f^a,$$ (7.70)

as in Eq. (9.42). There is also a single extra vacuum for $SU(N)$. For more complicated groups, there are several such vacua; we will rediscuss this issue in Chapter 9 — see Theorem 9.1 there. For example, there are 11 extra vacua for the E_8 gauge group. All together we have

$$\#_{\mathrm{vac}}[E_8] = h^\vee[E_8] + 11 = 30 + 11 = 41.$$ (7.71)

We observe here a slight mismatch with the fundamental Answer to the Question of Life, Universe and Everything, which is well known [131] to be 42.

Chapter 8

Vacuum structure of $4D$ supersymmetric theories III: chiral matter

In this chapter (see the review [132] for the abridged version), we will discuss supersymmetric gauge theories with chiral matter content where the number of left-handed and right-handed fermions is not equal. One of such (nonsupersymmetric) gauge theories is well known. It is the theory of electroweak interactions, which describes nature. But we are not studying in this book the real world, but rather a multitude of imaginary supersymmetric worlds...

8.1 Chiral SQED

The simplest version of a chiral gauge supersymmetric theory is chiral SQED. The ordinary QED and the ordinary SQED include a charged Dirac fermion. The Dirac bispinor field (1.52) includes a left-handed and a right-handed Weyl fermion fields or else a couple of left-handed fields with opposite electric charges [see Eq. (1.49)]. A chiral QED or SQED includes sets of left-handed and right-handed fields, which are inherently different. It is more convenient to represent them by a set of left-handed fields with different electric charges q_f.

However, there is one important restriction: the sum of the *cubes* of the charges should vanish. The matter is that each charged Weyl fermion generates a triple photon interaction M_{AAA} by the Feynman graph depicted in Fig. 8.1. This interaction breaks gauge invariance of the theory and its renormalizability. Clearly, the contribution of each fermion to M_{AAA} is proportional to q_f^3. The sum of all such contributions is proportional to $\sum_f q_f^3$, and it should vanish. Otherwise, the theory would be *anomalous* (gauge symmetry of classical action would be broken by quantum effects) and not kosher.[1]

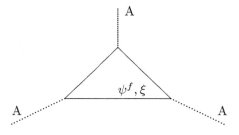

Fig. 8.1: Anomalous triangles in chiral SQED

The simplest version of chiral SQED includes 8 chiral superfields S^f carrying a positive charge e and a superfield T carrying a negative charge $-2e$. The Lagrangian of the model reads

$$\mathcal{L} = \frac{1}{4}\int d^4\theta \,(\overline{S}^f e^{eV} S^f + \overline{T}e^{-2eV}T) + \left(\frac{1}{8}\int d^2\theta\, W^\alpha W_\alpha + \text{c.c.}\right). \quad (8.1)$$

We keep the coupling constant e small.

The corresponding component Lagrangian includes, among other things, a term [cf. the last term in Eq. (3.34)] $-e^2(s^{*f}s^f - 2t^*t)^2/2$, where s^f and t are the scalar components of the multiplets S^f and T. This gives a classical vacuum valley of the type discussed in the previous chapter. Supersymmetry protects it — the corrections to the potential in all the orders of perturbation theory vanish. As a result, the spectrum of the theory does not have a gap: it is continuous starting right from zero.

In massless nonchiral theories we had the same situation, but this degeneracy could be lifted there if the matter fields were given a mass. In chiral theory, we cannot write a mass term, but we can add to the Lagrangian (8.1) a Yukawa term like

$$\mathcal{L}_{\text{Yukawa}} = \frac{h}{2}\int d^2\theta\,(S^f S^f T) + \text{c.c.} \quad (8.2)$$

---

[1]We are talking now about an *internal* anomaly when the vertices of the triangle are coupled to the dynamical vector fields. This should be distinguished from the *external* anomalies discussed in Chapters 6,7, when one of the vertices corresponded to external axial current. An external anomaly may break a certain global symmetry of the theory, but it does not ruin it, keeping its gauge invariance.

This gives a potential

$$V = h^2(4|s^f t|^2 + |s^f s^f|^2),\qquad(8.3)$$

which plays the same role as the mass term in nonchiral theories: it locks the valley and brings about a gap in the spectrum. Unfortunately, we cannot send in this case the interaction constant h to infinity, as we did with the mass parameter for the nonchiral SQED. In fact, we need to keep both h and e^2 small to avoid running into the Landau pole (because of which the chiral SQED, as well as the ordinary SQED and ordinary QED, are not, strictly speaking, well-defined quantum theories). Also here we do not have a nonperturbative instanton superpotential, which pushed for small masses the system far away along the valley in the case of nonchiral SQCD.

All this makes the analysis of the vacuum dynamics of chiral theories essentially more difficult than of nonchiral ones. Still, we will try to perform it. As we will see, one can choose the value of the Yukawa coupling in such a way that this dynamics is mainly determined by the gauge sector and the matter degrees of freedom can (with a proper care) be integrated over.

Following [133], we proceed in the same way as we did for the ordinary SQED in Sect. 6.2. We put the system in a finite box of size L, impose periodic boundary conditions on all the fields and expand the latter in Fourier series:

$$A_j(\boldsymbol{x}) = \sum_{\boldsymbol{n}} A_j^{(\boldsymbol{n})} e^{2\pi i \boldsymbol{n}\cdot\boldsymbol{x}/L}, \qquad \lambda_\alpha(\boldsymbol{x}) = \sum_{\boldsymbol{n}} \lambda_\alpha^{(\boldsymbol{n})} e^{2\pi i \boldsymbol{n}\cdot\boldsymbol{x}/L},$$

$$s^f(\boldsymbol{x}) = \sum_{\boldsymbol{n}} s^{f(\boldsymbol{n})} e^{2\pi i \boldsymbol{n}\cdot\boldsymbol{x}/L}, \qquad \psi_\alpha^f(\boldsymbol{x}) = \sum_{\boldsymbol{n}} \psi_\alpha^{f(\boldsymbol{n})} e^{2\pi i \boldsymbol{n}\cdot\boldsymbol{x}/L},$$

$$t(\boldsymbol{x}) = \sum_{\boldsymbol{n}} t^{(\boldsymbol{n})} e^{2\pi i \boldsymbol{n}\cdot\boldsymbol{x}/L}, \qquad \xi_\alpha(\boldsymbol{x}) = \sum_{\boldsymbol{n}} \xi_\alpha^{(\boldsymbol{n})} e^{2\pi i \boldsymbol{n}\cdot\boldsymbol{x}/L}.\qquad(8.4)$$

The slow variables $A_j^{(0)} \equiv c_j$ and their superpartners $\lambda_\alpha^{(0)} \equiv \eta_\alpha$ play a special role. Thinking in the Hamiltonian terms, we stay in the universe where the wave functions are invariant under the large gauge transformations [cf. Eq. (6.12)],

$$A_j \to A_j + \frac{2\pi n_j}{eL},$$
$$(s^f, \psi^f) \to (s^f, \psi^f)e^{-2\pi i \boldsymbol{n}\cdot\boldsymbol{x}/L}, \quad (t, \xi) \to (t, \xi)e^{4\pi i \boldsymbol{n}\cdot\boldsymbol{x}/L}.\qquad(8.5)$$

The zero mode c_j of the gauge potential lies in a dual torus, $c_j \in [0, \frac{2\pi}{eL})$, and a characteristic energy of this degree of freedom is $\sim e^2/L$. Most other degrees of freedom have a characteristic energy $\sim 1/L \gg e^2/L$ and are fast to be integrated out. We will then find the effective supercharges by averaging the full supercharges by the ground eigenstate of the fast Hamiltonian:[2]

$$\hat{Q}_\alpha^{\text{eff}} = \langle \hat{Q}_\alpha^{\text{slow}} \rangle_{\text{fast vacuum}}, \qquad \hat{\bar{Q}}^{\alpha\,\text{eff}} = (\hat{Q}_\alpha^{\text{eff}})^\dagger = \langle \hat{\bar{Q}}^{\alpha\,\text{slow}} \rangle_{\text{fast vacuum}}. \tag{8.6}$$

The effective Hamiltonian will then be restored as the anticommutator $\hat{H}^{\text{eff}} = \{\hat{Q}_\alpha^{\text{eff}}, \hat{\bar{Q}}^{\alpha\,\text{eff}}\}/2$.

However, in contrast to ordinary SQED, the variables c_j and their superpartners η_α are not the *only* slow variables that determine the low-energy dynamics of chiral SQED. There are also the variables describing the motion along the valley

$$\sum_{f=1}^{8} s^{*f} s^f - 2t^* t = 0. \tag{8.7}$$

When the Yukawa coupling h vanishes, the spectrum associated with this motion is continuous.

If h is nonzero but small, the spectrum is discrete, but the characteristic excitation energies and level spacings are also small. The scalar valley dynamics is described in this case by a WZ model with the quartic potential (8.3). By virial theorem, the characteristic kinetic and potential energies in the Hamiltonian should be of the same order, which means

$$\frac{1}{L^3|s|^2} \sim \frac{1}{L^3|t|^2} \sim h^2 L^3|s|^2|t|^2 \sim h^2 L^6|s|^4,$$

giving $|s|^2 \sim |t|^2 \sim h^{-2/3}/L^2$ and

$$E_{\text{char}}^{\text{WZ}} \sim \frac{h^{2/3}}{L}. \tag{8.8}$$

If $h \ll e^3$, this is much less than the characteristic energies in the sector of the theory associated with the slow gauge dynamics. We will be interested in the case where the low-energy dynamics is mainly determined by the gauge sector. To this end, we must keep $h \gg e^3$. Still, as we will see soon, h cannot be too large. The calculation below is valid in the region

$$e^3 \ll h \ll e. \tag{8.9}$$

What we should do is to take the fast Hamiltonian in the leading quadratic approximation, find its spectrum, treating the slow variables

---

[2]In contrast to what we had for ordinary SQED, we cannot now, even at the leading order, just to cross out the terms including fast variables in the full Hamiltonian. Well, we can do so for the terms including higher harmonics of A_j and λ_α, but not for the matter fields.

c_j, η_α as parameters, and determine the effective supercharges by calculating the averages of the full supercharges over the vacuum state of $\hat{H}^{\text{fast}}$. Take one of the multiplets of unit charge, set $L = 1$, neglect the Yukawa terms, and to begin with, keep only the zero Fourier modes of all the fields. We obtain a SQM model [134] with the supercharges[3]

$$\hat{Q}_\alpha = \left[\hat{P}_k(\sigma_k)_\alpha{}^\gamma + ies^*s\,\delta_\alpha^\gamma\right]\eta_\gamma + \sqrt{2}\left[-i\hat{\pi}^\dagger\delta_\alpha^\gamma + esc_k(\sigma_k)_\alpha{}^\gamma\right]\bar{\psi}_\gamma,$$

$$\hat{\bar{Q}}^\beta = \bar{\eta}^\delta\left[\hat{P}_k(\sigma_k)_\delta{}^\beta - ies^*s\,\delta_\delta^\beta\right] - \sqrt{2}\left[i\hat{\pi}\delta_\delta^\beta + es^*c_k(\sigma_k)_\delta{}^\beta\right]\psi^\delta \quad (8.10)$$

with $\hat{\pi} = -i\partial/\partial s$, $\hat{\pi}^\dagger = -i\partial/\partial s^*$. The anticommutator of the supercharges (8.10) has the form

$$\{\hat{Q}_\alpha, \hat{\bar{Q}}^\beta\} = 2\delta_\alpha{}^\beta\hat{H} - 2ec_k(\sigma_k)_\alpha{}^\beta\hat{G}, \quad (8.11)$$

where

$$\hat{H} = -\frac{\partial^2}{\partial s\partial s^*} + e^2c^2s^*s + ec_k\hat{\bar{\psi}}\sigma_k\psi$$

$$-\frac{1}{2}\triangle_c + \frac{e^2}{2}(s^*s)^2 + ie\sqrt{2}[(\hat{\bar{\psi}}\hat{\bar{\eta}})s - (\psi\eta)s^*] \quad (8.12)$$

is the Hamiltonian and the presence of the second term in the RHS of Eq. (8.11) with

$$\hat{G} = s^*\frac{\partial}{\partial s^*} - s\frac{\partial}{\partial s} + \psi_\alpha\hat{\bar{\psi}}^\alpha - 1 \quad (8.13)$$

tells us that we are dealing with a *gauge* SQM system. $\hat{G}$ is the generator of the Abelian gauge transformations. It commutes with the Hamiltonian. The physical states in the spectrum are the solutions of the Schrödinger equation $\hat{H}\Psi = E\Psi$ that satisfy the subsidiary condition $\hat{G}\Psi = 0$.

The Hamiltonian (8.12) matches the Lagrangian (3.34) [a part of it not involving the mass terms and the fields $t(x)$ and $\xi(x)$] deprived of the higher Fourier modes contribution. If we add the contribution of seven other matter fields (s^f, ψ^f) and of (t, ξ), taking into account as well higher Fourier harmonics, and evaluate the anticommutator $\{\hat{Q}_\alpha, \hat{\bar{Q}}^\alpha\}$, the full Hamiltonian of SQED in a finite box is reproduced.

The second terms in (8.10) are the fast supercharges and the first line in (8.12) is the fast Hamiltonian describing a variant of supersymmetric oscillator. The ground state of $\hat{H}^{\text{fast}}$, which the fast supercharges annihilate, has the unit fermion charge:

$$\Psi_0(s, s^*; \psi^\alpha) = w_\alpha(\mathbf{c})\psi^\alpha\exp\{-e|\mathbf{c}|ss^*\} \quad (8.14)$$

---

[3]Here $\bar{X}^\alpha = (X_\alpha)^\dagger$ for all the spinors. It follows, bearing in mind (0.2), that $\bar{X}_\alpha = -(X^\alpha)^\dagger$. And $(\sigma_k)_\alpha{}^\gamma$ are the ordinary Pauli matrices.

with

$$w_\alpha = \begin{pmatrix} -\sin\frac{\theta}{2}e^{-i\phi} \\ \cos\frac{\theta}{2} \end{pmatrix}. \tag{8.15}$$

Here θ and ϕ are the polar and azimuthal angles of the vector $\mathbf{c}$. The spinor (8.15) satisfies

$$\boldsymbol{n\sigma}w = -w$$

and coincides with the wave function of a spin $\frac{1}{2}$ particle with the spin directed along the negative z axis [cf. Eq. (1.47)].

The state (8.14) is gauge-invariant: $\hat{G}\Psi = 0$.

Our next task is to find the effective supercharges by averaging the slow supercharges [the first terms in (8.10)] over the fast vacuum (8.14). It is easy to derive

$$\langle s^* s \rangle_{\text{fast vacuum}} = \frac{1}{2e|\mathbf{c}|}. \tag{8.16}$$

But one has to also take into account the factor $w_\alpha\psi^\alpha$ in $\Psi_0(s, s^*; \psi^\alpha)$ with a nontrivial dependence on $\mathbf{c}$. As a result, $\langle\hat{P}_k\rangle_{\text{fast vacuum}}$ does not vanish, giving a "vector potential" $\mathcal{A}_k(\mathbf{c})$ which should be added to the operator $\hat{P}_k$ in the effective supercharge. We derive

$$\hat{Q}_\alpha^{\text{eff}} = \left[(\sigma_k)_\alpha{}^\beta(\hat{P}_k + \mathcal{A}_k) + i\delta_\alpha^\beta\mathcal{D}\right]\eta_\beta,$$

$$\hat{\bar{Q}}^{\alpha\,\text{eff}} = \bar{\eta}^\beta\left[(\sigma_k)_\beta{}^\alpha(\hat{P}_k + \mathcal{A}_k) - i\delta_\beta^\alpha\mathcal{D}\right], \tag{8.17}$$

$$\boxed{\hat{H}^{\text{eff}} = \frac{1}{2}(\hat{P}_k + \mathcal{A}_k)^2 + \frac{1}{2}\mathcal{D}^2 + \boldsymbol{\mathcal{H}}\cdot\bar{\eta}\boldsymbol{\sigma}\eta,} \tag{8.18}$$

where $\hat{P}_k = -i\partial/\partial c_k$, $\mathcal{D} = 1/(2|\mathbf{c}|)$ and

$$\boldsymbol{\mathcal{H}} = \boldsymbol{\nabla}\times\boldsymbol{\mathcal{A}} = \boldsymbol{\nabla}\mathcal{D}. \tag{8.19}$$

The vector function $\boldsymbol{\mathcal{A}}(\mathbf{c})$ depends on field variables rather than on spatial coordinates and has nothing to do with the physical vector potential $\boldsymbol{A}(\boldsymbol{x})$!

As was also the case for the ordinary SQED [see Eq. (6.25)], the effective Hamiltonian (8.18) has the benign Wilsonian nature as soon as $|\mathbf{c}|$ is not *too* small and

$$\kappa = \frac{1}{e|\mathbf{c}|^3} \ll 1. \tag{8.20}$$

But chiral SQED includes also the Yukawa term (8.2), and we want to make sure that the corresponding corrections to the Hamiltonian $\sim h^2(ss^*)^2$ are irrelevant if κ is small. We have to compare the characteristic value of these corrections with the gap in the spectrum of the fast Hamiltonian, which is of order $e|\boldsymbol{c}|$. Bearing in mind (8.16), we derive

$$\frac{h^2}{e^2|\boldsymbol{c}|^2} \ll e|\boldsymbol{c}|,$$

or $h^2\kappa \ll e^2$. We are on the safe side if

$$h \ll e,$$

and that is how the upper limit for the allowed range of h in (8.9) was derived.

Let us now take into account all other Fourier harmonics of all the matter fields $s^f(x), \psi^f(x), t(x)$ and $\xi(x)$ and restore the dependence on L. We obtain

$$\mathcal{D} = 4\sum_n \frac{1}{\left|\boldsymbol{c} - \frac{2\pi n}{eL}\right|} - \frac{1}{2}\sum_n \frac{1}{\left|\boldsymbol{c} - \frac{\pi n}{eL}\right|}. \tag{8.21}$$

The Hamiltonian (8.18) with a generic function $\mathcal{D}(\boldsymbol{c})$ coincides with the Hamiltonian written in the paper [135]. If $\mathcal{D} = -1/(2|\boldsymbol{c}|)$, the magnetic field

$$\mathcal{H} = \frac{\boldsymbol{c}}{2|\boldsymbol{c}|^3} \tag{8.22}$$

is the field of a magnetic monopole with the minimal magnetic charge[4] $g = 1/2$. This monopole singularity in the configuration space is known in other branches of physics by the name *Pancharatnam-Berry phase* [136].

In our case, we are dealing with a *cubic lattice* of monopoles. In fact, there are two superimposed cubic lattices: a lattice of monopoles of charge $1/2$ with edge length π/eL and the lattice of monopoles of charge -4 located at the nodes of a twice less dense lattice with edge length $2\pi/eL$. The net magnetic charge density in field space is zero — this is a corollary of the condition

$$\boxed{\sum_f q_f^3 = 0,} \tag{8.23}$$

which our chiral model enjoys.

---

[4]Note, however, the presence of the extra scalar potential $1/(8\boldsymbol{c}^2)$, which makes the problem supersymmetric.

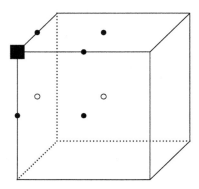

Fig. 8.2: A unit cubic cell of the monopole crystal in field space. The edge of the cube is $2\pi/(eL)$. The box marks the site with the monopole of charge $-7/2$. The circles mark the sites with the monopoles of charge $1/2$.

Now let us calculate the Witten index in this system. There are two ways to do so.

1. One can use the Cecotti-Girardello method and evaluate the integral (4.47) with the classical counterpart of the Hamiltonian (8.18). Integration over fermion variables reduces to the evaluation of the determinant

$$\det\left[\beta(\sigma_j)_\alpha{}^\beta\mathcal{H}_j\right] = -\beta^2\mathcal{H}^2 = -\beta^2(\boldsymbol{\nabla}\mathcal{D})^2. \tag{8.24}$$

When we also integrate with respect to the momenta, we obtain

$$I_W = \lim_{\beta\to0}\left[-\left(\frac{\beta}{8\pi^3}\right)^{1/2}\int d\boldsymbol{c}\,(\boldsymbol{\nabla}\mathcal{D})^2 e^{-\beta\mathcal{D}^2/2}\right], \tag{8.25}$$

where the integral is evaluated over the unit cell of our "crystal" depicted in Fig. 8.2. For small β, the integral is saturated by the regions adjacent to the singularities of $\mathcal{D}$ — the sites where the monopoles sit. The contribution of each such region is equal to $-|g|$, where g is the charge of the monopole. In our case, the unit cell contains a monopole of charge $-4 + 1/2 = -7/2$ and seven monopoles of charge $1/2$. The final result is

$$\boxed{I_W = -\frac{7}{2} - \frac{7}{2} = -7.} \tag{8.26}$$

This calculation strongly suggests the existence of seven fermion vacuum states in this model, but it does not *prove* it. The issue is that the CG method that reduces the functional integral for the index to an ordinary one has its applicability limits discussed on p. 79. First of all, it does not work for systems with continuous spectrum where the notion of index is ill-defined. In our case, the motion is finite and the spectrum is discrete, but we a facing another problem: the Hamiltonian (8.18) has singularities $\sim 1/|\boldsymbol{c} - \boldsymbol{c}_{\text{node}}|^2$ at the nodes of the lattice. The integral for the index is saturated by the regions of $\boldsymbol{c}$ at the vicinity of these singularities when $|\boldsymbol{c} - \boldsymbol{c}_{\text{node}}| \sim \sqrt{\beta}$. But if one looks at the original functional integral (4.43), one can observe that the characteristic values of the higher Fourier harmonics in the expansion

$$c_j(\tau) = \sum_n c_j^{(n)} e^{2\pi i n\tau/\beta} \tag{8.27}$$

are also of order $\sqrt{\beta}$, and one cannot be sure that the integral (8.25) not involving higher harmonics gives a correct value for the index.

Indeed, we will see in the next section that it *does* not [the value of the integral (4.47) is not integer] for non-Abelian chiral theories, For SQED it *does*, however. We know this because we can confirm the result (8.26) by a direct analysis of the Schrödinger equation.[5]

2. To do so, we deform the Hamiltonian (8.18) by adding a constant C to D. The condition (8.19) is still fulfilled, which assures sypersymmetry. The index does not change under such deformation.

Let C be positive and large, $C \gg 1/(eL)$. The wave functions of the low-lying states are now localized in the region of small $\mathcal{D}' = \mathcal{D}+C$, i.e. near the circles with positively charged monopoles. If C is large, the different circles do not "talk" to each other, and we can pick up one of them, say, the site in the center of the cube with $\boldsymbol{r} = \boldsymbol{c} - (\pi/eL)(1,1,1) = \boldsymbol{0}$. Then the problem reduces to the solution of the Schrödinger equation with the Hamiltonian

$$\hat{H} = \frac{1}{2}(\hat{\boldsymbol{P}} + \boldsymbol{A})^2 + \frac{1}{2}\left(C - \frac{1}{2r}\right)^2 + \frac{\boldsymbol{r}}{2r^3}\,\bar{\eta}\boldsymbol{\sigma}\eta\,. \tag{8.29}$$

Zero-energy states are possible only in the $F = 1$ sector, where the problem reduces to the motion of a fermion with gyromagnetic ratio $\gamma = 4$ in the

---

[5]The simplest example where the Hamiltonian involves singularities, but the phase space integral still gives the correct result for the index is Witten's model (4.53) with the superpotential

$$W(x) = -\ln x - \ln(a - x)\,. \tag{8.28}$$

In that case the motion is finite at the interval $x \in [0, a]$ and the potential has the singularities $\sim 1/x^2$ and $\sim 1/(x - a)^2$. (In the limit $a \to \infty$, we would arrive at the superconformal quantum mechanics of Ref. [62] with continuous spectrum.) There is only one bosonic zero-energy state with the wave function $\Psi(x, \psi) = x(a - x)$. Substituting the superpotential (8.28) in the integral like (4.54), we obtain $I_W = 1$ in accordance with the Schrödinger analysis.

field of a monopole and an additional spherically symmetric potential. The angular variables can then be separated, and the lowest harmonic has total angular momentum $j = |g| - 1/2 = 0$. The equation for the corresponding radial function $\chi(r)$ reads[6]

$$-\frac{1}{2r}\frac{\partial^2}{\partial r^2}(r\chi) + \frac{1}{2}\left(C^2 - \frac{C}{r} - \frac{1}{4r^2}\right)\chi = E\chi. \tag{8.30}$$

The zero-energy solution of (8.30) is readily found:

$$\chi(r) \sim \frac{1}{\sqrt{r}}e^{-Cr}. \tag{8.31}$$

In a particular gauge, where the "vector potentials" are chosen in the form

$$\mathcal{A}_1 = -\frac{y}{2r(r+z)} = -\frac{\sin\theta\sin\phi}{2r(1+\cos\theta)},$$

$$\mathcal{A}_2 = \frac{x}{2r(r+z)} = \frac{\sin\theta\cos\phi}{2r(1+\cos\theta)},$$

$$\mathcal{A}_3 = 0, \tag{8.32}$$

the full vacuum wave function has the form

$$\Psi(r, \theta, \phi; \eta_\alpha) = \frac{1}{\sqrt{r}}e^{-Cr}\left(\eta_1\cos\frac{\theta}{2} + \eta_2\sin\frac{\theta}{2}e^{-i\phi}\right). \tag{8.33}$$

The singularity of this function at $r = 0$ is benign — the integral $\int dr |\Psi|^2$ converges quite well. According to general theorems, Ψ is annihilated by the action of the supercharges (8.17). A dedicated reader may check it.

The unit cell in Fig. 8.2 contains seven sites with monopoles of charge $g = 1/2$, so that the spectrum contains seven fermionic zero-energy states.

If the constant C is large and negative, the wave function settles on the box site with $g = -7/2$. Then the ground state has the angular momentum $j = |g| - \frac{1}{2} = 3$, which produces a sevenfold angular degeneracy. And the equation for the radial wave function,

$$-\frac{1}{2r}\frac{\partial^2}{\partial r^2}(r\chi_7) + \frac{1}{2}\left(C^2 + \frac{7C}{r} + \frac{35}{4r^2}\right)\chi_7 = E\chi_7 \tag{8.34}$$

has only one zero-energy solution, as before:

$$\chi_7(r) = r^{5/2}e^{Cr}. \tag{8.35}$$

We calculated the index of the effective Hamiltonian (8.18). This Hamiltonian has a Wilsonian nature if the Yukawa constant lies in the

---

[6]We will derive it at the end of the section.

range (8.9) and if the conditions

$$\kappa_n = \frac{1}{e|Lc - \pi n/e|^3} \ll 1. \tag{8.36}$$

are all fulfilled so that we are not too close to the corners of the dual torus where the monopoles sit and where the BO corrections are large. Can we be sure that our result $I_W = -7$ is valid also for the original field theory (8.1) ?

Well, I can *swear* that it is, but I would not *bet* my head on that. The reason by which I still hesitate to do so is the limited range of the dynamic variables where the effective Hamiltonian (8.18) is Wilsonian. On the other hand, the original Witten's calculation of the index in SYM theories was also based on the effective Hamiltonian, which was only valid outside the corners of the dual torus. Hypothetically, these corners could bring about surprises for SYM theories, and we cannot exclude that they do so for chiral SQED. Of course, the Hamiltonian (6.53) is much simpler than (8.18) and it has no singularities. In addition, the result $I_W = h^\vee$ was independently confirmed by other methods discussed in Chapter 7.

For the chiral SQED, the Hamiltonian is more complicated, it has singularities in the corners, the necessity to introduce a Yukawa coupling is an extra complication, and we do not have an independent confirmation of the result (8.26).

Hopefully, we will be able to say more about this in the next edition of the book.

8.1.1   *Monopole harmonics*

For reader's benefit, we elucidate here some details concerning the Hamiltonian of a scalar and of a spinor charged particle interacting with a monopole. In particular, we will derive Eq. (8.30).

The Schrödinger equation for a scalar particle in a monopole field was solved by Igor Tamm[7] back in 1931 [137]. Let the electric charge of the particle be $e = 1$. The angular momentum operator that commutes with the Hamiltonian

$$\hat{H}_0 = \frac{1}{2}\left(\hat{P} + \mathcal{A}\right)^2, \tag{8.37}$$

---

[7] He was visiting Cambridge at that time and learned about Dirac's work directly from Dirac.

where $\boldsymbol{\mathcal{A}}$ is the vector potential of a monopole[8] with the minimal magnetic charge $g = 1/2$, has the form

$$\hat{\boldsymbol{L}}_0 = \boldsymbol{r} \times \left(\hat{\boldsymbol{P}} + \boldsymbol{\mathcal{A}} \right) + \frac{\boldsymbol{r}}{2r} . \tag{8.38}$$

The second term is the angular momentum of electromagnetic field, which is half-integer. The eigenvalues of $\hat{\boldsymbol{L}}_0^2$ are $l(l+1)$ with half-integer l. The eigenvalues of $\hat{L}_3^{(0)}$ are $m = -l, -l+1, \dots, l$. The corresponding eigenfunction are the so-called *monopole harmonics*.

We have

$$\hat{H}_0 = -\frac{1}{2r^2} \frac{\partial}{\partial r} \left(r^2 \frac{\partial}{\partial r} \right) + \frac{1}{2r^2} \left[\boldsymbol{r} \times (\hat{\boldsymbol{P}} + \boldsymbol{\mathcal{A}}) \right]^2$$

$$= -\frac{1}{2r^2} \frac{\partial}{\partial r} \left(r^2 \frac{\partial}{\partial r} \right) + \frac{\hat{\boldsymbol{L}}_0^2 - \frac{1}{4}}{2r^2} . \tag{8.39}$$

For the states with definite l, this gives

$$\hat{H}_0 = -\frac{1}{2r^2} \frac{\partial}{\partial r} \left(r^2 \frac{\partial}{\partial r} \right) + \frac{l(l+1) - \frac{1}{4}}{2r^2} . \tag{8.40}$$

The full Hamiltonian (8.18) commutes with

$$\hat{\boldsymbol{L}} = \hat{\boldsymbol{L}}_0 + \frac{1}{2} \hat{\bar{\eta}} \boldsymbol{\sigma} \eta. \tag{8.41}$$

The second term in (8.41) has a natural spin interpretation and the Hamiltonian (8.18), deprived of the potential term $\mathcal{D}^2/2$, may be interpreted as a Hamiltonian describing the interaction of a charged spin $1/2$ particle having the gyromagnetic ratio $\gamma = 4$ with a monopole. This problem was addressed in [138]. In the states with the net angular momentum $j = l - 1/2$, the term $-\boldsymbol{\mu} \cdot \boldsymbol{\mathcal{H}}$ in the Hamiltonian brings about an effective singular attractive potential $V_{\text{spin}}(r) = -\gamma/(8r^2)$. When $\gamma = 2$ (as is the case for the physical electron) and $j = 0$, this contribution exactly cancels the second term in the RHS of Eq. (8.40).

But in our case the gyromarnetic ratio is twice as large, and the interaction of the orbital and spin angular momentum with the monopole brings about in the s-wave the effective potential $-1/(4r^2)$. The Hamiltonian

$$\hat{H} = -\frac{1}{2} \triangle - \frac{\kappa}{r^2}$$

---

[8]Dirac and Tamm used an expression (8.32) for $\boldsymbol{\mathcal{A}}$ that involved a singularity at the *Dirac string* along the negative z axis, but later it was realized that its transverse projection $\boldsymbol{\mathcal{A}} - \boldsymbol{n}(\boldsymbol{\mathcal{A}} \cdot \boldsymbol{n})$ represents a *fiber bundle* on S^2, which can be described by nonsingular expressions on two overlapping charts [68].

with $\kappa > 1/8$ is known to exhibit a *falling on the center* phenomenon: the ground state is absent there and, which is worse, unitarity is lost [139].[9]

But the presence of the repulsive potential $\mathcal{D}^2/2$ changes the situation. We are interested only in the ground state Ψ_0 having zero total angular momentum. It satisfies the equations

$$\frac{\boldsymbol{r}}{r}\hat{\bar{\eta}}\boldsymbol{\sigma}\eta\,\Psi_0 = -\Psi_0,$$

$$\hat{\boldsymbol{L}}_0^2\,\Psi_0 = \frac{3}{4}\Psi_0 \tag{8.42}$$

and has the angular and Grassmann structure[10]

$$\Psi_0 \propto \eta_1 \cos\frac{\theta}{2} + \eta_2 \sin\frac{\theta}{2}e^{-i\phi} \tag{8.43}$$

The radial Schrödinger equation obtained after adding the term $\mathcal{D}^2/2$ in the Hamiltonian has the form of (8.30). This Hamiltonian still includes a singular attractive potential $-1/(8r^2)$, but $\kappa = 1/8$ is the border value — no falling on the center in this case!

8.2 A chiral $SU(3)$ model

The simplest such model includes, besides the vector multiplet, seven matter chiral multiplets in the fundamental group representation **3** and a multiplet in the representation $\bar{\mathbf{6}}$. In terms of superfields, the Lagrangian of the model is

$$\mathcal{L} = \left(\frac{1}{4}\mathrm{Tr}\int d^2\theta\,\hat{W}^\alpha\hat{W}_\alpha + \text{c.c.}\right)$$
$$+ \frac{1}{4}\int d^4\theta\left(\sum_{f=1}^{7}\overline{S}^{fj}e^{g\hat{V}}S_j^f + \Phi^{jk}e^{-g\hat{V}}\overline{\Phi}_{jk}\right) \tag{8.44}$$

with $j, k = 1, 2, 3$ and $\Phi^{jk} = \Phi^{kj}$.

---

[9]One can also consider the interaction of a W boson, which has spin 1, with the monopole [140]. The magnetic moment of W is larger than that of electron, and W boson falls on the monopole in s-wave. However, we hasten to say that this is true only for the Dirac monopole that has no size. For the 't Hooft-Polyakov monopole [141], which represents a soliton in certain non-Abelian gauge theory and has a small, but finite size, the W boson does not fall all the way down, but stops in the vicinity of the monopole core, forming with the monopole a bound state (called *dyon* [142]) that carries both magnetic and electric charges.

[10]The expressions multiplying $\eta_{1,2}$ in (8.43) have the meaning of the monopole harmonics $Y_{\frac{1}{2},\frac{1}{2},\pm\frac{1}{2}}$ in the notation of Ref. [68]. They are slightly different from the formulas written there due to the different sign convention in the Hamiltonian (8.18) that we adopted. Note that these expressions have the same form as in (8.15).

This theory is anomaly-free. Indeed, the contribution of the matter fermions in the non-Abelian anomalous triangle in Fig. 8.1 is equal in this case to

$$\langle A^a A^b A^c \rangle \propto 7 \operatorname{Tr}\{t^{(a}t^b t^{c)}\} - \operatorname{Tr}\{T^{(a}T^b T^{c)}\}, \tag{8.45}$$

where t^a and T^a are the generators in the triplet and sextet representations, correspondingly. The latter have the form

$$(T^a)^{kl}_{ij} = \frac{1}{2} \left[(t^a)^k_i \delta^l_j + (t^a)^l_i \delta^k_j + (t^a)^k_j \delta^l_i + (t^a)^l_j \delta^k_i \right], \tag{8.46}$$

where the factor $1/2$ stems from the requirement that T^a satisfy the same commutation relations, $[T^a, T^b] = i f^{abc} T^c$, as t^a.

For any $SU(N)$ group, one can derive

$$\operatorname{Tr}\{T^{(a}T^b T^{c)}\} = (N+4) \operatorname{Tr}\{t^{(a}t^b t^{c)}\}. \tag{8.47}$$

In our case, $N + 4 = 7$ and the net contribution to the anomaly vanishes.

The component Lagrangian of the model includes a quartic scalar potential, $V(s, \phi) = \frac{g^2}{2} \left(s^{*f} t^a s^f - \phi T^a \phi^* \right)^2$, which vanishes on the valley

$$s^{*f} t^a s^f - \phi T^a \phi^* = 0. \tag{8.48}$$

The motion along this valley brings about a continuous spectrum — in the same way as it did for chiral SQED. We can also lock this valley by including the Yukawa term in the Lagrangian:[11]

$$\mathcal{L}^{SU(3)}_{\text{Yukawa}} = \frac{h}{2} \int d^2\theta \, (S^f_j S^f_k \Phi^{jk}) + \text{c.c.}. \tag{8.49}$$

The corresponding scalar potential has the terms $\sim h^2 |s\phi|^2$ and $\sim h^2 |ss|^2$.

We repeat what we did for chiral SQED: put the theory in a small spatial box of length L, expand all the fields in the Fourier series, assume the effective coupling constant $g^2(L)$ to be small and assume that the Yukawa coupling lies in the range $g^3 \ll h \ll g$. Under these conditions, the low-energy spectrum of the theory is determined by the effective Hamiltonian in the gauge sector — it depends on zero Fourier modes of $A^a(x)$ belonging to the Cartan subalgebra of $SU(3)$ and their fermion superpartners, while

---

[11] When $h = 0$, the theory is asymptotically free: the effective coupling constant $g(\mu)$ runs to zero in the limit $\mu \to \infty$. But in the theory with a nonzero Yukawa coupling, the latter runs into a pole in this limit, so that this theory has the same status as QED and SQED: one can only treat it in the region of energies where both $g(\mu)$ and $h(\mu)$ are small.

the Yukawa terms and the whole dynamics associated with the valley (8.48) can be ignored. In the following, we will set $g = L = 1$.

The effective Hamiltonian can be found along the same lines as for the Abelian theory. We have 6 bosonic slow variables that can be presented as

$$\hat{c} = \frac{1}{2} \operatorname{diag}(a, b - a, -b) \tag{8.50}$$

with[12] $a_j, b_j \in [0, 4\pi)$.

Take a triplet s_j and concentrate on its zero Fourier mode. The fast Hamiltonian describing its interaction with the background gauge field $\hat{c}$ reads

$$\hat{H}_{\text{tripl}}^{\text{fast}} = -\frac{\partial^2}{\partial s_j^* \partial s_j} + s^*(\hat{c})^2 s$$

$$= -\frac{\partial^2}{\partial s_j^* \partial s_j} + \frac{1}{4} \left[a^2 s_1^* s_1 + (a - b)^2 s_2^* s_2 + b^2 s_3^* s_3 \right]. \tag{8.51}$$

The fast ground state wave function (its bosonic part) is

$$\Psi(s_j, s_j^*) = \exp\left\{ -\frac{|a|}{2} |s_1|^2 \right\} \exp\left\{ -\frac{|a - b|}{2} |s_2|^2 \right\} \exp\left\{ -\frac{|b|}{2} |s_3|^2 \right\}. \tag{8.52}$$

It follows:

$$\langle |s_1|^2 \rangle = \frac{1}{|a|}, \qquad \langle |s_2|^2 \rangle = \frac{1}{|a - b|}, \qquad \langle |s_3|^2 \rangle = \frac{1}{|b|}. \tag{8.53}$$

The slow supercharge [cf. Eq. (8.10)] is

$$\hat{Q}_\alpha = \left[(\sigma_k)_\alpha{}^\gamma \hat{P}_k^a + i\delta_\alpha^\gamma s_f^* t^a s_f \right] \eta_\gamma^a. \tag{8.54}$$

Averaging this over the fast bosonic vacuum, we see the following contribution to the effective supercharge:

$$\frac{i\eta_\alpha^3}{2} [\langle |s_1|^2 \rangle - \langle |s_2|^2 \rangle] + \frac{i\eta_\alpha^8}{2\sqrt{3}} [\langle |s_1|^2 \rangle + \langle |s_2|^2 \rangle - 2\langle |s_3|^2 \rangle] = \tag{8.55}$$

$$\frac{i\eta_\alpha^{(a)}}{2} \left[\frac{1}{|a|} - \frac{1}{|a - b|} \right] + \frac{i\eta_\alpha^{(b)}}{2} \left[\frac{1}{|a - b|} - \frac{1}{|b|} \right] \equiv i \left[\eta^{(a)} \mathcal{D}_{(a)} + i\eta^{(b)} \mathcal{D}_{(b)} \right],$$

where[13]

$$\eta_\alpha^{(a)} = \eta_\alpha^3 + \frac{1}{\sqrt{3}} \eta_\alpha^8, \qquad \eta_\alpha^{(b)} = \frac{2}{\sqrt{3}} \eta_\alpha^8. \tag{8.57}$$

---

[12]It is the rhombus in Fig. 6.1. For the reasons to become clear at the end of this section, we do not care now about the Weyl alcove.

[13]That means that

$$\hat{\eta}_\alpha = \eta_\alpha^a t^a = \frac{1}{2} \operatorname{diag}(\eta_\alpha^{(a)}, \eta_\alpha^{(b)} - \eta_\alpha^{(a)}, -\eta_\alpha^{(b)}). \tag{8.56}$$

Here $\mathcal{D}_{(a)}$ and $\mathcal{D}_{(b)}$ are the contributions due to the zero Fourier mode of the triplet to the induced D-terms. If we add the contributions of the other matter Fourier modes with the averages $\langle |s_1^{(n)}|^2 \rangle = 1/|a - 4\pi n|$ etc. and multiply the result by 7 (we have 7 triplets), we derive the expressions [cf. the first term in (8.21)]:

$$\mathcal{D}_{(a)} = \frac{7}{2} \sum_n \left[\frac{1}{|a_{2n}|} - \frac{1}{|(a - b)_{2n}|} \right],$$

$$\mathcal{D})_{(b)} = \frac{7}{2} \sum_n \left[\frac{1}{|(a - b)_{2n}|} - \frac{1}{|b_{2n}|} \right], \tag{8.58}$$

where $a_m = a - 2\pi m$, $b_m = b - 2\pi m$.

Consider now the contribution of the antisextet. The part of the fast Hamiltonian including the zero modes of $\phi_{ij}(\boldsymbol{x})$ is

$$\hat{H}_{\text{sext}}^{\text{fast}} = \hat{\bar{\pi}}^{ij} \hat{\pi}_{ij} + \phi^{ij} \boldsymbol{\mathcal{X}}^2 \phi_{ij}^*, \tag{8.59}$$

where

$$\boldsymbol{\mathcal{X}} = A^3 T^3 + A^8 T^8 = aX + (b - a)Y - bZ$$

with

$$X_{ij}^{kl} = \frac{1}{4}[\delta_{i1}\delta^{k1}\delta_j^l + \delta_{i1}\delta^{l1}\delta_j^k + \delta_{j1}\delta^{k1}\delta_i^l + \delta_{j1}\delta^{l1}\delta_i^k],$$

$$Y_{ij}^{kl} = \frac{1}{4}[\delta_{i2}\delta^{k2}\delta_j^l + \delta_{i2}\delta^{l2}\delta_j^k + \delta_{j2}\delta^{k2}\delta_i^l + \delta_{j2}\delta^{l2}\delta_i^k],$$

$$Z_{ij}^{kl} = \frac{1}{4}[\delta_{i3}\delta^{k3}\delta_j^l + \delta_{i3}\delta^{l3}\delta_j^k + \delta_{j3}\delta^{k3}\delta_i^l + \delta_{j3}\delta^{l3}\delta_i^k]. \tag{8.60}$$

This gives

$$\begin{aligned}
\hat{H}_{\text{sext}}^{\text{fast}} = {} & |\pi_{11}|^2 + |\pi_{22}|^2 + |\pi_{33}|^2 + 2(|\pi_{12}|^2 + |\pi_{13}|^2 + |\pi_{23}|^2) \\
& + a^2|\phi_{11}|^2 + (a - b)^2|\phi_{22}|^2 + b^2|\phi_{33}|^2 \\
& + \frac{b^2}{2}|\phi_{12}|^2 + \frac{(a - b)^2}{2}|\phi_{13}|^2 + \frac{a^2}{2}|\phi_{23}|^2.
\end{aligned} \tag{8.61}$$

One finds the following scalar averages:

$$\langle |\phi^{11}|^2 \rangle = \frac{1}{2|a|}, \qquad \langle |\phi^{22}|^2 \rangle = \frac{1}{2|a - b|}, \qquad \langle |\phi^{33}|^2 \rangle = \frac{1}{2|b|},$$

$$\langle |\phi^{23}|^2 \rangle = \frac{1}{|a|}, \qquad \langle |\phi^{12}|^2 \rangle = \frac{1}{|b|}, \qquad \langle |\phi^{13}|^2 \rangle = \frac{1}{|a - b|}. \tag{8.62}$$

The slow supercharge acquires the contribution

$$\begin{aligned}
\Delta^{\text{antisextet}} \hat{Q}_\alpha = {} & -i\left(\eta_\alpha^3 \phi T^3 \phi^* + \eta_\alpha^8 \phi T^8 \phi^* \right) = i\eta_\alpha^{(a)}(|\phi_{23}|^2 - |\phi_{11}|^2 \\
& + |\phi_{22}|^2 - |\phi_{13}|^2) + i\eta_\alpha^{(b)}(|\phi_{13}|^2 - |\phi_{22}|^2 + |\phi_{33}|^2 - |\phi_{12}|^2).
\end{aligned} \tag{8.63}$$

Averaging it over the fast vacuum, we derive the contribution

$$
\frac{i\eta_\alpha^{(a)}}{2}\left[\frac{1}{|a|}-\frac{1}{|a-b|}\right] + \frac{i\eta_\alpha^{(b)}}{2}\left[\frac{1}{|a-b|}-\frac{1}{|b|}\right] \tag{8.64}
$$

to the effective supercharge, which happens to coincide with the contribution of a triplet. Consider now the contribution of the higher modes $\phi^{ij}_{(n)}$ in the fast Hamiltonian. An analysis similar to what we did for the triplets brings about the averages

$$
\langle|\phi^{11}_{(n)}|^2\rangle = \frac{1}{2|a_n|}, \qquad \langle|\phi^{22}_{(n)}|^2\rangle = \frac{1}{2|(a-b)_n|}, \qquad \langle|\phi^{33}_{(n)}|^2\rangle = \frac{1}{2|b_n|},
$$
$$
\langle|\phi^{23}_{(n)}|^2\rangle = \frac{1}{|a_{2n}|}, \qquad \langle|\phi^{12}_{(n)}|^2\rangle = \frac{1}{|b_{2n}|}, \qquad \langle|\phi^{13}_{(n)}|^2\rangle = \frac{1}{|(a-b)_{2n}|}. \tag{8.65}
$$

Note the different periodicity patterns for the diagonal components $\phi^{11}, \phi^{22}, \phi^{33}$ vs. the mixed components $\phi^{12}, \phi^{13}, \phi^{23}$. In contrast to the mixed components, the diagonal ones carry double charges with respect to T^3 and/or $T^* = \frac{1}{2}(\sqrt{3}\,T^8 - T^3)$ and play the same role as the field t played in chiral SQED. For a gauge transformation when, say, a_x is shifted by 4π, the sextet matter fields $\phi^{11}(x), \phi^{22}(x), \phi^{33}(x)$ have to "make two turns" being multiplied by $e^{4\pi ix}$ to compensate this shift, in the same way as the fields of double charge did in the Abelian model. And the fields $\phi^{12}, \phi^{13}, \phi^{23}$ make only "one turn" being multiplied by $e^{2\pi ix}$.

Plugging (8.65) into the fast vacuum averages of the slow supercharge and adding the triplet contribution (8.58), we derive:

$$
\mathcal{D}_{(a)} = \frac{9}{2}\sum_n\left[\frac{1}{|a_{2n}|}-\frac{1}{|(a-b)_{2n}|}\right] - \frac{1}{2}\sum_n\left[\frac{1}{|a_n|}-\frac{1}{|(a-b)_n|}\right],
$$
$$
\mathcal{D}_{(b)} = \frac{9}{2}\sum_n\left[\frac{1}{|(a-b)_{2n}|}-\frac{1}{|b_{2n}|}\right] - \frac{1}{2}\sum_n\left[\frac{1}{|(a-b)_n|}-\frac{1}{|b_n|}\right]. \tag{8.66}
$$

The second terms are the contributions of the sextet components $\phi^{11}, \phi^{22}, \phi^{33}$. The corresponding lattice is twice as dense as the lattice for the triplets and for the components $\phi^{12}, \phi^{13}, \phi^{23}$. The reason is the same as for chiral SQED.

As in the Abelian model, the effective supercharges acquire besides the induced $\mathcal{D}$ terms the "vector potential" contributions coming from the averages $\langle\partial/\partial a\rangle$ and $\langle\partial/\partial b\rangle$ over the fermion factor in the fast ground state wave function. As a result, we derive the following expressions for the effective supercharges:

$$
\begin{aligned}
\hat{Q}_\alpha^{\text{eff}} &= \eta_\beta^{(a)} \left[(\sigma_k)_\alpha{}^\beta (\hat{P}_k^{(a)} + \mathcal{A}_k^{(a)}) + i\delta_\alpha^\beta \mathcal{D}_{(a)} \right] \\
&+ \eta_\beta^{(b)} \left[(\sigma_k)_\alpha{}^\beta (\hat{P}_k^{(b)} + \mathcal{A}_k^{(b)}) + i\delta_\alpha^\beta \mathcal{D}_{(b)} \right] , \\
\hat{\bar{Q}}^{\alpha\,\text{eff}} &= \hat{\bar{\eta}}^{(a)\beta} \left[(\sigma_k)_\beta{}^\alpha (\hat{P}_k^{(a)} + \mathcal{A}_k^{(a)}) - i\delta_\beta^\alpha \mathcal{D}_{(a)} \right] \\
&+ \hat{\bar{\eta}}^{(b)\beta} \left[(\sigma_k)_\beta{}^\alpha (\hat{P}_k^{(b)} + \mathcal{A}_k^{(b)}) - i\delta_\beta^\alpha \mathcal{D}_{(b)} \right] ,
\end{aligned} \tag{8.67}
$$

where $\hat{P}_k^{(a)} = -i\partial/\partial a^k$, $\hat{P}_k^{(b)} = -i\partial/\partial b^k$ and $\boldsymbol{\mathcal{A}}^{(a,b)}$ are related to $\mathcal{D}_{(a,b)}$ so that

$$
\begin{aligned}
\mathcal{H}_j^{(a)} &\equiv \varepsilon_{jkl}\, \partial_k^a \mathcal{A}_l^{(a)} = \partial_j^a \mathcal{D}_{(a)} , \\
\mathcal{H}_j^{(b)} &\equiv \varepsilon_{jkl}\, \partial_k^b \mathcal{A}_l^{(b)} = \partial_j^b \mathcal{D}_{(b)} , \\
\mathcal{H}_j^{(ab)} &\equiv \varepsilon_{jkl}\, \partial_k^a \mathcal{A}_l^{(b)} = \varepsilon_{jkl}\, \partial_k^b \mathcal{A}_l^{(a)} = \partial_j^a \mathcal{D}_{(b)} = \partial_j^b \mathcal{D}_{(a)} .
\end{aligned} \tag{8.68}
$$

Bearing in mind that

$$
\{\hat{\bar{\eta}}^{(a)\beta}, \eta_\alpha^{(a)}\} = \{\hat{\bar{\eta}}^{(b)\beta}, \eta_\alpha^{(b)}\} = \frac{4}{3}\delta_\alpha^\beta, \quad \{\hat{\bar{\eta}}^{(a)\beta}, \eta_\alpha^{(b)}\} = \{\hat{\bar{\eta}}^{(b)\beta}, \eta_\alpha^{(a)}\} = \frac{2}{3}\delta_\alpha^\beta, \tag{8.69}
$$

we derive the effective Hamiltonian:[14]

$$
\begin{aligned}
\hat{H}^{\text{eff}} =& \frac{2}{3} \left[(\hat{P}_k^{(a)} + \mathcal{A}_k^{(a)})^2 + (\hat{P}_k^{(b)} + \mathcal{A}_k^{(b)})^2 + (\hat{P}_k^{(a)} + \mathcal{A}_k^{(a)})(\hat{P}_k^{(b)} + \mathcal{A}_k^{(b)}) \right. \\
&+ \left. \mathcal{D}_{(a)}^2 + \mathcal{D}_{(b)}^2 + \mathcal{D}_{(a)}\mathcal{D}_{(b)} \right] \\
&+ \boldsymbol{\mathcal{H}}^{(a)} \cdot \hat{\bar{\eta}}^{(a)} \boldsymbol{\sigma} \eta^{(a)} + \boldsymbol{\mathcal{H}}^{(b)} \cdot \hat{\bar{\eta}}^{(b)} \boldsymbol{\sigma} \eta^{(b)} + \boldsymbol{\mathcal{H}}^{(ab)} \\
&\cdot (\hat{\bar{\eta}}^{(a)} \boldsymbol{\sigma} \eta^{(b)} + \hat{\bar{\eta}}^b \boldsymbol{\sigma} \eta^{(a)}) .
\end{aligned}
$$

$$
\tag{8.70}
$$

---

[14]The effective supercharges (8.67) and the Hamiltonian (8.70) were derived in Ref. [143]. Unfortunately, the expressions for the induced $\mathcal{D}$ terms quoted in that paper involved a mistake. As a result, the value of the phase space integral for the Witten index was also not correct. But we will confirm in the calculations below the main conclusion of that work: the phase space integral (4.47) evaluated for the Hamiltonian (8.70) has a fractional value, so that the CG method to evaluate the index does not work in this case.

8.2.1 Phase space integral

The Hamiltonian (8.70) is more complicated than (8.18), and in this case it is difficult to find the vacuum states explicitly, as we did for SQED. Hence we shall use the CG method and evaluate the integral (4.47) for the classical version of the Hamiltonian (8.70). This calculation is more complicated than in the Abelian case, but is still feasible.

Note first of all that the integral (4.47) was written in the assumption that the bosonic and fermion variables form a canonical basis with the Poisson brackets (5.14). In our case, it is so for bosonic variables, but not for the fermion variables whose Poisson brackets are nontrivial, as follows from (8.69). The Jacobian of the transformation from the canonical variables $\eta_\alpha^3, \eta_\alpha^8, \bar{\eta}^{\alpha 3}, \bar{\eta}^{\alpha 8}$ to $\eta_\alpha^{(a)}, \eta_\alpha^{(b)}, \bar{\eta}^{\alpha(a)}, \bar{\eta}^{\alpha(b)}$ gives the factor $(2/\sqrt{3})^4 = 16/9$. Integrating over the fermions, we derive

$$
I_W = \lim_{\beta \to 0} \frac{16}{9} \beta^4 \int \frac{da\, dP^{(a)}\, db\, dP^{(b)}}{(2\pi)^6} \begin{vmatrix} \mathcal{H}^{(a)}\sigma, & \mathcal{H}^{(ab)}\sigma \\ \mathcal{H}^{(ab)}\sigma, & \mathcal{H}^{(b)}\sigma \end{vmatrix}
$$
$$
\cdot \exp\left\{ -\frac{2}{3}\beta \left[(\boldsymbol{P}^{(a)} + \boldsymbol{\mathcal{A}}_k^{(a)})^2 + (\boldsymbol{P}^{(b)} + \boldsymbol{\mathcal{A}}_k^{(b)})^2 \right.\right.
$$
$$
\left.\left. + (\boldsymbol{P}^{(a)} + \boldsymbol{\mathcal{A}}_k^{(a)})(\boldsymbol{P}^{(b)} + \boldsymbol{\mathcal{A}}_k^{(b)}) + \mathcal{D}_{(a)}^2 + \mathcal{D}_{(b)}^2 + \mathcal{D}_{(a)}\mathcal{D}_{(b)} \right] \right\}. \quad (8.71)
$$

Calculating the determinant and integrating over momenta, we obtain

$$
\boxed{\begin{aligned}
I_W = \lim_{\beta \to 0} \frac{\beta}{4\pi^3 \sqrt{3}} \int dadb & \left[\|\mathcal{H}^{(ab)}\|^4 + 2\|\mathcal{H}^{(ab)}\|^2 (\mathcal{H}^{(a)} \cdot \mathcal{H}^{(b)}) \right.\\
& \left. + \|\mathcal{H}^{(a)}\|^2 \|\mathcal{H}^{(b)}\|^2 - 4(\mathcal{H}^{(a)} \cdot \mathcal{H}^{(ab)})(\mathcal{H}^{(b)} \cdot \mathcal{H}^{(ab)}) \right]\\
& \cdot \exp\left\{ -\frac{2}{3}\beta \left[\mathcal{D}_{(a)}^2 + \mathcal{D}_{(b)}^2 + \mathcal{D}_{(a)}\mathcal{D}_{(b)} \right] \right\}.
\end{aligned}}
\quad (8.72)
$$

The integral runs over the fundamental cell $a_j, b_j \in [0, 4\pi)$ of our "crystal" (recall that the gauge invariant slow wave functions are periodic in the space of weights with the period 4π for each component of $\boldsymbol{a}$ and $\boldsymbol{b}$). If β is small, the integral is saturated by the regions near the singularities of $\mathcal{D}_{(a)}$ and $\mathcal{D}_{(b)}$. There are 64 such regions in the fundamental cell:

(i) its corner $\boldsymbol{a} \sim \boldsymbol{b} \sim 0$;

(ii) 42 symmetric regions of the type $\boldsymbol{a} \sim \boldsymbol{a}_0 = 2\pi(1, 0, 0), \boldsymbol{b} \sim \boldsymbol{b}_0 = 2\pi(0, 1, 0)$ (so that all three vectors $\boldsymbol{a}_0, \boldsymbol{b}_0, \boldsymbol{a}_0 - \boldsymbol{b}_0$ are nonzero);

(iii) 21 asymmetric regions of the type $\boldsymbol{a} \sim 2\pi(1, 0, 0), \boldsymbol{b} \sim \boldsymbol{0}$ where one of these vectors is equal to zero.

Consider first the region around the corner. The angular integration gives

$$dadb = 8\pi^2 \rho_a d\rho_a \rho_b d\rho_b \rho_{ab} d\rho_{ab}$$ (8.73)

with $\rho_a = |a|, \rho_b = |b|, \rho_{ab} = |a - b|$. The induced $\mathcal{D}$ terms are

$$\mathcal{D}_{(a)} = 4\left(\frac{1}{\rho_a} - \frac{1}{\rho_{ab}}\right), \qquad \mathcal{D}_{(b)} = 4\left(\frac{1}{\rho_{ab}} - \frac{1}{\rho_b}\right).$$ (8.74)

Substituting this into (8.72) and bearing in mind (8.68), we obtain the following expression for the corner contribution:[15]

$$I_W(\text{corner}) = \frac{32}{\pi\sqrt{3}} \iiint_0^\infty \frac{d\rho_a d\rho_b d\rho_{ab}}{\rho_a^3 \rho_b^3 \rho_{ab}^3} \left[\rho_a^4 + \rho_b^4 + \rho_{ab}^4 + \right.$$
$$\rho_a \rho_b(\rho_{ab}^2 - \rho_a^2 - \rho_b^2) + \rho_a \rho_{ab}(\rho_b^2 - \rho_a^2 - \rho_{ab}^2) + \rho_b \rho_{ab}(\rho_a^2 - \rho_b^2 - \rho_{ab}^2)]$$
$$\cdot \exp\left[-\frac{2}{3}\left(\frac{1}{\rho_a^2} + \frac{1}{\rho_b^2} + \frac{1}{\rho_{ab}^2} - \frac{1}{\rho_a\rho_b} - \frac{1}{\rho_a\rho_{ab}} - \frac{1}{\rho_b\rho_{ab}}\right)\right]$$ (8.75)

with the constraint $|\rho_a - \rho_b| \le \rho_{ab} \le \rho_a + \rho_b$.

It is convenient to make a variable change

$$\rho_a = \mu\rho, \qquad \rho_b = \nu\rho, \qquad \rho_{ab} = (1 - \mu - \nu)\rho,$$ (8.76)

after which the integral over ρ can be easily done, and we derive

$$I_W(\text{corner}) = \frac{8\sqrt{3}}{\pi} \int_0^{1/2} d\mu \int_{1/2-\mu}^{1/2} d\nu \, \frac{A(\mu, \nu)}{\mu\nu\kappa B(\mu, \nu)},$$ (8.77)

where $\kappa = 1 - \mu - \nu$ and

$$A(\mu, \nu) = \mu^4 + \nu^4 + \kappa^4 + \mu\nu\kappa - \mu\nu(\mu^2 + \nu^2) - \mu\kappa(\mu^2 + \kappa^2) - \nu\kappa(\nu^2 + \kappa^2),$$
$$B(\mu, \nu) = \mu^2\nu^2 + \mu^2\kappa^2 + \nu^2\kappa^2 - \mu\nu\kappa.$$ (8.78)

Mathematica calculated for us the integral (8.78) numerically, and this calculation produced an integer answer:

$$I_W^{\text{corn}} = 16.$$ (8.79)

So far so good. Consider now the contributions of 42 symmetric regions. The corresponding $\mathcal{D}$ terms have the same form as in Eq. (8.74) with $\rho_a = |a - a_0|$ etc., but with the factor $-1/2$ rather than 4. Also the "magnetic fields" $\mathcal{H}^{(a)}, \mathcal{H}^{(b)}$ and $\mathcal{H}^{(ab)}$ entering the preexponential in

---

[15]The integral does not depend on β and we have set $\beta = 1/16$.

(8.72) are 8 times smaller than for the corner. As a result, the contribution of a symmetric region in the index is equal to $16/64 = 1/4$. And the contribution of all such regions is

$$I_W^{\text{sym}} = \frac{42}{4} = 10\frac{1}{2}. \tag{8.80}$$

It is *fractional*.

Consider now the 21 asymmetric regions. For example, a region around the singularity at $\boldsymbol{a}_0 = 2\pi(1,0,0), \boldsymbol{b}_0 = \boldsymbol{0}$. The $\mathcal{D}$ terms are

$$\mathcal{D}_{(a)} = \frac{1}{2}\left(\frac{1}{\rho_{ab}} - \frac{1}{\rho_a}\right), \qquad \mathcal{D}_{(b)} = -\frac{1}{2\rho_{ab}} - \frac{4}{\rho_b} \tag{8.81}$$

with $\rho_a = |\boldsymbol{a} - \boldsymbol{a}_0|, \rho_b = |\boldsymbol{b}|,$ and $\rho_{ab} = |\boldsymbol{a} - \boldsymbol{b} - \boldsymbol{a}_0|$. The contribution of this region to the index is[16]

$$I_W^{\text{asym}} = \frac{1}{2\pi\sqrt{3}} \iiint \frac{d\rho_a d\rho_b d\rho_{ab}}{\rho_a^3 \rho_b^3 \rho_{ab}^3} \left[64(\rho_a^4 + \rho_{ab}^4) + \rho_b^4 \right.$$
$$8\rho_a\rho_b(\rho_a^2 + \rho_b^2 - \rho_{ab}^2) + 64\rho_a\rho_{ab}(\rho_b^2 - \rho_a^2 - \rho_{ab}^2) + 8\rho_b\rho_{ab}(\rho_b^2 + \rho_{ab}^2 - \rho_a^2)]$$
$$\cdot \exp\left[-\frac{2}{3}\left(\frac{1}{\rho_a^2} + \frac{1}{\rho_{ab}^2} + \frac{64}{\rho_b^2} - \frac{1}{\rho_a\rho_{ab}} + \frac{8}{\rho_a\rho_b} + \frac{8}{\rho_b\rho_{ab}}\right)\right]. \tag{8.82}$$

A numerical calculation of this integral[17] gives the value 3. The contribution of all 21 asymmetric regions is

$$I_W^{\text{asym}} = 63. \tag{8.84}$$

---

[16]We have now set $\beta = 4$.

[17]The fact that a numerical calculation of these complicated integrals gives nicely looking simple fractions or even integers looks as a small mathematical miracle, of which I have no explanation. Moreover, consider a set of integrals

$$I_q = \frac{1}{2\pi\sqrt{3}} \iiint \frac{d\rho_a d\rho_b d\rho_{ab}}{\rho_a^3 \rho_b^3 \rho_{ab}^3} \left[q^2(\rho_a^4 + \rho_{ab}^4) + \rho_b^4 \right.$$
$$q\rho_a\rho_b(\rho_{ab}^2 - \rho_a^2 - \rho_b^2) + q^2\rho_a\rho_{ab}(\rho_b^2 - \rho_a^2 - \rho_{ab}^2) + q\rho_b\rho_{ab}(\rho_a^2 - \rho_b^2 - \rho_{ab}^2)]$$
$$\cdot \exp\left[-\frac{2}{3}\left(\frac{1}{\rho_a^2} + \frac{1}{\rho_{ab}^2} + \frac{q^2}{\rho_b^2} - \frac{1}{\rho_a\rho_{ab}} - \frac{q}{\rho_a\rho_b} - \frac{q}{\rho_b\rho_{ab}}\right)\right] \tag{8.83}$$

with integer q. The "experimental" values of these integrals are

$$I_{q\leq 0} = \frac{1-q}{3}, \qquad I(1) = \frac{1}{4}, \qquad I_{q>1} = \frac{2(q-1)}{3}.$$

I wonder if one can derive these amusing results analytically.

Adding all the contributions, we obtain

$$I_W = 89\frac{1}{2}. \tag{8.85}$$

Actually, the notation I_W is not correct here, because the Witten index cannot be fractional in a system with discrete spectrum (the range of bosonic dynamic variables $\boldsymbol{a}$ and $\boldsymbol{b}$ in the Hamiltonian (8.70) is finite, and the spectrum should be discrete). We have not calculated the index, but instead just quoted the value of the integral (8.71). The fractional value of the integral is not a paradox by the reasons clarified on p. 181: the effective Hamiltonian (8.70) has singularities such that the CG method by which the index was evaluated above may not work. Unfortunately, we cannot say why it works well for chiral SQED, for Witten's model with the superpotential (8.28), but does not work for chiral SQCD.[18]

8.3   The minimal $SU(5)$ model

The simplest chiral anomaly-free $SU(5)$ supersymmetric model includes a quintet S_j and an antidecuplet $\Phi^{[jk]}$. The generators in the antisymmetric representation of an $SU(N)$ group read

$$(T^a)^{kl}_{ij} = \frac{1}{2}\left[(t^a)^k_i\delta^l_j - (t^a)^l_i\delta^k_j - (t^a)^k_j\delta^l_i + (t^a)^l_j\delta^k_i\right]. \tag{8.86}$$

One can derive that

$$\mathrm{Tr}\{T^{(a}T^bT^{c)}\} = (N-4)\,\mathrm{Tr}\{t^{(a}t^bt^{c)}\}, \tag{8.87}$$

and, for $N = 5$, the contributions of the quintet and antidecuplet in the anomalous triangle indeed cancel.

The Lagrangian reads

$$\mathcal{L} = \left(\frac{1}{4}\mathrm{Tr}\int d^2\theta\,\hat{W}^\alpha\hat{W}_\alpha + \mathrm{c.c.}\right) + \frac{1}{4}\int d^4\theta\left(\overline{S}e^{g\hat{V}}S + \Phi e^{-g\hat{V}}\overline{\Phi}\right).$$

$$\tag{8.88}$$

An important distinction of this model compared to the models considered in the previous sections is the *absence* of the scalar valley. In other

---

[18] Were the result of the calculation of the phase space integral integer, we would have to recall that not all the eigenfunctions of the Hamiltonian (8.70) contribute in the index of the original theory, but only the functions invariant under Weyl permutations: $\boldsymbol{a} \leftrightarrow \boldsymbol{b} - \boldsymbol{a}$, etc. But the fractional result makes the discussion of this subtle question pointless.

words, the equation system

$$\phi^{ij}(T^a)_{ij}{}^{kl}\phi^*_{kl} - s^{*j}(t^a)_j{}^k s_k = 0 \tag{8.89}$$

with T^a written in Eq. (8.86) has no solutions.

One can be convinced in that quite explicitly.[19] Introduce the matrices $A_k{}^j = s_k s^{*j}$ and $B_k{}^j = \phi^*_{kl}\phi^{lj}$. Then (8.89) boils down to $\mathrm{Tr}\{(A+2B)t^a\} = 0$, i.e. $A + 2B = C \cdot \mathbb{1}$. By a gauge rotation, ϕ^{ij} can be brought to the following canonical form:

$$\phi = \begin{pmatrix} 0 & x & 0 & 0 & 0 \\ -x & 0 & 0 & 0 & 0 \\ 0 & 0 & 0 & y & 0 \\ 0 & 0 & -y & 0 & 0 \\ 0 & 0 & 0 & 0 & 0 \end{pmatrix}. \tag{8.90}$$

This gives

$$B = \phi^*\phi = -\mathrm{diag}\left(|x|^2, |x|^2, |y|^2, |y|^2, 0\right).$$

Hence

$$s_j s^{*k} = \mathrm{diag}\left(C + 2|x|^2, C + 2|x|^2, C + 2|y|^2, C + 2|y|^2, C\right). \tag{8.91}$$

It easy to see that this relation, treated as an equation for s_j, has no solution.

Thus, we do not need to introduce here a Yukawa coupling to lock a nonexistent scalar valley (besides, we are not *able* to do so: in this case, one cannot write a trilinear gauge-invariant term in the superfield Lagrangian). The theory only has the gauge coupling and is asymptotically free. To study its vacuum structure, it is sufficient to study the gauge branch of classical vacua.

Take a constant Abelian background having the form

$$\frac{1}{2}\,\mathrm{diag}(a, b - a, c - b, d - c, -d). \tag{8.92}$$

The fast Hamiltonian describing its interaction with the zero Fourier modes of the scalar quintet components is

$$\begin{aligned} H_{\mathrm{quint}}^{\mathrm{fast}} &= -\frac{\partial^2}{\partial s^{*j}\partial s_j} + \frac{1}{4}\,\big[a^2|s_1|^2 \\ &+ (b-a)^2|s_2|^2 + (c-b)^2|s_3|^2 + (d-c)^2|s_4|^2 + d^2|s_5|^2\big]. \end{aligned} \tag{8.93}$$

[19]See Appendix A of the review [110].

The scalar averages are

$$\langle |s_1|^2 \rangle = \frac{1}{|\boldsymbol{a}|}, \qquad \langle |s_2|^2 \rangle = \frac{1}{|\boldsymbol{b} - \boldsymbol{a}|}, \qquad \langle |s_3|^2 \rangle = \frac{1}{|\boldsymbol{c} - \boldsymbol{b}|},$$

$$\langle |s_4|^2 \rangle = \frac{1}{|\boldsymbol{d} - \boldsymbol{c}|}, \qquad \langle |s_5|^2 \rangle = \frac{1}{|\boldsymbol{d}|}. \tag{8.94}$$

The fast Hamiltonian describing the interaction of the gauge background with the zero Fourier modes of the scalar decuplet components is

$$
\begin{aligned}
H_{\text{dec}}^{\text{fast}} &= 2\left[|\pi_{12}|^2 + \ldots + |\pi_{45}|^2 \right] + \frac{1}{2}\left[\boldsymbol{b}^2 |\phi^{12}|^2 + (\boldsymbol{a} + \boldsymbol{c} - \boldsymbol{b})^2 |\phi^{13}|^2 \right. \\
&\quad + (\boldsymbol{a} + \boldsymbol{d} - \boldsymbol{c})^2 |\phi^{14}|^2 + (\boldsymbol{a} - \boldsymbol{d})^2 |\phi^{15}|^2 + (\boldsymbol{a} - \boldsymbol{c})^2 |\phi^{23}|^2 \\
&\quad + (\boldsymbol{b} - \boldsymbol{a} + \boldsymbol{d} - \boldsymbol{c})^2 |\phi^{24}|^2 + (\boldsymbol{b} - \boldsymbol{a} - \boldsymbol{d})^2 |\phi^{25}|^2 + (\boldsymbol{d} - \boldsymbol{b})^2 |\phi^{34}|^2 \\
&\quad \left. + (\boldsymbol{c} - \boldsymbol{b} - \boldsymbol{d})^2 |\phi^{35}|^2 + \boldsymbol{c}^2 |\phi^{45}|^2 \right].
\end{aligned}
\tag{8.95}
$$

The scalar averages are

$$\langle |\phi^{12}|^2 \rangle = \frac{1}{|\boldsymbol{b}|}, \quad \langle |\phi^{23}|^2 \rangle = \frac{1}{|\boldsymbol{a} - \boldsymbol{c}|}, \quad \langle |\phi^{15}|^2 \rangle = \frac{1}{|\boldsymbol{a} - \boldsymbol{d}|}, \quad \langle |\phi^{45}|^2 \rangle = \frac{1}{|\boldsymbol{c}|},$$

$$\langle |\phi^{13}|^2 \rangle = \frac{1}{|\boldsymbol{a} + \boldsymbol{c} - \boldsymbol{b}|}, \quad \langle |\phi^{14}|^2 \rangle = \frac{1}{|\boldsymbol{a} + \boldsymbol{d} - \boldsymbol{c}|}, \quad \langle |\phi^{24}|^2 \rangle = \frac{1}{|\boldsymbol{b} - \boldsymbol{a} + \boldsymbol{d} - \boldsymbol{c}|},$$

$$\langle |\phi^{25}|^2 \rangle = \frac{1}{|\boldsymbol{b} - \boldsymbol{a} - \boldsymbol{d}|}, \quad \langle |\phi^{34}|^2 \rangle = \frac{1}{|\boldsymbol{b} - \boldsymbol{d}|}, \quad \langle |\phi^{35}|^2 \rangle = \frac{1}{|\boldsymbol{c} - \boldsymbol{b} - \boldsymbol{d}|}.$$

$$\tag{8.96}$$

Taking into account also higher Fourier harmonics of the matter fields and proceeding in the same way as we did for the $SU(3)$ model, we can find the effective supercharges and Hamiltonian [143].

$$
\begin{aligned}
\hat{Q}_\alpha^{\text{eff}} &= \eta_\beta^{(a)} \left[(\sigma_k)_\alpha{}^\beta (\hat{P}_k^{(a)} + \mathcal{A}_k^{(a)}) + i\delta_\alpha^\beta \mathcal{D}_{(a)} \right] \\
&\quad + \eta_\beta^{(b)} \left[(\sigma_k)_\alpha{}^\beta (\hat{P}_k^{(b)} + \mathcal{A}_k^{(b)}) + i\delta_\alpha^\beta \mathcal{D}_{(b)} \right], \\
&\quad + \eta_\beta^{(c)} \left[(\sigma_k)_\alpha{}^\beta (\hat{P}_k^{(c)} + \mathcal{A}_k^{(c)}) + i\delta_\alpha^\beta \mathcal{D}_{(c)} \right] \\
&\quad + \eta_\beta^{(d)} \left[(\sigma_k)_\alpha{}^\beta (\hat{P}_k^{(d)} + \mathcal{A}_k^{(d)}) + i\delta_\alpha^\beta \mathcal{D}_{(d)} \right]
\end{aligned}
\tag{8.97}
$$

and similarly for $\hat{\bar{Q}}^{\alpha\,\text{eff}}$ expressed via $\bar{\eta}^{(a)\alpha}$ etc. Here

$$
\begin{aligned}
\eta_\alpha^{(a)} &= \eta_\alpha^3 + \frac{\eta_\alpha^8}{\sqrt{3}} + \frac{\eta_\alpha^{15}}{\sqrt{6}} + \frac{\eta_\alpha^{24}}{\sqrt{10}}, \\
\eta_\alpha^{(b)} &= 2\left(\frac{\eta_\alpha^8}{\sqrt{3}} + \frac{\eta_\alpha^{15}}{\sqrt{6}} + \frac{\eta_\alpha^{24}}{\sqrt{10}}\right) \\
\eta_\alpha^{(c)} &= 3\left(\frac{\eta_\alpha^{15}}{\sqrt{6}} + \frac{\eta_\alpha^{24}}{\sqrt{10}}\right), \\
\eta_\alpha^{(d)} &= \frac{4\eta_\alpha^{24}}{\sqrt{10}}.
\end{aligned}
\tag{8.98}
$$

The fermion variables have the following anticommutators:

$$
\{\bar{\eta}^{(a)\beta}, \eta_\alpha^{(a)}\} = \{\bar{\eta}^{(d)\beta}, \eta_\alpha^{(d)}\} = \frac{8}{5}\delta_\alpha^\beta, \quad \{\bar{\eta}^{(b)\beta}, \eta_\alpha^{(b)}\} = \{\bar{\eta}^{(c)\beta}, \eta_\alpha^{(c)}\} = \frac{12}{5}\delta_\alpha^\beta,
$$

$$
\{\bar{\eta}^{(a)\beta}, \eta_\alpha^{(b)}\} = \{\bar{\eta}^{(b)\beta}, \eta_\alpha^{(a)}\} = \{\bar{\eta}^{(c)\beta}, \eta_\alpha^{(d)}\} = \{\bar{\eta}^{(d)\beta}, \eta_\alpha^{(c)}\} = \frac{6}{5}\delta_\alpha^\beta,
$$

$$
\{\bar{\eta}^{(a)\beta}, \eta_\alpha^{(c)}\} = \{\bar{\eta}^{(c)\beta}, \eta_\alpha^{(a)}\} = \{\bar{\eta}^{(b)\beta}, \eta_\alpha^{(d)}\} = \{\bar{\eta}^{(d)\beta}, \eta_\alpha^{(b)}\} = \frac{4}{5}\delta_\alpha^\beta,
$$

$$
\{\bar{\eta}^{(a)\beta}, \eta_\alpha^{(d)}\} = \{\bar{\eta}^{(d)\beta}, \eta_\alpha^{(a)}\} = \frac{2}{5}\delta_\alpha^\beta \quad \{\bar{\eta}^{(b)\beta}, \eta_\alpha^{(c)}\} = \{\bar{\eta}^{(c)\beta}, \eta_\alpha^{(b)}\} = \frac{8}{5}\delta_\alpha^\beta.
$$

$$
\tag{8.99}
$$

The supercharges $\hat{Q}^{\text{eff}}$ and $\hat{\bar{Q}}^{\alpha\,\text{eff}}$ look similar to (8.67), but they now include extra dynamic variables (c, d and their superpartners) and represent sums of four terms rather than two. The induced $\mathcal{D}$ terms are

$$
\begin{aligned}
\mathcal{D}_{(a)} &= \frac{1}{2}\left(\langle|s_1|^2\rangle - \langle|s_2|^2\rangle\right. \\
&\quad - \langle|\phi_{13}|^2\rangle - \langle|\phi_{14}|^2\rangle - \langle|\phi_{15}|^2\rangle + \langle|\phi_{23}|^2\rangle + \langle|\phi_{24}|^2\rangle + \langle|\phi_{25}|^2\rangle\left.\right), \\
\mathcal{D}_{(b)} &= \frac{1}{2}\left(\langle|s_2|^2\rangle - \langle|s_3|^2\rangle\right. \\
&\quad - \langle|\phi_{21}|^2\rangle - \langle|\phi_{24}|^2\rangle - \langle|\phi_{25}|^2\rangle + \langle|\phi_{31}|^2\rangle + \langle|\phi_{34}|^2\rangle + \langle|\phi_{35}|^2\rangle\left.\right), \\
\mathcal{D}_{(c)} &= \frac{1}{2}\left(\langle|s_3|^2\rangle - \langle|s_4|^2\rangle\right. \\
&\quad - \langle|\phi_{31}|^2\rangle - \langle|\phi_{32}|^2\rangle - \langle|\phi_{35}|^2\rangle + \langle|\phi_{41}|^2\rangle + \langle|\phi_{42}|^2\rangle + \langle|\phi_{45}|^2\rangle\left.\right), \\
\mathcal{D}_{(d)} &= \frac{1}{2}\left(\langle|s_4|^2\rangle - \langle|s_5|^2\rangle\right. \\
&\quad - \left(\langle|\phi_{41}|^2\rangle - \langle|\phi_{42}|^2\rangle - \langle|\phi_{43}|^2\rangle + \langle|\phi_{51}|^2\rangle + \langle|\phi_{52}|^2\rangle + \langle|\phi_{53}|^2\rangle\right)\left.\right),
\end{aligned}
\tag{8.100}
$$

which gives (when we take into account the contributions of all the Fourier harmonics)

$$
\begin{aligned}
\mathcal{D}_{(a)} = \frac{1}{2}\sum_n \Bigg[& \frac{1}{|a_{2n}|} + \frac{1}{|(a-c)_{2n}|} + \frac{1}{|(b-a+d-c)_{2n}|} + \frac{1}{|(b-a-d)_{2n}|} \\
& - \frac{1}{|(a-b)_{2n}|} - \frac{1}{|(a+c-b)_{2n}|} - \frac{1}{|(a+d-c)_{2n}|} - \frac{1}{|(a-d)_{2n}|} \Bigg], \\
\mathcal{D}_{(b)} = \frac{1}{2}\sum_{2n} \Bigg[& \frac{1}{|(a-b)_{2n}|} + \frac{1}{|(b-d)_{2n}|} + \frac{1}{|(a+c-b)_{2n}|} + \frac{1}{|(b-c+d)_{2n}|} \\
& - \frac{1}{|b_{2n}|} - \frac{1}{|(b-c)_{2n}|} - \frac{1}{|(b-a-d)_{2n}|} - \frac{1}{|(b-a+d-c)_{2n}|} \Bigg], \\
\mathcal{D}_{(c)} = \frac{1}{2}\sum_n \Bigg[& \frac{1}{|c_{2n}|} + \frac{1}{|(b-c)_{2n}|} + \frac{1}{|(a-c+d)_{2n}|} + \frac{1}{|(b-a+d-c)_{2n}|} \\
& - \frac{1}{|(d-c)_{2n}|} - \frac{1}{|(a-c)_{2n}|} - \frac{1}{|(a+c-b)_{2n}|} - \frac{1}{|(c-b-d)_{2n}|} \Bigg], \\
\mathcal{D}_{(d)} = \frac{1}{2}\sum_{2n} \Bigg[& \frac{1}{|(d-c)_{2n}|} + \frac{1}{|(d-a)_{2n}|} + \frac{1}{|(c-b-d)_{2n}|} + \frac{1}{|(b-a-d)_{2n}|} \\
& - \frac{1}{|d_{2n}|} - \frac{1}{|(b-d)_{2n}|} - \frac{1}{|(a+d-c)_{2n}|} - \frac{1}{|(b-a+d-c)_{2n}|} \Bigg].
\end{aligned}
$$

$$\tag{8.101}$$

In this case, the elementary cell of our 12-dimensional crystal has only one node. The induced vector potentials are related to $\mathcal{D}_{(a,b,c,d)}$ by the relations (8.68) to which one should add similar relations involving $\mathcal{D}_{(c,d)}$ and $\boldsymbol{A}^{(c,d)}$:

$$
\begin{aligned}
\mathcal{H}_j^{(c)} &\equiv \varepsilon_{jkl}\,\partial_k^c \mathcal{A}_l^{(c)} = \partial_j^c \mathcal{D}_{(c)}, \\
\mathcal{H}_j^{(ad)} &\equiv \varepsilon_{jkl}\,\partial_k^a \mathcal{A}_l^{(d)} = \varepsilon_{jkl}\,\partial_k^d \mathcal{A}_l^{(a)} = \partial_j^a \mathcal{D}_{(d)} = \partial_j^d \mathcal{D}_{(a)},
\end{aligned}
$$

$$\tag{8.102}$$

etc. Calculating the anticommutator $\{\hat{\bar{Q}}^{\alpha\,\mathrm{eff}}, \hat{Q}_\alpha^{\mathrm{eff}}\}$, one can derive an expression for the effective Hamiltonian. It represents a rather obvious generalization of (8.70). For example, bearing in mind (8.99), one can write a contribution

$$
\boxed{
\begin{aligned}
\hat{H}_{\mathcal{D}}^{\mathrm{eff}} = \frac{1}{5}\Big[& 4\left(\mathcal{D}_{(a)}^2 + \mathcal{D}_{(d)}^2\right) + 6\left(\mathcal{D}_{(b)}^2 + \mathcal{D}_{(c)}^2\right) + 6\left(\mathcal{D}_{(a)}\mathcal{D}_{(b)} + \mathcal{D}_{(c)}\mathcal{D}_{(d)}\right) \\
& + 4\left(\mathcal{D}_{(a)}\mathcal{D}_{(c)} + \mathcal{D}_{(b)}\mathcal{D}_{(d)}\right) + 2\mathcal{D}_{(a)}\mathcal{D}_{(d)} + 8\mathcal{D}_{(b)}\mathcal{D}_{(c)} \Big].
\end{aligned}
}
$$

$$\tag{8.103}$$

Another contribution represents an analogous quadratic form of $\hat{P}_k^{(a,b,c,d)}$ + $\mathcal{A}_k^{(a,b,c,d)}$. And there are 10 bifermion terms including $\mathcal{H}^{(a)}, \mathcal{H}^{(bc)}$, etc.

One can also write, and probably, by applying certain efforts and using a clever numerical method,[20] calculate numerically the phase space integral (4.47) for this model in an attempt to evaluate the index of the effective Hamiltonian. Bearing in mind the absence of the scalar vacuum valleys and the presence of only one relevant node in the elementary cell (these are certain conceptual simplifications), it might coincide with the index of the original field theory (8.88). But it also cannot be excluded that this complicated calculation would give a meaningless fractional result, as a similar calculation did for $SU(3)$...

8.4 Spontaneous supersymmetry breaking

The main physical question to which Witten index analysis helps to answer is whether supersymmetry is broken spontaneously in a given theory. In the gauge theories discussed so far, supersymmetry was not broken. But there are models involving chiral matter multiplets where supersymmetry *is* broken. We are sure of that because the breaking occurs in the *weak coupling regime* where everything is under control!

The simplest such model (the so-called *3-2 model*) is based on the gauge group $SU(3) \times SU(2)$ [123] (see also the reviews [110, 111, 144]). It includes the following chiral matter multiplets: a multiplet $Q_{j=1,2,3;f=1,2}$ in the representation $(\mathbf{3}, \mathbf{2})$, two multiplets U^j and D^j in the representation $(\bar{\mathbf{3}}, \mathbf{1})$, and a multiplet L_f in the representation $(\mathbf{1}, \mathbf{2})$.

This matter content is similar to the fermion matter content of one generation of the Standard Model.[21] The fields Q (their fermion components) may be associated with left quarks, the fields U and D with right antiquarks $\bar{u}_R$ and $\bar{d}_R$, and the fields L with left electron and neutrino. There is no $U(1)$ group and no "right electron". The model is anomaly-free in the same way as the Standard Model is.

---

[20] The problem boils down to calculating a rather complicated 10-dimensional integral, which is a nontrivial task even for Mathematica.

[21] Of course, we do not mean that this model has some phenomenological significance; it probably does not. But for a physicist familiar with the SM (and I am afraid that we lost our reader mathematician, if any, somewhere in the middle of Chapter 6), this remark may have some mnemonic value.

The component Lagrangian of the model includes the quartic scalar interaction terms [cf. Eqs. (3.34), (3.44)]:

$$V \sim (\mathcal{D}^{a=1,\ldots,8})^2 + (\tilde{\mathcal{D}}^{A=1,2,3})^2 \qquad (8.104)$$

with

$$\begin{aligned}
\mathcal{D}^a &= (q_{jf})^*(t^a)_j{}^k q_{kf} - u^j(t^a)_j{}^k u_k^* - d^j(t^a)_j{}^k d_k^*, \\
\tilde{\mathcal{D}}^A &= (q_{jf})^*(t^A)_f{}^g q_{jg} + (l_f)^*(t^A)_f{}^g l_g,
\end{aligned} \qquad (8.105)$$

where t^a are the generators of $SU(3)$ and t^A are the generators of $SU(2)$.

The important observation is that this model admits a vacuum valley — there are classical field configurations for which the $\mathcal{D}$ terms (8.105) and the potential (8.104) vanish. Up to gauge rotations, a general solution to the equations $\mathcal{D}^a = \tilde{\mathcal{D}}^A = 0$ is

$$q_{jf} = \begin{pmatrix} \tau_1 & 0 \\ 0 & \tau_2 \\ 0 & 0 \end{pmatrix}, \qquad l_f = \left(0, \sqrt{|\tau_1|^2 - |\tau_2|^2}\right),$$

$$u^j = (\tau_1, 0, 0), \qquad d^j = (0, \tau_2, 0), \qquad (8.106)$$

where $\tau_{1,2}$ are complex parameters ($|\tau_1| \geq |\tau_2|$).

We have already dealt with the vacuum valleys — for nonchiral models, the valleys (lifted by the instanton superpotential and the mass terms) were discussed in Chapter 7 and for chiral models they were discussed in the first half of this chapter. For the simplest nonchiral $SU(2)$ model, the valley equations were written in Eq. (7.15). But the solution to these equations,

$$s_j = \begin{pmatrix} \tau \\ 0 \end{pmatrix}, \qquad t^j = (\tau, 0) \qquad (8.107)$$

was not stable due to the presence of the nonperturbative instanton-generated superpotential (7.11).

The same happens in the 3-2 model. The instantons generate the following superpotential:

$$\mathcal{W}_{3\text{-}2}^{\mathrm{inst}} = \frac{\Lambda^7}{Q_{jf}Q_{kg}U^j D^k \varepsilon^{fg}}. \qquad (8.108)$$

As was the case in the nonchiral SQCD, the valley solution (8.106) is not stable under the scale transformations $\tau \to \lambda\tau$, and we are facing the phenomenon of run-away vacuum. In nonchiral SQCD, the vacuum was stabilized by giving a mass to the matter fields. In the chiral 3-2 model, we cannot do so, but we can instead include a Yukawa term

$$\mathcal{W}_{\mathrm{Yukawa}} = h\,\varepsilon^{fg}Q_{jf}U^j L_g, \qquad (8.109)$$

which plays the same role — it locks the valley and fixes the values of $\tau_{1,2}$ at

$$\tau_{1,2} = c_{1,2}\frac{\Lambda}{h^{1/7}}, \tag{8.110}$$

where $c_{1,2}$ are some particular numerical constant, irrelevant for us.

And here comes an essential difference between the chiral and nonchiral models. In the nonchiral massive SQCD, supersymmetry is not broken and the energy of the vacuum states is zero. But in the 3-2 model the vacuum energy density $\mathcal{E}$ is not zero. To estimate it, it suffices for us to substitute the characteristic value (8.110) of the scalar fields in the quartic WZ potential:

$$\boxed{\mathcal{E} \sim \left(\frac{\partial \mathcal{W}_{\text{Yukawa}}}{\partial \phi}\right)^2 \sim h^2\phi_{\text{char}}^4 = c_3 h^{10/7}\Lambda^4.} \tag{8.111}$$

As the breaking occurs in the weak coupling regime, the numerical factor c_3 can also be evaluated [123].

The fact that supersymmetry is broken in this case can be explained as follows. Both in nonchiral SQCD and in the 3-2 model, instanton effects bring about a nonzero gluino condensate $\langle\lambda\lambda\rangle$. In nonchiral SQCD, the gluino condensate was related to the scalar v.e.v., $\langle\lambda\lambda\rangle \propto m\tau^2$, in such a way that the vacuum average of the RHS of the Konishi identity (7.17) vanished. This implied that the average of the anticommutator in the LHS of (7.17) vanished too, which meant that supersymmetry was not broken.

But there is no mass in the 3-2 model, and nonvanishing $\langle\lambda\lambda\rangle$ means that the anticommutator in the left hand side of the corresponding Konishi identity (which is valid for all gauge supersymmetric theories including matter multiplets) has a nonzero v.e.v. Hence $\hat{Q}_\alpha|\text{vac}\rangle \neq 0$ and supersymmetry is spontaneously broken. The coupling constants in this theory are small and everything (including the vacuum energy) can be perturbatively calculated.

Spontaneous supersymmetry breaking may also occur in other models if:

• The model admits a vacuum valley.
• Instanton-generated superpotential pushes the system along the valley.
• The model is chiral and the mass term cannot be written. To lock the valley, one has to introduce the Yukawa terms.

In Sects. 8.1, 8.2, 8.3, we considered three different chiral models, but neither of them satisfied all the conditions above.

- The chiral SQED involves a valley, the equation (8.7) has a plenty of solutions, but there are no instantons and no instanton-generated superpotential.
- The minimal chiral $SU(3)$ model with 7 triplets and an antisextet has a valley, has instantons, but the latter fail to generate the required superpotential [123].
- The minimal chiral $SU(5)$ model has no valleys.

The next in complexity model, where supersymmetry is spontaneously broken in the weak coupling regime, is the $SU(5)$ chiral model with *two* quintets $S^{f=1,2}$ and two antidecuplets Φ^f. In this case, the expression for $\mathcal{D}^a$ includes twice as much variables as in the minimal $SU(5)$ model and solutions to the equations $\mathcal{D}^a = 0$ can be found though their explicit form is not simple [145]. The instanton superpotential *is* generated in this valley, it reads

$$\mathcal{W}^{\text{inst}}_{SU(5)} = \frac{\Lambda^{11}}{B^{fg}B_{fg}}, \tag{8.112}$$

where $B_{fg} = \varepsilon_{ff'}\varepsilon_{gg'}B^{f'g'}$ and $B^{fg} = \varepsilon_{ijklp}(\Phi^f)^{ij}(\Phi^h)^{kl}(\Phi_h)^{pq}S^g_q$. To lock the valley, one introduces the Yukawa term $\propto (\Phi^1)^{ij}S^f_iS^g_j\varepsilon_{fg}$.

Many more models with spontaneous SUSY breaking have been studied in the literature. We address the reader to the original paper [123], to the reviews [110,111] and to many papers written afterwards for further details.

Chapter 9

Maximally supersymmetric gauge quantum mechanics

As we have seen in Chapter 6, the spectrum of pure SYM theory placed in a finite box of size L is discrete and there are $I_W = h^\vee$ ground zero-energy states separated from the excited states by a gap $\sim g^2/L$. But what happens in the limit $L = 0$ when a finite box is replaced by a single point and we are dealing with supersymmetric Yang-Mills quantum mechanics obtained from the field theory by dimensional reduction?

This SQM system is characterized by the Hamiltonian (3.59) and constraints (3.60). The dynamical variables A_j^a can acquire any value; there is a classical vacuum valley along the directions where $f^{abc}A_i^a A_j^b = 0$, which is not lifted in a supersymmetric system by quantum effects [47]; the motion is infinite and the spectrum is continuous [84]. A naive definition of Witten index does not work in this case.

The same physical picture holds in the extended $\mathcal{N} = 8$ gauge SQM system obtained by dimensional reduction from 6-dimensional SYM theory (3.72) (or from $\mathcal{N} = 2$ four-dimensional SYM theory). But the maximally supersymmetric $\mathcal{N} = 16$ gauge SQM system obtained by dimensional reduction from the maximally supersymmetric $\mathcal{N} = 4$ four-dimensional SYM theory (or from 10-dimensional SYM theory) has a different rather interesting behavior.[1] The continuous spectrum starting from zero is still there, but on top of that the system possesses some number of *normalizable* states (in mathematical language, the spectrum has a pure point component) whose energy exactly equals to zero.

---

[1]$\mathcal{N} = 16$ is the maximal value for a *gauge* SQM. In recent [146], a nontrivial nongauge SQM system with an arbitrary number of supercharges was constructed. Another nontrivial *weak* SQM system with an arbitrary number of supercharges will be discussed in Chapter 11. Also the SQM system obtained by dimensional reduction of the $4D$ $\mathcal{N} = 8$ supergravity has $\mathcal{N} = 32$ supersymmetries, which is more than $\mathcal{N} = 16$. But this is a system with *gauged* rather than rigid supersymmetry, where the Hamiltonian is one of the constraints. Such systems are beyond the scope of our book.

There are also other simpler QM systems having this property.

- Consider a motion in 5-dimensional space with the following spherically symmetric potential:

$$V(r) = -\frac{15a^2}{2(r^2 + a^2)^2}. \tag{9.1}$$

We look for spherically symmetric solutions to the Schrödinger equation,

$$\left[-\frac{1}{2}\triangle + V(r) \right] \Psi(r) =$$

$$-\frac{1}{2}\left[\frac{1}{r^2}\frac{\partial^2}{\partial r^2}[r^2\Psi(r)] - \frac{2}{r^2}\Psi(r) \right] + V(r)\Psi(r) = 0. \tag{9.2}$$

The potential falls out at infinity, and the spectrum includes a continuum component. But on top of that, it has a *normalized* zero energy solution:

$$\Psi_0(r) = \frac{C}{(r^2 + a^2)^{3/2}}. \tag{9.3}$$

The integral $\int |\Psi_0(r)|^2 r^4 dr$ converges at infinity.

At large r, the wave function (9.3) falls down $\propto r^{-3}$. Indeed, $\Psi(r) = 1/r^3$ is a formal solution to the free Schrödinger equation $\triangle\Psi = 0$. The existence of such an asymptotic solution with normalizable behavior at infinity is a *necessary* condition for a normalizable zero mode to exist.[2] But of course, it is not a sufficient condition. For example, a normalized (not only at infinity, but also at zero) zero-energy state does not exist for repulsive potentials.

- The second example was discussed in Chapter 5 (see p. 109) and concerned the Dirac spectrum on the background of BPST instanton [72]. The instanton is a non-Abelian Euclidean field configuration living in $\mathbb{R}^4$. The field falls out at infinity and the spectrum is continuous. But on top of that, one has an exact zero mode. As was explained in Chapter 5 (see Theorem 5.5), the operators $\not{D}(1 \pm \gamma^5)$ obey the supersymmetry algebra, and the index of the Dirac operator can be interpreted as the Witten index of a certain SQM system. In the instanton background, this index is equal to one (and thus can be well defined in spite of the presence of continuous spectrum). Being supersymmetric, this instanton system is rather similar in behaviour to the maximal SQM to be discussed in the following.

---

[2]Note that this condition is not fulfilled for free 3-dimensional motion!

The Hamiltonian of the maximal SQM system was written in Eq. (3.64). For the reader's convenience, we reproduce it here:

$$\hat{H} = \frac{1}{2}(\hat{E}_I^a)^2 + \frac{g^2}{4}(f^{abc}A_I^a A_J^b)^2 + \frac{ig}{2}f^{abc}\hat{\lambda}_\alpha^a(\Gamma_I)_{\alpha\beta}\hat{\lambda}_\beta^b A_I^c \qquad (9.4)$$

with $I, J = 1, \ldots, 9$ and $\alpha = 1, \ldots, 16$; $\hat{E}_I^a = -i\partial/\partial A_I^a$. The matrices Γ_I satisfy the Clifford algebra, $\{\Gamma_I, \Gamma_J\}_{\alpha\beta} = 2\delta_{IJ}\delta_{\alpha\beta}$. The same concerns the operators $\hat{\lambda}_\alpha^a : \{\hat{\lambda}_\alpha^a, \hat{\lambda}_\beta^b\} = \delta^{ab}\delta_{\alpha\beta}$. The latter can be represented as

$$\hat{\lambda}_1^a = \frac{1}{\sqrt{2}}\left(\mu_1^a + \frac{\partial}{\partial\mu_1^a}\right), \qquad \hat{\lambda}_9^a = \frac{i}{\sqrt{2}}\left(\frac{\partial}{\partial\mu_1^a} - \mu_1^a\right),$$

$$\vdots \qquad\qquad\qquad \vdots$$

$$\hat{\lambda}_8^a = \frac{1}{\sqrt{2}}\left(\mu_8^a + \frac{\partial}{\partial\mu_8^a}\right), \qquad \hat{\lambda}_{16}^a = \frac{i}{\sqrt{2}}\left(\frac{\partial}{\partial\mu_8^a} - \mu_8^a\right), \quad (9.5)$$

where $\mu_{\tilde{\alpha}=1,\ldots,8}^a$ are holomorphic fermion variables. The wave functions depend on them as well as on the bosonic variables A_I^a. The variables A_I^a and $\mu_{\tilde{\alpha}}$ are dimensionless in this description, and the same concerns the coupling constant g^2 and the energies — eigenvalues of the Hamiltonian (9.4). The characteristic values of A following from the virial theorem are $A_{\text{char}} \sim g^{-1/3}$. The characteristic energies of excited states are[3] $\mathcal{E}_{\text{char}} \sim g^{2/3}$. Not all possible wave functions $\Psi(A, \mu)$ are admissible, but only those which obey the gauge constraint

$$\hat{G}^a\Psi \equiv f^{abc}\left(A_I^b\hat{E}_I^c - \frac{i}{2}\hat{\lambda}_\alpha^b\hat{\lambda}_\alpha^c\right)\Psi = 0. \qquad (9.6)$$

The Hamiltonian (9.4) is invariant under $SO(9)$ rotations. A specific difference between this model and the $\mathcal{N} = 4$ and $\mathcal{N} = 8$ gauge SQM models, is the fact that the holomorphic variables $\mu_{\tilde{\alpha}}$ do not form a representation of $SO(9)$. A related fact is that the Hamiltonian does not commute with the fermion charge operator,

$$\hat{F} = \mu_{\tilde{\alpha}}^a\frac{\partial}{\partial\mu_{\tilde{\alpha}}^a}, \qquad (9.7)$$

though it commutes with the operator $(-1)^{\hat{F}}$, as is required by supersymmetry.

Consider the simplest $SU(2)$ model. In this case, there are at least three ways to be convinced in the presence of a normalized zero-energy state in the spectrum.

---

[3] E_{char} also involves a factor growing with the rank of the group — see Sect. 9.4.

9.1   Phase space integral

The first two methods work well for the $SU(2)$ gauge group — otherwise the calculations become pretty difficult.

To begin with, let us close our eyes at the presence of the continuous low-energy spectrum and try to calculate the index of the gauge SQM system like we usually do for benign systems including a gap between the vacuum and excited states. Namely, we use the CG method and replace the functional integral for the index by the finite-dimensional phase space integral (4.47). However, in this case it is not so straightforward. We are dealing with a gauge system and, before calculating the integral, we have to implement the gauge constraints $G^a = 0$. The easiest way to do so is to insert in the integrand the factor[4]

$$\prod_a \delta(G^a) \;=\; \prod_a \frac{1}{2\pi} \int dA_0^a \, e^{iA_0^b G^b} \,. \tag{9.8}$$

Thereby the variables A_0^a, which were present in the role of Lagrange multipliers in the SYM Lagrangian, are restored in the phase space integral. After integrating over the momenta E_I^a and the fermion variables, we arrive at the integral

$$I = \frac{1}{8\pi^2} \left(\frac{\beta g^2}{2\pi}\right)^{3(\mathcal{N}+2)/4} C_\mathcal{N} \cdot$$

$$\int \prod_{aM} dA_M^a \, \det\left(iA_M^a \Gamma_M \varepsilon^{abc}\right) \exp\left\{-\frac{\beta g^2}{4} \varepsilon^{abe} \varepsilon^{cde} A_M^a A_N^b A_M^c A_N^d\right\}. \tag{9.9}$$

We are mainly interested in the maximal $\mathcal{N} = 16$ supersymmetric model where $M = 0, 1, \ldots, 9$, $\Gamma_0 = i$ and $C_{16} = 1/2$ (this factor multiplies the determinant in the $\mathcal{N} = 16$ case due to the reality of fermion variables λ_α^a), but the result (9.9) also applies to $\mathcal{N} = 4$ and $\mathcal{N} = 8$ gauge SQM's with $M = 0, 1, 2, 3$ and $M = 0, \ldots, 5$. For $\mathcal{N} = 4$, Γ_M should be replaced by $\sigma_\mu^{\text{Eucl}} = (i, \boldsymbol{\sigma})$ and, for $\mathcal{N} = 8$, by the Euclidean 4×4 gamma matrices talking to complex $SO(5)$ fermions. $C_\mathcal{N} = 1$ in both cases.

Let us calculate these integrals. It is convenient to use the *polar representation*. In the simplest $\mathcal{N} = 4$ case, it reads

$$A_\mu^a \;=\; V^{ab} \Lambda_\nu^b U_{\nu\mu} \,, \tag{9.10}$$

---

[4]This is how this calculation was performed in the papers [57]. An alternative method consists in resolving the gauge constraints *explicitly*, expressing the Hamiltonian via the gauge-invariant variables and calculating the integral (4.47) for *that* Hamiltonian — see Ref. [84]b (where I unfortunately missed the factor 1/8) and the Appendix in Ref. [61] (where this mistake was corrected).

where $V^{ab} \in SO(3), U_{\nu\mu} \in SO(4)$ and Λ^b_ν is a rectangular quasidiagonal matrix:

$$\Lambda^b_\nu = \begin{pmatrix} a & 0 & 0 & 0 \\ 0 & b & 0 & 0 \\ 0 & 0 & c & 0 \end{pmatrix}. \tag{9.11}$$

Such a representation is well known in nuclear physics [152], where four 3-vectors A^a_μ may be interpreted as four nucleon coordinates in a ^{5}He (or ^{5}Li) nucleus in the center-of-mass system. The polar representation was applied to Yang-Mills theory in Refs. [153].

For $\mathcal{N} = 8$ and $\mathcal{N} = 16$ theories, we will use a similar representation, with Λ^b_M possesing additional columns with zero elements: two such columns for $\mathcal{N} = 8$ (the mathematics is the same as for the ^{7}Li nucleus) and six extra columns for $\mathcal{N} = 16$ (like for ^{11}B). The integrands in (9.9) do not depend on angles. The angular integration can be done using the formula [152]

$$\int \prod_{a=1}^{3} \prod_{M=1}^{D} dA^a_M = \kappa_D \int (abc)^{D-3}(a^2 - b^2)(a^2 - c^2)(b^2 - c^2) \, da\,db\,dc, \tag{9.12}$$

where $D \geq 3$; a, b, c are restricted to lie in the range $\{\Omega_0 : a \geq b \geq c \geq 0\}$; and

$$\kappa_D = \frac{16\pi^{(3D+1)/2}}{\Gamma\left(\frac{D-2}{2}\right)\Gamma\left(\frac{D-1}{2}\right)\Gamma\left(\frac{D}{2}\right)}. \tag{9.13}$$

In particular,

$$\kappa_3 = 32\,\pi^4, \qquad \kappa_4 = 32\,\pi^6, \qquad \kappa_5 = \frac{128\,\pi^7}{3},$$

$$\kappa_6 = \frac{32\,\pi^9}{3}, \qquad \kappa_9 = \frac{1024\,\pi^{13}}{5\cdot 9!!}, \qquad \kappa_{10} = \frac{16\,\pi^{15}}{9!!}. \tag{9.14}$$

The determinant factors in the integrands in (9.9) read

$$\det_{\mathcal{N}=4} = 4(abc)^2, \quad \det_{\mathcal{N}=8} = 16(abc)^4, \quad \frac{1}{2}\det_{\mathcal{N}=16} = 256(abc)^8. \tag{9.15}$$

The exponent is

$$-\frac{\beta g^2}{4}\left[(A^a)^2(A^b)^2 - (A^a \cdot A^b)^2\right] = -\frac{\beta g^2}{2}(a^2b^2 + a^2c^2 + b^2c^2). \tag{9.16}$$

Consider first the simplest case $\mathcal{N} = 4$. To do the integral, we use Eq. (9.12) with $D = 4$ and pose $\beta g^2/2 = 1$ (the integral does not depend on β or g). This gives

$$I_4 = \frac{16}{\sqrt{\pi}} \cdot$$

$$\int_{\Omega_0} dadbdc \,(abc)^3 (a^2 - b^2)(a^2 - c^2)(b^2 - c^2) e^{-a^2 b^2 - a^2 c^2 - b^2 c^2}. \quad (9.17)$$

Next we introduce the variables

$$A = bc, \qquad B = ac, \qquad C = ab, \qquad\qquad (9.18)$$

in which terms the integral acquires the form

$$I_4 = \frac{8}{\sqrt{\pi}} \int_{\Omega_0} dAdBdC \,(A^2 - B^2)(A^2 - C^2)(B^2 - C^2) e^{-A^2 - B^2 - C^2}. (9.19)$$

Then we use again (9.12) with $D = 3$ in the "opposite direction" by treating A, B, C as the elements of a diagonal matrix $\tilde{\Lambda}_j^a$ and introducing nine variables $F_j^a = V^{ab} \tilde{\Lambda}_k^b U_{kj}$, with V and U being two $SO(3)$ matrices. We obtain

$$I_4 \;=\; \frac{1}{4\pi^{9/2}} \int_{-\infty}^{\infty} \prod_{aj} dF_j^a \, \exp\{-(F_j^a)^2\} \;=\; \frac{1}{4}. \qquad (9.20)$$

The cases $\mathcal{N} = 8$ and $\mathcal{N} = 16$ are treated analogously. We derive

$$I_8 \;=\; \frac{1}{4}, \qquad I_{16} \;=\; \frac{5}{4}. \qquad\qquad (9.21)$$

One can write a universal formula

$$\boxed{I_{\mathcal{N}} \;=\; 2^{(\mathcal{N}-8)/2} \frac{\Gamma\left(\frac{\mathcal{N}}{4} - \frac{1}{2}\right)}{\sqrt{\pi}\,\Gamma\left(\frac{\mathcal{N}}{4} + 1\right)} \cdot} \qquad (9.22)$$

The values of the index are fractional reflecting the presence of continuous spectrum.

9.1.1 *Deficit term*

Now the idea is to *subtract* from these values the continuum contribution, so you are left with *integer* values of the index counting the *normalizable* zero-energy states. This possible to do. As we will shortly see, the continuum contribution is 1/4 in all three models giving the "genuine" index $I_W = 0$ for $\mathcal{N} = 4, 8$ and $I_W = 1$ for the maximal SQM.

To evaluate the continuum contribution to be subtracted (it is sometimes called the *deficit term*, and the result (9.22) is *principal term*), we should recall the physical origin of the continuous spectrum. It appears due to an infinite motion along the classical valley at

$$\varepsilon^{abc} A_I^b A_J^c = 0,\qquad (9.23)$$

which is not lifted by quantum effects due to supersymmetry. This infinite motion is associated with large values of $|A^a|$. Thus, all that we have to do is to single out the contribution of small $\varepsilon^{abc} A_I^b A_J^c$ but large $|A^a|$ in the integral (9.9) and subtract it!

In the vicinity of the bottom of the valley, far enough from the origin, the Hamiltonian (9.4) simplifies. As was explained in Chapter 6, for a field right at the bottom, we can choose the gauge $A_I^a = c_I \delta^{a3}$ so that c_I are the slow variables describing the motion along the valley. We may then represent $A_I^a = c_I \delta^{a3} + a_I^{\tilde{a}}$, where $a_I^{\tilde{a}=1,2}$ are the transverse components of the gauge potential, which are much smaller than c_I for large enough $|c|$. Then we may write the fast Hamiltonian including c_I as parameters and the components of $a_I^{\tilde{a}}$ orthogonal to c_I (the projections $a_I^{\tilde{a}} c_I$ amount to gauge transformations which do not cost energy) and their superpartners as dynamical variables. In the leading BO approximation, the fast Hamiltonian reads

$$\hat{H}^{\text{fast}} = \left(\delta_{IJ} - \frac{c_I c_J}{c^2}\right)\left(-\frac{1}{2}\frac{\partial^2}{\partial a_I^{\tilde{a}} \partial a_J^{\tilde{a}}} + \frac{g^2 c^2}{2} a_I^{\tilde{a}} a_J^{\tilde{a}}\right) + \frac{ig}{2} c_I \varepsilon^{\tilde{a}\tilde{b}} \hat{\lambda}^{\tilde{a}} \Gamma_I \hat{\lambda}^{\tilde{b}}. \qquad (9.24)$$

It is basically a supersymmetric oscillator describing transverse fluctuations around the bottom. The ground state of this oscillator has zero energy, and that is why the spectrum of the effective Hamiltonian describing the motion along the valley is continuous.[5] Averaging the full supercharges (3.65) over the ground state of the fast Hamiltonian (9.24) and calculating the anticommutator $\{\hat{Q}_\alpha^{\text{eff}}, \hat{\bar{Q}}^{\alpha\,\text{eff}}\}$, we may find the effective slow Hamiltonian. It has a very simple form [cf. Eq. (6.53)]:

$$\hat{H}^{\text{eff}} = \frac{1}{2}\hat{E}_I \hat{E}_I \qquad (9.25)$$

---

[5] Incidentally, the Hamiltonian (9.4) for the group $SU(N)$ in the large N limit coincides with the mass operator of 2+1 supermembranes embedded in (9+1) - dimensional space [147, 148]. The fact that the spectrum of (9.4) is continuous means that the supermembrane mass spectrum is continuous [149, 150]. Realization of this fact quenched early attempts to build up a supermembrane theory (where supermembranes would be treated as fundamental objects).

On the other hand, there were suggestions that this Hamiltonian represents a certain limit of 11-dimensional M theory, which is considered as one of the candidates for the Holy Grail Theory of Everything [151].

with $\hat{E}_I = -i\partial/\partial c_I$. To find a contribution of continuous spectrum to the phase space integral for the index, one has to calculate the integral

$$\int \prod_I \frac{dE_I dc_I}{2\pi} \prod_{\tilde{\alpha}} d\eta_{\tilde{\alpha}} d\bar{\eta}_{\tilde{\alpha}} \exp\left\{-\beta H^{\text{eff}}\right\} \tag{9.26}$$

with $\eta_{\tilde{\alpha}} = \mu_{\tilde{\alpha}}^3$.

"But it is zero!", says the confused and a little bit agressive reader. The Hamiltonian (9.25) does not depend on fermion variables, and the Grassmann integration in (9.26) gives zero! And where is one quarter that you promised??

Well, give me a break...

Aha! I've got it! The effective Hamiltonian (9.25) is correct, but not all its eigenstates are physical states that should be taken into account in the supertrace $\text{Tr}\{(-1)^{\hat{F}} e^{-\beta \hat{H}^{\text{eff}}}\}$. The physical states satisfy the condition of Weyl invariance (recall its discussion in Chapter 6):

$$\Psi(-c_I, -\eta_{\tilde{\alpha}}) = \Psi(c_I, \eta_{\tilde{\alpha}}), \tag{9.27}$$

and we have to implement this condition in our calculation of the path integral. The supertrace projected on the states invariant under the action of the Weyl symmetry can be presented as an integral

$$I_W^{\text{cont}} = \frac{1}{2} \int \prod_I dc_I \prod_{\tilde{\alpha}} d\eta_{\tilde{\alpha}} d\bar{\eta}_{\tilde{\alpha}} \exp\{-\eta_{\tilde{\alpha}}\bar{\eta}_{\tilde{\alpha}}\} \cdot$$
$$[\mathcal{K}(c_I, \eta_{\tilde{\alpha}}; c_I, \bar{\eta}_{\tilde{\alpha}}; -i\beta) + \mathcal{K}(-c_I, -\eta_{\tilde{\alpha}}; c_I, \bar{\eta}_{\tilde{\alpha}}; -i\beta)], \tag{9.28}$$

where $\mathcal{K}(\cdots)$ is the evolution kernel [cf. Eq. (4.41)]. In our case (free motion!) the kernel has a simple form[6]

$$\mathcal{K}(c_I', \eta_{\tilde{\alpha}}'; c_I, \bar{\eta}_{\tilde{\alpha}}; -i\beta) = (2\pi\beta)^{-9/2} \exp\left\{\eta_{\tilde{\alpha}}'\bar{\eta}_{\tilde{\alpha}} - \frac{(c_I' - c_I)^2}{2\beta}\right\}. \tag{9.29}$$

Substituting it in (9.28), we find that: *(i)* the first term is the supertrace for the system with no constraints imposed, and it is zero; *(ii)* the second term is not zero, giving [57, 154]

$$I_W^{\text{cont}} = \frac{1}{2}\frac{2^8}{2^9} = \frac{1}{4}. \tag{9.30}$$

The factor 2^8 in the numerator comes from the fermion integrals and the factor 2^9 in the denominator from bosonic integrals.

---

[6]The product of nine free bosonic evolution kernels (4.10) with $\Delta t = -i\beta$ is multiplied by the product of eight Grassmann kernels (4.38) with $H = 0$.

Basically, a nonzero contribution in the integral for the index is due to asymmetry between bosonic and fermionic states in the continuous spectrum brought about by the Weyl invariance requirement. For example, the bosonic state $\Psi(c_I, \eta_{\tilde{\alpha}}) = 1$ is Weyl invariant, while its fermionic counterparts like $\Psi(c_I, \eta_{\tilde{\alpha}}) = \eta_1$ etc. are not.

Note that the result (9.30) holds universally for $\mathcal{N} = 16$ theory and also for $\mathcal{N} = 8$ and $\mathcal{N} = 4$ theories. Indeed, for $\mathcal{N} = 4$, we have just 2 complex fermionic variables and 3 bosonic variables, while for $\mathcal{N} = 8$, we have 4 complex fermionic variables and 5 bosonic variables. We derive:

$$\boxed{I_W^{\text{cont}} = \frac{1}{2}\frac{2^8}{2^9} = \frac{1}{2}\frac{2^4}{2^5} = \frac{1}{2}\frac{2^2}{2^3} = \frac{1}{4}} \tag{9.31}$$

for any $\mathcal{N}$. Subtracting this from the values (9.22), we conclude that the proper Witten index is zero for the $\mathcal{N} = 4$ and $\mathcal{N} = 8$ theories and $I_W = 1$ for the $\mathcal{N} = 16$ theory. In other words, the $\mathcal{N} = 4$ and $\mathcal{N} = 8$ systems do not have normalizable zero-energy states in their spectra, whereas the maximal gauge SQM has one.

9.2 Asymptotics along the valley

For the toy model with the spherically symmetric potential in $\mathbb{R}^5$ that falls down at infinity, we mentioned that a necessary condition for the existence of a normalized bound state is the existence of a normalized asymptotic zero-energy solution to the free Schrödinger equation at large r. A similar statement can be done for the gauge SQM models in interest: if the asymptotic normalized solution to the effective valley Hamiltonian exists for large $|\boldsymbol{A}|$, this is a strong indication that such solution also exists for the full Hamiltonian.

Such asymptotic solution exists, indeed, for the maximally supersymmetric $SU(2)$ model [58–60], and it does not exist for $\mathcal{N} = 4$ and $\mathcal{N} = 8$ models.

Let us first prove the latter statement. For $\mathcal{N} = 4$, it is obvious: the effective Hamiltonian for the $SU(2)$ theory is just a 3-dimensional Laplacian, which does not have normalized eigenstates.

In $\mathcal{N} = 8$ theory, the effective wave function depends on a 5-dimensional vector c_I and the holomorphic Grassmann $SO(5)$ spinor $\eta_{\alpha=1,2,3,4}$. A 16-component wave function is decomposed into two $SO(5)$ singlets with the fermion numbers $F = 0$ and $F = 4$, two spinors with $F = 1$ and $F = 3$ and a vector and a singlet in the $F = 2$ sector. The vacuum function must be annihilated by the effective supercharges,

$$\hat{Q}_\alpha^{\text{eff}} = -i\frac{\partial}{\partial c_I}(\gamma_I)_\alpha{}^\beta\eta_\beta, \qquad \hat{\bar{Q}}^{\alpha\,\text{eff}} = (\hat{Q}_\alpha^{\text{eff}})^\dagger = -i\frac{\partial}{\partial c_I}\hat{\bar{\eta}}^\beta(\gamma_I)_\beta{}^\alpha, \quad (9.32)$$

where γ_I are the 5-dimensional gamma matrices.

And one *can* construct such a function! It has the form

$$\Psi^{\text{slow}}(c_I,\eta_\alpha) \overset{?}{=} \partial_I\frac{1}{|\boldsymbol{c}|^3}\,\eta_\alpha C^{\alpha\epsilon}(\gamma_K)_\epsilon{}^\beta\eta_\beta, \qquad (9.33)$$

where C is a charge conjugation matrix lifting spinor indices.[7] It is easy to see that

$$\hat{\bar{Q}}^{\alpha\,\text{eff}}\Psi = -i\partial_I\partial_K\frac{1}{|\boldsymbol{c}|^3}\,\hat{\bar{\eta}}^\beta(\gamma_I)_\beta{}^\alpha\eta_\gamma(C\gamma_K)^{\gamma\delta}\eta_\delta = 2i\partial_I\partial_K\frac{1}{|\boldsymbol{c}|^3}\,\eta_\gamma(C\gamma_K\gamma_I)^{\gamma\alpha}$$

$$\propto \triangle^{d=5}\frac{1}{|\boldsymbol{c}|^3} = 0. \qquad (9.34)$$

The result of the action of the supercharge $\hat{Q}_\alpha^{\text{eff}}$ on the function (9.33) is also zero.

However, our best try (9.33) *is* not an admissible wave function because it does not satisfy the requirement of Weyl invariance (9.27). Throwing it away, we are left with nothing!

Consider now the $\mathcal{N} = 16$ theory. The effective wave function $\Psi^{\text{slow}}(c_I,\eta_{\tilde{\alpha}})$ has now $2^8 = 256$ components. A half of them are bosonic and the other half, fermionic. As was mentioned, in contrast to $\mathcal{N} = 4$ and $\mathcal{N} = 8$ models, the variables $\eta_{\tilde{\alpha}}$ do not constitute a reprepresentation of $SO(9)$, but the components of the wave function *do*. The fermion components belong to the irreducible **128**-plet of $SO(9)$. This is a kind of Rarita-Schwinger spin-vector $\mathbf{128}_{I\alpha}$ ($I = 1,\ldots,9$; $\alpha = 1,\ldots,16$) satisfying the constraints $(\Gamma_I)_{\alpha\beta}\mathbf{128}_{I\beta} = 0$. The bosonic components split in two irreducible representations **44** + **84**. The first one is the traceless symmetric tensor $\mathbf{44}_{(IJ)}$ and the second one is the antisymmetric tensor $\mathbf{84}_{[IJK]}$.

Let us pick up the symmetric $\mathbf{44}_{(IJ)}$-plet[8] and construct the following function

$$\boxed{\Psi^{\text{slow}}(c_I,\eta_{\tilde{\alpha}}) = \mathbf{44}_{IJ}(\eta_{\tilde{\alpha}})\,\partial_I\partial_J\frac{1}{|\boldsymbol{c}|^7}.} \qquad (9.35)$$

---

[7]One of many possible particular choices for γ_I and C is:

$$\gamma_{i=1,2,3} = \sigma_i \otimes \sigma_2, \qquad \gamma_4 = \mathbb{1} \otimes \sigma_1, \qquad \gamma_5 = \mathbb{1} \otimes \sigma_3, \qquad C = \sigma_2 \otimes \mathbb{1}.$$

Note that both C and $C\gamma_I$ are antisymmetric.

[8]Its explicit structure in terms of $\eta_{\tilde{\alpha}}$ is rather nontrivial [59]: it involves the component without η factors, the component $\prod_{\tilde{\alpha}=1}^8 \eta_{\tilde{\alpha}}$ and *some* components with the products of four η factors.

It is an $SO(9)$ singlet. It falls out rapidly and is normalizable at infinity. Acting on it with the supercharge $\hat{Q}_\alpha^{\text{eff}} = -i(\Gamma_K)_{\alpha\beta}\hat{\lambda}_\beta\partial_K$, we obtain

$$\hat{Q}_\alpha^{\text{eff}}\Psi^{\text{slow}}(c_I, \eta_{\tilde{\alpha}}) \; \propto \; (\Gamma_K)_{\alpha\beta}\hat{\lambda}_\beta \, \mathbf{44}_{IJ}(\eta_{\tilde{\alpha}}) \, \partial_I\partial_J\partial_K \frac{1}{|\boldsymbol{c}|^7} \, . \qquad (9.36)$$

The function in (9.36) is odd in $\eta_{\tilde{\alpha}}$ and must represent our Rarita-Schwinger **128**-plet. We can symmetrize $(\Gamma_K)_{\alpha\beta}\hat{\lambda}_\beta \, \mathbf{44}_{IJ}$ over (IJK) and write

$$(\Gamma_{(K})_{\alpha\beta}\hat{\lambda}_\beta \, \mathbf{44}_{IJ)}(\eta_{\tilde{\alpha}}) \; = \; \delta_{IK}\mathbf{128}_{J\alpha} + \delta_{JK}\mathbf{128}_{I\alpha} + \delta_{IJ}\mathbf{128}_{K\alpha} \, . \qquad (9.37)$$

Substituting it in (9.36), we obtain zero due to the property

$$\triangle^{d=9}\frac{1}{|\boldsymbol{c}|^7} \; = \; 0 \, .$$

In contrast to (9.33), the function (9.35) is Weyl-even and admissible.

The asymptotic analysis of the solutions to Schrödinger equation and the explicit construction of the asymptotic wave function (9.35) gives a vivid understanding of the low-energy dynamics of the maximal SQM systems — in what respects it is similar and in what respects different from simpler $\mathcal{N} = 4$ and $\mathcal{N} = 8$ systems. However, as was mentioned above, the existence of an asymptotic solution does not guarantee the existence of a normalized solution for the full Hamiltonian. In addition, this method works well for $SU(2)$, but it is difficult to generalize it to other groups. We are only aware of an explicit construction of the asymptotic wave function for $SU(3)$ [155].

The method based on the calculation of the phase space integral is less direct, but it allows generalizations to some groups of low rank. The main complication of this method is not so much a necessity to subtract the continuum contribution, since in many cases this can be done [61], but the actual calculation of the *principal* contribution — of multidimensional phase space integrals for higher groups. They were evaluated numerically [156], but not analytically.

9.3   Mass deformation

There exists a simple universal method to determine the existence of normalized vacuum states. It is based on *mass deformation* [157] of the Hamiltonian (9.4). For $SU(2)$, one can derive the existence of a normalized state in a very simple way, but it can also be easily generalized to $SU(N)$ and, using some known mathematical results, to all classical Lie groups [61].

As was mentioned back in Chapter 3, the SQM model (9.4) can be obtained by dimensional reduction of the $\mathcal{N} = 4$ four-dimensional SYM theory with the Lagrangian (3.52). We now deform the latter by adding a mass term to the superpotential:

$$\mathcal{W}^M = \frac{g}{3\sqrt{2}} \varepsilon_{fgh} f^{abc} \Phi_f^a \Phi_g^b \Phi_h^c - \frac{M}{2} \Phi_f^a \Phi_f^a. \tag{9.38}$$

The theory thus obtained is still supersymmetric, though it is now $\mathcal{N} = 1$ rather than $\mathcal{N} = 4$. The Witten indices in the original theory and in the deformed theory coincide.

Having done that, we may also ignore the spatial dependence of the fields to obtain a deformed SQM model, which keeps 4 out of 16 conserved Hermitian supercharges intact. We will discuss in the following the deformed $4D$ field theory and deformed SQM in parallel.

In the deformed model, the classical potential reads

$$V = \left| \frac{g}{\sqrt{2}} \varepsilon_{fgh} f^{abc} \phi_g^b \phi_h^c - M\phi_f^a \right|^2 + \left| \frac{g}{\sqrt{2}} f^{abc} \phi_f^b \phi_f^{*c} \right|^2. \tag{9.39}$$

The classical vacuum fields satisfy the equations

$$\varepsilon_{fgh} f^{abc} \phi_g^b \phi_h^c = \frac{M\sqrt{2}}{g} \phi_f^a, \qquad f^{abc} \phi_f^b \phi_f^{*c} = 0 \tag{9.40}$$

or

$$\varepsilon_{fgh} [\hat{\phi}_g, \hat{\phi}_h] = \frac{iM\sqrt{2}}{g} \hat{\phi}_f, \qquad [\hat{\phi}_f, \hat{\phi}_f^\dagger] = 0 \tag{9.41}$$

with $\hat{\phi}_f = \phi_f^a t^a$.

Consider first the $SU(2)$ case. Besides the obvious solution $\phi = 0$, the system (9.40) enjoys a nonzero solution:

$$\boxed{\phi_f^a = \frac{M}{g\sqrt{2}} \delta_f^a.} \tag{9.42}$$

It is unique up to an overall gauge rotation (7.70).

The appearance of the Higgs average (9.42) breaks down the gauge invariance completely: all gauge fields and their superpartners acquire mass of order M. The same applies to the matter fields, regardless of whether we are at the vicinity of the trivial or nontrivial classical vacuum.

When mass is large, the state (9.42) is separated from the sector $\langle \phi \rangle_{\text{vac}} = 0$ by a high barrier. In the limit $M \to \infty$, this barrier becomes unpenetrable, and if, in the morning, we wake up in the sector with $\langle \phi \rangle_{\text{vac}} \sim 0$, we are still going to stay there at the end of the day. The presence of heavy matter fields would not be felt and the dynamics would be the same as in $\mathcal{N} = 1$ 4–dimensional SYM theory [or in $\mathcal{N} = 4$ gauge SQM, if we are talking about the dimensionally reduced (0+1) model].

On the other hand, if $M \lesssim g^{2/3}$, the barrier disappears and the new vacuum state overlaps essentially with the conventional vacuum sector. In the limit $M \to 0$, the state (9.42) goes over into the celebrated localized supersymmetric vacuum state of the hamiltonian (9.4).

Let us emphasize that the final conclusion of the existence of a localized supersymmetric state is valid irrespectively of whether we are thinking in the language of the SQM system or in the language of the associated 4–dimensional field theory.

In the former case, both undeformed and deformed model involve the continuous spectrum associated with the gauge potentials $A_{j=1,2,3}^a$. For large M, the main support of the localized state wave function is well separated from the states from continuum, but in the limit $M \to 0$, these wave functions dwell in the same region of A, and that makes a dynamical analysis difficult. Still, the true index, the number of the *normalized* vacuum states, does not depend on M and, for the $SU(2)$ gauge group, it is equal to 1.

If we are thinking in terms of 4–dimensional field theory, the most convenient way to treat it, as we know, is to put it in a finite spatial volume. That makes the spectrum discrete and removes all the uncertainties connected with nonzero "deficit term" discussed in Sect. 9.1. Going from quantum mechanics to field theory defined in the box gives us, if you like, a convenient *infrared regularization* that makes the motion finite and respects supersymmetry.

In the field theory with large mass, we have one extra state at large values of Higgs average, and also two conventional vacuum states of $\mathcal{N} = 1$ SYM theory coming from the region $\phi \sim 0$ where the heavy matter fields decouple. The total index is $I_W = 3$, and, as was already mentioned at the end of Chapter 7, this holds also for $\mathcal{N} = 4$, $D = 4$ SYM $SU(2)$ theory when the mass term is deformed away.

Let us now discuss more complicated gauge groups. We have to solve again the equation system (9.40), but with an additional requirement: the Higgs averages of the scalar fields should break the gauge invariance *completely* and give mass to *all* gauge fields (otherwise, the wave functions would smear out along the flat directions corresponding to the remaining massless fields, and the state would not be localized.)

In mathematical language, that means that we are looking for the triples $\hat{\phi}_f$ belonging to the complex Lie algebra $\mathfrak{g}$ (the fields ϕ_f^a are complex being the lowest components of the chiral superfields Φ_f^a), which satisfy the relations (9.41) and whose centralizer is trivial (i.e. there is no such $g \in \mathfrak{g}$ that $[g, \phi_f] = 0$ for all $f = 1, 2, 3$). This problem is reduced [130] to the mathematical problem of classifying the embeddings $su(2) \subset \mathfrak{g}$ with the trivial centralizer factorized over the action of the complexified group $\mathcal{G}$.

The latter problem was addressed and solved by mathematicians in the papers [158]. Using these results, one can count the number of normalized vacuum states for all classical Lie groups [61]. The result is the following.[9]

Theorem 9.1. *The number of inequivalent by conjugation solutions of Eq.(9.40) and hence the number $\#_{\text{vac}}[\mathcal{G}]$ of different normalized supersymmetric vacua in the theory (9.4) with gauge group $\mathcal{G}$ is*[10]

(a) $\#_{\text{vac}}[SU(N)] = 1$.

(b) $\#_{\text{vac}}[Sp(2r)]$ *coincides with the number of partitions of r into inequal parts.*[11]

(c) $\#_{\text{vac}}[SO(N)]$ *coincides with the number of partitions of N into inequal odd parts.*[12]

(d) $\#_{\text{vac}}[G_2] = 2$, $\#_{\text{vac}}[F_4] = 4$, $\#_{\text{vac}}[E_6] = 3$, $\#_{\text{vac}}[E_7] = 6$, and $\#_{\text{vac}}[E_8] = 11$.

We are not in a position to give the proof of this theorem in our book, but we will give here some illustrations, directing the reader to Ref. [61] and to the cited mathematical papers for more details.

• The simplest case is $\mathcal{G} = SU(N)$, where there is a single such embedding and a single vacuum state. It is sufficient to write down the generators of $SU(2)$ in the representation with the spin $j = (N-1)/2$ and treat them as the elements of the $su(N)$ algebra in the fundamental representation.

---

[9]The numbers of vacua listed in this theorem refers to the SQM systems (9.4). Adding these numbers to the numbers of vacuum states in the pure $\mathcal{N} = 1$ SYM theory derived in Chapter 6 and coinciding with the dual Coxeter numbers of the gauge groups, we may evaluate the Witten indices of the extended $\mathcal{N} = 4$ SYM theories — see the very end of Chapter 7.

[10]Please do not mix up N characterizing the group with $\mathcal{N}$ characterizing the supersymmetry algebra.

[11]For $r = 1$ or $r = 2$, we have only one universal solution (9.46). For $r = 3 = 2 + 1$, we have two inequivalent solutions, the same for $r = 4$, for $r = 5 = 4 + 1 = 3 + 2$, we have three solutions, there are four solutions for $r = 6 = 5 + 1 = 4 + 2 = 3 + 2 + 1$, etc.

[12]The additional solutions appear starting from $N = 8 = 7 + 1 = 5 + 3$ and $N = 9 = 5 + 3 + 1$.

For example, for $su(3)$, the non–trivial triple is

$$2\hat{\phi}_3 = \frac{M}{g\sqrt{2}}h = \frac{M}{g\sqrt{2}}\begin{pmatrix} 1 & 0 & 0 \\ 0 & 0 & 0 \\ 0 & 0 & -1 \end{pmatrix},$$

$$\hat{\phi}_1 + i\hat{\phi}_2 = \frac{M}{g\sqrt{2}}e = \frac{M}{g\sqrt{2}}\begin{pmatrix} 0 & 1 & 0 \\ 0 & 0 & 1 \\ 0 & 0 & 0 \end{pmatrix},$$

$$\hat{\phi}_1 - i\hat{\phi}_2 = \frac{M}{g\sqrt{2}}f = \frac{M}{g\sqrt{2}}\begin{pmatrix} 0 & 0 & 0 \\ 1 & 0 & 0 \\ 0 & 1 & 0 \end{pmatrix}. \tag{9.43}$$

Such a triple is unique up to congugation. We see that h belongs to the Cartan subalgebra of $\mathcal{G}$, e belongs to the subalgebra $\mathfrak{g}_+$ spanned by the positive root vectors and f belongs to the subalgebra $\mathfrak{g}_-$ spanned by the negative root vectors. The relations

$$[h, e] = e, \qquad [h, f] = -f, \qquad [e, f] = h \tag{9.44}$$

hold. In other words, our triple realizes an embedding $su(2) \subset \mathfrak{g} = su(3)$ with a trivial centralizer. Let us call such embeddings (and such triples) *distinguished*.

To prepare for generalization to other algebras, we express the embedding (9.43) in the Chevalley basis spelled out in Appendix A:

$$h = \alpha^\vee + \beta^\vee, \qquad e = E_\alpha + E_\beta, \qquad f = E_{-\alpha} + E_{-\beta}. \tag{9.45}$$

A similar embedding van be constructed for any simple algebra. Choose an element ρ of the Cartan subalgebra $\mathfrak{h}$ such that $[\rho, E_{\alpha_j}] = E_{\alpha_j}$ for all positive simple roots α_j. It represents a linear combination $\rho = \sum_{j=1}^{r} b_j \alpha_j^\vee$ of the simple coroots with positive coefficients[13] b_j. Then a generalization of the embedding (9.45) to an arbitrary algebra is

$$\boxed{h = \rho, \qquad e = \sum_{j=1}^{r} \sqrt{b_j}\, E_{\alpha_j}, \qquad f = \sum_{j=1}^{r} \sqrt{b_j}\, E_{-\alpha_j}.} \tag{9.46}$$

The properties (9.44) hold — this follows from the definition of ρ and from the normalization $[E_\alpha, E_{-\alpha}] = \alpha^\vee$.

Theorem 9.2. *The triple (9.46) is distinguished.*

---

[13] In particular, $\rho = \alpha^\vee + \beta^\vee$ for $su(3)$, $\rho = \frac{3}{2}(\alpha^\vee + \gamma^\vee) + 2\beta^\vee$ for $su(4)$, $\rho = 2(\alpha^\vee + \delta^\vee) + 3(\beta^\vee + \gamma^\vee)$ for $su(5)$, etc.

Proof. Indeed, the centralizer of ρ is the Cartan subalgebra $\mathfrak{h}$. Take any nonzero element $x \in \mathfrak{h}$ and show that $[x, e] \neq 0$. Represent x as a linear combination of simple coroots, $x = \sum_k c_k \alpha_k^\vee$. Then

$$[x, e] = \sum_{jk} c_k \sqrt{b_j}\, \alpha_j(\alpha_k^\vee) E_{\alpha_j} = \sum_{jk} c_k \sqrt{b_j} A_{jk} E_{\alpha_j}, \qquad (9.47)$$

where A_{jk} is the Cartan matrix (A.12). Bearing in mind that $b_j \neq 0$, this can be zero only if $\forall j : \{A_{jk}c_k = 0\}$, i.e. if c_k is an eigenvector of A with a zero eigenvalue. But A has a nonzero determinant (see a remark on p. 285) and does not have zero eigenvalues. $\qquad\square$

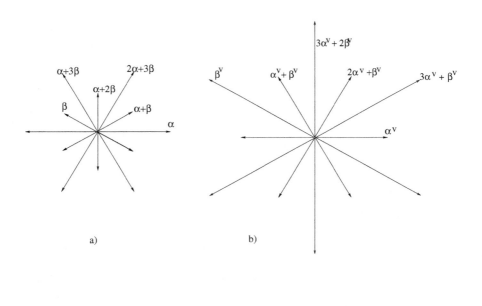

Fig. 9.1: *a)* Root system, *b)* coroot system, *c)* the marked Dynkin diagram for the algebra g_2

For $su(N)$, the solution (9.46) is unique up to conjugation, but for almost all other groups, it is not so. There are other solutions. The general theory that allows to construct such solutions is layed out in Ref. [61]. We will not do so here, but will discuss in some detail the simplest nontrivial example: the algebra g_2. In that case, besides the universal embedding (9.46), which is

$$h = 5\alpha^\vee + 3\beta^\vee, \qquad e, f = \sqrt{5}E_{\pm\alpha} + \sqrt{3}E_{\pm\beta}, \qquad (9.48)$$

another distinguished embedding exists.

The system of roots of g_2 is depicted in Fig. 9.1a, the system of corresponding coroots[14] in Fig. 9.1b, and the *marked* Dynkin diagram[15] in Fig. 9.1c.

The second distiguished triple reads[16]

$$\boxed{h = 2\alpha^\vee + \beta^\vee, \quad e = E_\alpha + E_{\alpha+3\beta}, \quad f = E_{-\alpha} + E_{-(\alpha+3\beta)}.} \qquad (9.49)$$

To show that it is a triple (9.44) and not just a triple but a *distiguished* triple, we need to know some basic commutators. To begin with, we write

$$[\alpha^\vee, E_\alpha] = 2E_\alpha, \qquad [\beta^\vee, E_\alpha] = -3E_\alpha,$$
$$[\alpha^\vee, E_\beta] = -E_\beta, \qquad [\beta^\vee, E_\beta] = 2E_\beta. \qquad (9.50)$$

[The coefficients in (9.50) are nothing but the elements of the G_2 Cartan matrix in Eq. (A.13).] Then, using the commutators (A.4),

$$[E_\alpha, E_\beta] = E_{\alpha+\beta}, \qquad [E_{\alpha+\beta}, E_\beta] = 2E_{\alpha+2\beta}, \quad \text{etc.},$$

and the Jacobi identities, one can derive:

$$[\alpha^\vee, E_{\alpha+\beta}] = E_{\alpha+\beta}, \quad [\alpha^\vee, E_{\alpha+2\beta}] = 0, \quad [\alpha^\vee, E_{\alpha+3\beta}] = -E_{\alpha+3\beta},$$
$$[\alpha^\vee, E_{2\alpha+3\beta}] = E_{2\alpha+3\beta}, \quad [\beta^\vee, E_{\alpha+\beta}] = -E_{\alpha+\beta}, \quad [\beta^\vee, E_{\alpha+2\beta}] = E_{\alpha+2\beta},$$
$$[\beta^\vee, E_{\alpha+3\beta}] = 3E_{\alpha+3\beta}, \quad [\beta^\vee, E_{2\alpha+3\beta}] = 0. \qquad (9.51)$$

It follows that the centralizer of h is the subalgebra $\mathfrak{g}_0$ with the basis $\alpha^\vee, \beta^\vee, E_\beta$, and[17] $E_{-\beta}$. On the other hand,

$$[h, E_\alpha] = E_\alpha, \quad [h, E_{\alpha+\beta}] = E_{\alpha+\beta}, \quad [h, E_{\alpha+2\beta}] = E_{\alpha+2\beta},$$
$$[h, E_{\alpha+3\beta}] = E_{\alpha+3\beta} \quad \text{and} \quad [h, E_{2\alpha+3\beta}] = 2E_{2\alpha+3\beta}. \qquad (9.52)$$

That is the meaning of the marking in Fig. 9.1c : h is chosen in such a way that the root vectors, corresponding to the roots whose simple root

---

[14]The long coroots $\beta^\vee, 3\alpha^\vee + \beta^\vee$ and $3\alpha^\vee + 2\beta^\vee$ in Fig. 9.1b correspond to short roots: $(\alpha + \beta)^\vee = [E_{\alpha+\beta}, E_{-(\alpha+\beta)}] = 3\alpha^\vee + \beta^\vee$ and $(\alpha + 2\beta)^\vee = 3\alpha^\vee + 2\beta^\vee$. The short coroots $\alpha^\vee, \alpha^\vee + \beta^\vee$ and $2\alpha^\vee + \beta^\vee$ correspond to long roots: $(2\alpha + 3\beta)^\vee = 2\alpha^\vee + \beta^\vee$ and $(\alpha + 3\beta)^\vee = \alpha^\vee + \beta^\vee$.

[15]The marking is the blob in the circle describing the long simple root α. The meaning of this marking will be shortly explained.

[16]It has also many other avatars related to (9.49) by conjugation.

[17]The fact that h commutes with $E_{\pm\beta}$, whereas $[h, E_{\pm\alpha}] = \pm E_{\pm\alpha}$ means that h is a fundamental coweight ω_α of the root α.

expansions include the marked simple root α, do not commute with h, whereas the root vectors E_β and $E_{-\beta}$, corresponding to the unmarked root β, do.

Now, the element e does not commute with any element of $\mathfrak{g}_0$. Indeed, it follows from (9.50) and (9.51) that e does not commute with the elements of $\mathfrak{h}$. Neither does it commute with E_β or $E_{-\beta}$ due to the nonzero commutators $[E_\alpha, E_\beta] = E_{\alpha+\beta}$ and $[E_{\alpha+3\beta}, E_{-\beta}] = E_{\alpha+2\beta}$. The same concerns f, and the triple (9.49) is indeed distinguished.

As is seen from (9.52), the element h realizes a $\mathbb{Z}$-*gradation* of the algebra $\mathfrak{g}_2$. The latter can be subdivided in five linear spaces:

$\mathfrak{g}_0$ with the basis $\alpha^\vee, \beta^\vee, E_\beta$, and $E_{-\beta}$;

$\mathfrak{g}_1$ with the basis $E_\alpha, E_{\alpha+\beta}, E_{\alpha+2\beta}$, and $E_{\alpha+3\beta}$;

$\mathfrak{g}_{-1}$ with the basis $E_{-\alpha}, E_{-(\alpha+\beta)}, E_{-(\alpha+2\beta)}$, and $E_{-(\alpha+3\beta)}$;

$\mathfrak{g}_2$ with the basis $E_{2\alpha+3\beta}$;

$\mathfrak{g}_{-2}$ with the basis $E_{-(2\alpha+3\beta)}$.   (9.53)

Then $[h, x] = jx$ if $x \in \mathfrak{g}_j$. (The commutators of h with the negative root vectors have the same form as in (9.52), but with a negative sign.) In addition: *(i)* $\mathfrak{g}_1 \oplus \mathfrak{g}_2$ coincides with the subalgebra $\mathfrak{g}_+$ of all positive root vectors and $\mathfrak{g}_{-1} \oplus \mathfrak{g}_{-2}$ is the subalgebra $\mathfrak{g}_-$ of all negative root vectors; *(ii)* the dimensions of $\mathfrak{g}_0$, $\mathfrak{g}_1$ and $\mathfrak{g}_{-1}$ coincide.

These are actually the underlying reasons, unravelled in [158], by which the triple (9.49) is distinguished and is not equivalent to the universal triple (9.46).[18] Based on that, all the distinguished triples for all the algebras were classified in these mathematical works.

The next in complexity example is $sp(6)$ [for $sp(4) \simeq so(5)$, there is only the universal distinguished triple (9.46)]. Consider the marked Dynkin diagram in Fig. 9.2

Take the element
$$h = \omega_\alpha + \omega_\gamma = \frac{3}{2}\alpha^\vee + 2\beta^\vee + \frac{5}{2}\gamma^\vee . \qquad (9.54)$$

---

[18]Both properties mentioned above are important. For example, the element $\alpha^\vee$ in $su(3)$ realizes the gradation: $\mathfrak{g}_0 = \mathfrak{h}$; $\mathfrak{g}_1$ spanned by $E_{\alpha+\beta}$ and $E_{-\beta}$; $\mathfrak{g}_{-1}$ spanned by $E_{-(\alpha+\beta)}$ and E_β; $\mathfrak{g}_2 \propto E_\alpha$; $\mathfrak{g}_{-2} \propto E_{-\alpha}$. The dimensions of $\mathfrak{g}_0$, $\mathfrak{g}_1$, and $\mathfrak{g}_{-1}$ coincide. But $\mathfrak{g}_1$ includes a negative root vector, while $\mathfrak{g}_{-1}$ includes a positive root vector. And that is why the exceptional triple associated with this gradation,

$$h = \begin{pmatrix} 1 & 0 & 0 \\ 0 & -1 & 0 \\ 0 & 0 & 0 \end{pmatrix}, \quad e = \begin{pmatrix} 0 & 0 & 1 \\ 0 & 0 & 0 \\ 0 & -1 & 0 \end{pmatrix}, \quad f = \begin{pmatrix} 0 & 0 & 0 \\ 0 & 0 & -1 \\ 1 & 0 & 0 \end{pmatrix},$$

is equivalent to (9.43) by conjugation.

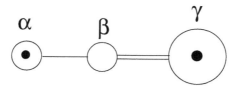

Fig. 9.2: Nontrivial distinguished marked Dynkin diagram for the algebra $sp(6)$

Its centralizer is the subalgebra $\mathfrak{g}_0$ with the basis $\alpha^\vee$, $\beta^\vee$, $\gamma^\vee$, E_β, $E_{-\beta}$. It realizes the following $\mathbb{Z}$-gradation of $sp(6)$ (such as $[h, \mathfrak{g}_j] = j\mathfrak{g}_j$):

$\qquad \mathfrak{g}_0$ with the basis $\alpha^\vee$, $\beta^\vee$, $\gamma^\vee$, E_β, $E_{-\beta}$,

$\qquad \mathfrak{g}_1$ with the basis E_α, $E_{\alpha+\beta}$, E_γ, $E_{\beta+\gamma}$, $E_{2\beta+\gamma}$,

$\qquad \mathfrak{g}_2$ with the basis $E_{\alpha+\beta+\gamma}$, $E_{\alpha+2\beta+\gamma}$,

$\qquad \mathfrak{g}_3$ with the basis $E_{2\alpha+2\beta+\gamma}$, $\qquad\qquad$ (9.55)

and the linear spaces $\mathfrak{g}_{-1}$, $\mathfrak{g}_{-2}$, $\mathfrak{g}_{-3}$ spanned by the corresponding negative root vectors. The condition $\dim \mathfrak{g} = \dim \mathfrak{g}_1$ is satisfied, the condition $\mathfrak{g}_1 \oplus \mathfrak{g}_2 \oplus \mathfrak{g}_3 = \mathfrak{g}_+$ is also satisfied, and hence the marking in Fig. 9.2 is distinguished.

The elements e and f in the corresponding distinguished triple may be chosen as

$$ e = \sqrt{\frac{3}{2}}\, E_\alpha + \frac{1}{\sqrt{2}}\, E_\gamma + \sqrt{2}\, E_{2\beta+\gamma} $$

$$ f = \sqrt{\frac{3}{2}}\, E_{-\alpha} + \frac{1}{\sqrt{2}}\, E_{-\gamma} + \sqrt{2}\, E_{-(2\beta+\gamma)}\,. \qquad (9.56) $$

All other distinguished triples in $sp(6)$ are equivalent by conjugation either to the universal triple (9.46) or to the triple (9.54), (9.56).

More examples and more discussion can be found in [61].

The last remark is that our counting of distinguished triples for the symplectic and orthogonal groups giving the number of normalized vacuum states in the spectrum of the Hamiltonian (9.4) has got an unexpected interpetation in [159]. In this paper, our result was reproduced using the D-brane language and ideology.

9.4 Thermodynamics and duality

The title of this book is *Witten index*. It counts the number of vacuum states, and vacuum dynamics of different supersymmetric quantum mechanical and field theory systems is our principal question of interest. There are plenty of other interesting dynamical questions that can be asked for the theories studied in the previous chapters, but we have not done so.

However, in this particular section, we decided to change our habits and discuss a very interesting question which concerns not the vacuum states of the Hamiltonian (9.4) (which were studied in the first three sections of the chapter), but its *excited* states. This discussion will preceded by a digression. For now I want to talk (very briefly and without any justification for the statements that I want to make) about *Maldacena duality* [160].

Consider a $4D$ $\mathcal{N} = 4$ SYM theory with the $SU(N)$ group in the '*t Hooft limit* when N tends to infinity, the coupling constant g^2 to zero, but the product

$$\lambda = g^2 N \tag{9.57}$$

is kept fixed. In such a theory, all physical quantities depend only on λ. If $\lambda \ll 1$, everything can be calculated perturbatively as a series in λ. For large λ, perturbation theory does not work, but Maldacena's conjecture was that the properties of this theory at $\lambda \gg 1$ can be derived by studying classical solutions of $10D$ supergravity on a special $AdS \times S^5$ background.

This conjecture has not been proven yet, but it was verified in several nontrivial cases where an *exact* solution was known. Arguably, the most lucid example is the circular supersymmetric Wilson loop [161]. For large N, its vacuum average can be perturbatively evaluated in any order in λ. The sum of the perturbative series is

$$\langle W \rangle_{\text{circle}} = \frac{2I_1(\sqrt{\lambda})}{\sqrt{\lambda}} . \tag{9.58}$$

On the other hand, the same quantity can be calculated at large λ on the supergravity side. The result,

$$\langle W \rangle_{\text{circle}} = \sqrt{\frac{2}{\pi}} e^{\sqrt{\lambda}} \frac{1}{\lambda^{3/4}} \left[1 - \frac{3}{8\sqrt{\lambda}} + \cdots \right], \tag{9.59}$$

coincides exactly with the asymptotics and preasymptotics of (9.58).

Duality relationships can be established and duality predictions can be made, however, not only for $4D$ theory, but also for its low-dimensional "sisters" obtained by dimensional reduction. In particular, by studying a certain charged black hole solution in $10D$ supergravity, one can evaluate the average internal energy of the SQM system described by the Hamiltonian (9.4). The prediction is [162]:

$$\boxed{\left\langle \frac{\mathcal{E}}{N^2} \right\rangle_{T \ll \lambda^{1/3}} \approx 7.41 \lambda^{1/3} \left(\frac{T}{\lambda^{1/3}} \right)^{14/5} \left[1 + O\left(\frac{T}{\lambda^{1/3}} \right)^{9/5} \right] .} \tag{9.60}$$

A question arises whether this rather nontrivial critical behaviour can be understood in terms of the dynamics of the system (9.4) without going through the duality glass to the supergravity side. Even though the system (9.4) is complicated, it is just a QM system with large, but finite (for finite N) number of degrees of freedom. The analysis of its dynamics at strong coupling is *a priori* a much more simple task than the analysis of a strongly coupled field theory. In Refs. [163], a numerical analysis of the system (9.4) was performed. The results are in a good agreement with (9.60).

Can one understand it also analytically (staying firmly on the SQM side)? Our answer to this question is positive.

9.4.1 *Nonsupersymmetric preliminaries*

As a warm-up, consider the harmonic oscillator, $\hat{H} = (\hat{p}^2 + \omega^2 x^2)/2$, at finite temperature. The partition function is

$$Z = \sum_{n=0}^{\infty} \exp\left\{-\beta\omega\left(n + \frac{1}{2}\right)\right\} = \frac{1}{2\sinh\frac{\beta\omega}{2}} . \tag{9.61}$$

At large temperatures $T = \beta^{-1}$,

$$Z_{T\gg\omega} \approx \frac{T}{\omega} . \tag{9.62}$$

The latter result can also be obtained semiclassically

$$Z_{\text{high T}} \approx Z_{\text{semicl}} = \int \frac{dpdx}{2\pi} \exp\left\{-\frac{\beta}{2}(p^2 + \omega^2 x^2)\right\} = \frac{T}{\omega} . \tag{9.63}$$

The mean energy is

$$\langle\mathcal{E}\rangle_T = -\frac{\partial}{\partial\beta}\ln Z = \frac{\omega}{2\tanh\frac{\beta\omega}{2}} . \tag{9.64}$$

At low temperatures, $\langle\mathcal{E}\rangle_T \approx \omega/2 + O(e^{-\beta\omega})$, while at high temperatures,

$$\langle\mathcal{E}\rangle_{T\gg\omega} \approx T . \tag{9.65}$$

The behaviour $\langle\mathcal{E}\rangle_{\text{high T}} \propto T$ is characteristic not only for the oscillator, but for *any* reasonable QM system. Basically, it is the analog of the Stefan-Boltzmann law in zero spatial dimensions. Consider e.g. the Hamiltonian $\hat{H} = \hat{p}^2/2$ describing free one-dimensional motion. The partition function is

$$Z = \int \frac{dpdx}{2\pi} e^{-\beta p^2/2} = L\sqrt{\frac{T}{2\pi}} , \tag{9.66}$$

where L is the length of the box. In the limit $L \to \infty$, it is infinite, but the average energy $\langle \mathcal{E} \rangle_T$ defined as in (9.64) does not depend on the infinite factor L (playing the role of the infrared regulator) and is equal to $T/2$. The factor multiplying T in the high–temperature estimate for $\langle \mathcal{E} \rangle_T$ is just the number #$_{\text{d.f.}}$ of degrees of freedom for a finite motion and #$_{\text{d.f.}}/2$ for an infinite motion (no potential energy there).

Consider now a nonsupersymmetric $su(N)$ matrix model described by the Hamiltonian

$$\hat{H} = \frac{1}{2}(\hat{E}_I^a)^2 + \frac{g^2}{4}(f^{abc}A_I^a A_J^b)^2. \tag{9.67}$$

There are classical valleys (9.23), but they are locked by quantum corrections so that the spectrum is discrete. To estimate the energy $\mathcal{E}_0$ of the ground state, take the simplest variational Ansatz:

$$\Psi_0 \propto \exp\left\{-\alpha(A_I^a)^2\right\}. \tag{9.68}$$

Then, comparing the characteristic potential and kinetic energies, we find

$$\alpha \sim \lambda^{1/3}, \qquad \mathcal{E}_0 \sim N^2\lambda^{1/3} \sim N^2\alpha. \tag{9.69}$$

In other words, the system has only one natural energy scale $\alpha \sim \lambda^{1/3}$ with λ defined in (9.57). It also defines a characteristic size of the wave functions:

$$\left(A^2\right)_{\text{char}} \sim \frac{1}{\mathcal{E}_{\text{char}}} \sim \frac{1}{\alpha}. \tag{9.70}$$

The gap between the first excited state and the vacuum state is also of order $\mathcal{E}_{\text{char}}$.

The mean internal energy follows the same pattern as for the harmonic oscillator: it is equal to $\mathcal{E}_0$ plus exponentially small corrections at $T \ll \alpha$ and is of order N^2T when $T \gtrsim \alpha$. This pattern was confirmed by numerical calculations [164].

9.4.2   *Supersymmetric models*

In contrast to bosonic YM models, in all SYM models ($\mathcal{N} = 4$, $\mathcal{N} = 8$, and $\mathcal{N} = 16$), the spectrum has a continuous component associated with the flat directions (9.23). The number of continuous degrees of freedom is equal to

$$\frac{(\mathcal{N} + 2)(N - 1)}{2}$$

and grows $\propto N$ with N. The partition function of these models involves the contribution

$$Z_{\text{cont}} = \left(\frac{T}{\mu}\right)^{(\mathcal{N}+2)(N-1)/4}, \tag{9.71}$$

where μ is the infrared cutoff.

But the spectrum also has a discrete component with the localized states associated with finite classical trajectories. In $\mathcal{N} = 4$ and $\mathcal{N} = 8$ theories, these states are expected to have the characteristic energies of order α, and their contribution to the partition function behaves in the same way as the partition function of the purely bosonic YM system: it is exponentially suppressed at $T \ll \alpha$ and is of order

$$Z_{\text{discrete}} \sim \left(\frac{T}{\alpha}\right)^{cN^2} \quad \text{if} \quad T \gtrsim \alpha. \tag{9.72}$$

Consider the sum $Z = Z_{\text{cont}} + Z_{\text{discrete}}$. In the limit $\mu \to 0$ at fixed N, the continuum contribution dominates both at low and at high temperatures and the mean internal energy is just

$$\langle \mathcal{E} \rangle_T = \frac{(\mathcal{N}+2)(N-1)}{4} T, \tag{9.73}$$

which has nothing to do with (9.60).

But if μ is fixed and we send N to infinity, the continuum contribution still dominates at low temperatures, but at high temperatures Z_{discrete} grows much faster than Z_{cont} and dominates. Suppose that we managed to filter out the continuum contribution.[19] Then the mean internal energy is predicted to behave in the same way as for the bosonic system: it grows $\propto N^2 T$ at $T \gtrsim \alpha$ and is frozen at $\mathcal{E}_0$ at $T \ll \alpha$. And this prediction was confirmed in numerical experiment [165].

As we know from the previous discussion, the $\mathcal{N} = 16$ model has one important feature that distinguishes it from the models with lower $\mathcal{N}$: the existence of a normalized state with zero energy, so that $\mathcal{E}_0 = 0$. But this alone does not affect thermodynamics so much. To understand in the

---

[19]This is *exactly* what was effectively done in numerical calculations. The functional integral for Z was evaluated by Metropolis algorithm with initial values of all components A_I^a chosen to be of the same order and not very large. It was then observed that, for small N, this configuration is unstable, so that the field variables tend to grow along the flat directions. However, for larger N, the system penetrates the valley only after a considerable number of iterations. The larger is N and/or T, the more stable is the system. An effective barrier is erected. For large N, one can thus evaluate the averages *before* the system penetrates through this barrier and escapes along the valley.

Hamiltonian language the dependence (9.60), we are making the following conjecture [166]:

In the limit of large N, the $\mathcal{N} = 16$ system enjoys, on top of the scale α, another much smaller scale $\sim N^{-5/9}\alpha$, and the spectrum includes many localized states having this characteristic energy and greater, but smaller than α. These states contribute in the partition function. As a result, the mean thermal energy can exhibit the behaviour (9.60) in the intermediate temperature region,

$$N^{-5/9}\alpha < T < \alpha. \tag{9.74}$$

Let us explain where this new energy scale may come from. In the leading BO approximation, the effective valley Hamiltonian is

$$\hat{H}^{\text{eff}} = \frac{1}{2}\hat{\boldsymbol{E}}^a\hat{\boldsymbol{E}}^a, \tag{9.75}$$

where $a = 1,\ldots,N-1$. What are the subleading terms? In $\mathcal{N} = 4$ and $\mathcal{N} = 8$ theories, a nontrivial metric in the space of A_f^a appears, these corrections becoming important when $g|A|^3 \sim N$ [86]. In $\mathcal{N} = 16$ theory, there is no corrections to the metric[20] and the leading corrections to the Hamiltonian are of order $\hat{E}^4$. The exact form of these corrections was found for $SU(2)$ in [169] and for $SU(N)$ in [170]:

$$\boxed{\hat{H}_{\text{eff}} = \sum_{k=1}^{N}(\hat{\boldsymbol{E}}_k)^2 + \frac{15}{16}\sum_{k>l}^{N}\frac{[(\hat{\boldsymbol{E}}_k - \hat{\boldsymbol{E}}_l)^2]^2}{g^3|\boldsymbol{A}_k - \boldsymbol{A}_l|^7} + \ldots,} \tag{9.76}$$

where $\boldsymbol{A}_k$ and $\hat{\boldsymbol{E}}_k$ are the entries of the corresponding diagonal $su(N)$ matrices. The second term in (9.76) represents the sum over all positive roots of $su(N)$. For large N, it involves $\sim N^2$ terms, while the first term has N terms. The estimate for A_{char} is obtained from the condition

$$\frac{N}{A_{\text{char}}^2} \sim \frac{N^2}{g^3 A_{\text{char}}^{11}}.$$

We have

$$A_{\text{char}}^2 \sim \frac{N^{2/9}}{g^{2/3}} \sim \frac{N^{5/9}}{\alpha}. \tag{9.77}$$

In Ref. [170], also two-loop corrections to the effective Hamiltonian were evaluated. They are estimated as $\sim N^3 E^6/(g^6 A^{14})$ and are of the same

---

[20]This result obtained first in [167] has much in common with $4D$ nonrenormalization theorems [168].

order as the leading term at the scale (9.77). One can conjecture that the corrections at the k-loop level have the order

$$\hat{H}_{\text{eff}}^k \sim \frac{(NE^2)^{k+1}}{(g^3 A^7)^k} \sim \frac{N^{k+1}}{g^{3k} A^{9k+2}},$$

so that *all* the terms in the effective Hamiltonian have the same order at this scale.

We see that the estimated size of the vacuum wave function turns out to be essentially *larger* than the characteristic size (9.70) of bosonic eigenstates. The large characteristic size (9.77) suggests the existence of the energy scale

$$\boxed{\mathcal{E}_{\text{char}}^{\text{new}} \sim N^{-5/9} \alpha,} \qquad (9.78)$$

which is considerably *smaller* than the principal energy scale α.

An assumption that a rich spectrum of the localized states with the energies lying in the interval (9.74) exists allows one to explain the result (9.60). Indeed:

- The dependence (9.60) implies that, at $T \sim \lambda^{1/3} = \alpha$, $\langle \mathcal{E} \rangle_T \sim N^2 \alpha$. This is natural as the physics at this large scale is the same as in the pure YM model or in $\mathcal{N} = 4$ and $\mathcal{N} = 8$ SYM models: the system has $\sim N^2$ degrees of freedom and each of them contributes $\sim \alpha$ in $\langle \mathcal{E} \rangle_T$. At still larger temperatures, the partition function behaves as in Eq.(9.72), which gives $\langle \mathcal{E} \rangle_T \sim N^2 T$.
- At the lower edge of the interval (9.74), $\langle \mathcal{E} \rangle_T \sim N^{4/9} \alpha \sim N \mathcal{E}_{\text{char}}^{\text{new}}$. This can be understood by noticing that, at still smaller energies, the contribution of both "old" and "new" discrete spectrum states becomes exponentially suppressed and the continuum contribution takes over. At still lower temperatures, $\langle \mathcal{E} \rangle_T \sim NT$ as in Eq.(9.73).[21]
- The law Eq.(9.60) is a natural power interpolation between the estimates for $\langle \mathcal{E} \rangle_T$ at the two edges of (9.74). It implies that the partition function grows in this intermediate region with temperature as

$$Z(T) \propto \exp\left\{ N^2 \left(\frac{T}{\alpha} \right)^{9/5} \right\}. \qquad (9.79)$$

Bearing in mind that

$$Z(T) \sim \int_{N^{-5/9}\alpha}^{\alpha} \rho(\mathcal{E}) e^{-\mathcal{E}/T} d\mathcal{E}, \qquad (9.80)$$

---

[21] This works especially well if the infrared cutoff μ is assumed to be of order $\mathcal{E}_{\text{char}}^{\text{new}}$. Do not ask me for a justification!

where $\rho(\mathcal{E})$ is the spectral density of localized states, we deduce that

$$\rho(\mathcal{E}) \ \propto \ \exp\left\{N^2\left(\frac{\mathcal{E}}{\alpha}\right)^{9/5}\right\}. \tag{9.81}$$

To summarize, the full gauge SQM prediction for the temperature dependence of the mean internal energy is

$$\langle\mathcal{E}\rangle_T \sim NT \qquad \text{if} \qquad T \lesssim N^{-5/9}\alpha,$$

$$\langle\mathcal{E}\rangle_T \sim N^2\alpha\left(\frac{T}{\alpha}\right)^{14/5} \qquad \text{if} \qquad N^{-5/9}\alpha \lesssim T \lesssim \alpha,$$

$$\langle\mathcal{E}\rangle_T \sim N^2 T \qquad \text{if} \qquad T \gtrsim \alpha. \tag{9.82}$$

We conclude this section and this chapter with two remarks.

(1) Remarkably, the scale (9.78) also shows up on the supergravity side. It is the temperature (the inverse of the compactified circle) where the *black string* solution in classical $10D$ supergravity becomes unstable[22] and transforms to the black hole [172].

(2) On the other hand, one also sees another scale on the supergravity side:

$$\mathcal{E}_{\text{string}} \sim N^{-10/21}\alpha. \tag{9.83}$$

At $T \lesssim \mathcal{E}_{\text{string}}$, the corrections due to string loops become large [173] and invalidate the prediction (9.60). Numerically, $\frac{10}{21}$ is rather close to $\frac{5}{9}$, but still string corrections that one cannot control set in when the temperature has not reached yet the low edge of the interval (9.74). Unfortunately, we see no trace of the scale (9.83) on the maximal SQM side.

---

[22]This effect is known as *Gregory-Laflamme instability* [171].

Chapter 10

Three-dimensional gauge theories

10.1 Setting the problem

Consider the Lagrangian (3.75) of $3D$ $\mathcal{N} = 1$ supersymmetric Yang-Mills-Chern-Simons theory, which we rewrite here for convenience:

$$
\boxed{
\begin{aligned}
\mathcal{L} &= \frac{1}{g^2} \left[-\frac{1}{4} F^a_{\mu\nu} F^{a\mu\nu} + \frac{i}{2} \lambda^{a\alpha} (\gamma^\mu)_{\alpha\beta} \nabla_\mu \lambda^{a\beta} \right] \\
&\quad + \frac{\kappa}{2} \left[\varepsilon^{\mu\nu\rho} \left(A^a_\mu \partial_\nu A^a_\rho + \frac{1}{3} f^{abc} A^a_\mu A^b_\nu A^c_\rho \right) + i \lambda^{a\alpha} \lambda^a_\alpha \right]
\end{aligned}
}
\tag{10.1}
$$

with $\varepsilon^{012} = 1$ and real and symmetric

$$
(\gamma^\mu)_{\alpha\beta} = (\mathbb{1}, \sigma^1, \sigma^3)_{\alpha\beta} . \tag{10.2}
$$

Here $A^a_\mu t^a$ and $\lambda^a t^a$ belong to a simple Lie algebra $\mathfrak{g}$. We will be mostly interested in the case $\mathfrak{g} = su(N)$. The first line in Eq. (10.1) is the SYM Lagrangian in three dimensions. The coupling constant g^2 carries the dimension of mass. In the second line, we see a supersymmetric version of the Chern-Simons Lagrangian. The constant κ is dimensionless.

Taken in isolation, the Chern-Simons Lagrangian describes a *topological* theory [174]. That means that the Chern-Simons action represents a topological invariant (3.79). And the canonical Hamiltonian corresponding to the pure CS Lagrangian *vanishes*. Such a theory does not have classical dynamics, and, in the spectrum of its quantum version, there is only a finite number of states annihilated by the gauge constraints,

$$
\hat{G}^a \Psi = 0 . \tag{10.3}
$$

However, the presence of the Yang-Mills term in the Lagrangian makes the theory dynamical and rather nontrivial. The physical boson and fermion excitations in this theory carry the mass $m = \kappa g^2$. The parity transformation flips the sign of mass, so that a theory with a given nonzero κ breaks parity. As was mentioned in Chapter 3.7, if we require the invariance of the exponential e^{iS} of the classical action under a large (non-contractible) gauge transformation, κ should be quantized:

$$\kappa = \frac{k}{4\pi} \tag{10.4}$$

with integer or sometimes half-integer (see below) k.

The Witten index in this theory was evaluated in Ref. [175]. For the $SU(N)$ gauge group, the result is

$$\boxed{I(k, N) = [\operatorname{sgn}(k)]^{N-1} \binom{|k| + N/2 - 1}{N - 1} \,.} \tag{10.5}$$

This is valid for $|k| \geq N/2$. For $|k| < N/2$, the index vanishes and supersymmetry is probably broken. In the simplest $SU(2)$ case, the index is just

$$I(k, 2) = k \,. \tag{10.6}$$

For $SU(3)$, it is

$$I(k, 3) = \frac{k^2 - 1/4}{2} \,, \tag{10.7}$$

We can now notice that, for the index to be integer, the level k should be half-integer rather than integer for $SU(3)$ and for all unitary groups with odd N. The explanation is that, in these cases, the large gauge transformation mentioned above not only shifts the classical action, but also contributes the extra factor (-1) in the functional integral due to the modification of the fermion determinant [45, 46].

The result (10.5) was derived in [175] by the following reasoning. Consider the theory in a *large* spatial box, $g^2 L \gg 1$. Consider then the functional integral for the index and mentally perform a Gaussian integration over fermionic variables. This gives an effective bosonic action that involves the Chern-Simons term, the Yang-Mills term, and other higher-derivative gauge-invariant terms. The fermion loop renormalizes the coefficient of the CS term:[1]

$$k \to k - \frac{N}{2} \,. \tag{10.8}$$

---

[1]This relation (to be derived later) holds for $k > 0$. In what follows, k will be assumed to be positive by default, although the results for negative k will also be mentioned.

We are interested in the vacuum states, and their dynamics depends on the term with the lowest number of derivatives in the effective Lagrangian, i.e. the CS term, the effects due to the YM term and still higher derivative terms being suppressed at small energies and a large spatial volume. Basically, the spectrum of vacuum states coincides with the full spectrum in the topological pure CS theory. The latter was determined by canonical quantization of the CS theory and by explicitly solving the Gauss law constraints (10.3) [176, 177].

The index (10.5) is then determined as the number of states in pure CS theory with the shift (10.8). For example, in the $SU(2)$ case, the number of CS states is $k + 1$, which gives (10.6) after the shift.

10.2 SYMCS theory in a small box

The method described above is somewhat indirect. It is interesting to reproduce the result (10.5) using the same direct method as the one applied for $4D$ SYM theories in Ref. [47] and laid out in Chapter 6. This was done in Ref. [178] (see also a minireview [179]). We choose the spatial box to be *small* rather than large, $g^2 L \ll 1$, and study the dynamics of the corresponding Born-Oppenheimer Hamiltonian.

Take the simplest $SU(2)$ theory and impose the periodic boundary conditions for all the fields. Then, similar to what we had in four dimensions, the slow variables in the effective BO Hamiltonian are the zero Fourier modes of the gauge potential lying in the Cartan subalgebra of $su(2)$ and of its holomorphic superpartner:

$$c_{j=1,2} = A_j^{(0)3}, \qquad \eta = \lambda_1^{(0)3} - i\lambda_2^{(0)3}. \tag{10.9}$$

The variables c_j lie on the dual torus,

$$c_j \in \left[0, \frac{4\pi}{L}\right) : \tag{10.10}$$

the configurations c_j and $c_j + 4\pi/L$ are equivalent under a contractible gauge transformation.

10.2.1 *Tree approximation*

10.2.1.1 $SU(2)$

In four dimensions, the leading order BO Hamiltonian was very simple being given by Eq. (6.53). In the $SU(2)$ case, it was just the Laplacian. And the vacuum wave functions (6.54) did not depend on c_j. In three dimensions, the presence of the CS term makes the problem more complicated. The effective supercharges and Hamiltonian have the following form [180]:[2]

$$\hat{Q}^{\text{eff}} = \frac{g}{L}\eta(\hat{P}_- + \mathcal{A}_-)$$

$$\hat{\bar{Q}}^{\text{eff}} = \frac{g}{L}\hat{\bar{\eta}}(\hat{P}_+ + \mathcal{A}_+), \tag{10.11}$$

$$\hat{H}^{\text{eff}} = \frac{g^2}{2L^2}\left[\left(\hat{P}_j + \mathcal{A}_j\right)^2 + \mathcal{B}(\eta\hat{\bar{\eta}} - \hat{\bar{\eta}}\eta)\right], \tag{10.12}$$

where

$$\mathcal{A}_j = -\frac{\kappa L^2}{2}\epsilon_{jk}c_k, \tag{10.13}$$

$\hat{P}_\pm = \hat{P}_1 \pm i\hat{P}_2 = -i\left(\frac{\partial}{\partial c_1} \pm i\frac{\partial}{\partial c_2}\right)$, $\hat{\bar{\eta}} = \frac{\partial}{\partial \eta}$, $\mathcal{A}_\pm = \mathcal{A}_1 \pm i\mathcal{A}_2$, and $\mathcal{B} = \frac{\partial \mathcal{A}_2}{\partial c_1} - \frac{\partial \mathcal{A}_1}{\partial c_2} = \kappa L^2$. The bosonic part of the Hamiltonian describes the motion on the dual torus in the presence of a uniform magnetic field. The quantization (10.4) of κ means that the flux of the magnetic field on the dual torus[3] is equal to

$$\Phi = \mathcal{B}\left(\frac{4\pi}{L}\right)^2 = 4\pi k. \tag{10.14}$$

According to what we learned in Chapter 4 [see Eq. (4.61)], the Witten index of the Hamiltonian (10.12) is equal to $\Phi/(2\pi) = 2k$. However, it is not a correct answer for the original field theory for two reasons:

(1) Only *Weyl-invariant* wave functions are admissible.
(2) In contrast to what we had in four dimensions, it is not sufficient in this case to restrict oneself by the analysis of the tree BO Hamiltonian, but it is necessary to take into account one-loop effects.

We will explain here how the Weyl invariance requirement (following from the gauge invariance of the original theory) affects the index, and come to grips with the loops in Sect. 10.2.2.

A distinguishing feature of the 3D problem compared to 4D problem is that the wave functions are not just invariant under the shifts along the cycles of the dual torus, but acquire extra phase factors,

$$\Psi(X + 1, Y) = e^{-2\pi ikY}\Psi(X, Y),$$

$$\Psi(X, Y + 1) = e^{2\pi ikX}\Psi(X, Y), \tag{10.15}$$

where $X = c_1 L/(4\pi)$ and $Y = c_2 L/(4\pi)$.

---

[2]In this paper, one should read only the first part including the tree-level analysis. The assertion in the second part that not only the fermion, but also the gluon loops contribute to the renormalization of k was not correct. It was corrected in [178].

[3]It has nothing to do, of course, with the physical magnetic field.

Let us explain where these factors come from. As was mentioned, the shifts $X \to X + 1$ and $Y \to Y + 1$ represent contractible gauge transformations. In the YM and SYM theories, wave functions are invariant under such transformations. But the YMCS theory is special in this respect. Indeed, the Gauss law constraint in the YMCS theory involve not just $D_j \Pi_j^a$, but rather [181]

$$G^a = \frac{\delta \mathcal{L}}{\delta A_0^a} = D_j \Pi_j^a + \frac{\kappa}{2} \epsilon_{jk} \partial_j A_k^a , \qquad (10.16)$$

where $\Pi_j^a = F_{0j}^a / g^2 + (\kappa/2)\epsilon_{jk} A_k^a$ are the canonical momenta. The wave functions $\Psi[A_j^a(\boldsymbol{x})]$ (the dependence on fermion variables is irrelevant here) are annihilated by the action of $\hat{G}^a$. The second term in (10.16) gives rise to the following shift of $\Psi[A]$ associated with an infinitesimal gauge transformation[4] $\delta A_j^a(\mathbf{x}) = D_j \alpha^a(\mathbf{x})$:

$$\delta \Psi[A] = \int d^2 x \, D_j \alpha^a(\boldsymbol{x}) i \hat{\Pi}_j^a \Psi[A] = -i \int d^2 x \, \alpha^a(\boldsymbol{x}) D_j \hat{\Pi}_j^a \Psi[A]$$

$$= \frac{i\kappa}{2} \epsilon_{jk} \int d^2 x \, \alpha^a(\boldsymbol{x}) \partial_j A_k^a \Psi[A] = -\frac{i\kappa}{2} \int d^2 x \, \epsilon_{jk} \partial_j \alpha^a(\boldsymbol{x}) A_k^a(\boldsymbol{x}) \Psi[A].$$
$$(10.17)$$

For the finite contractible gauge transformations $\alpha^a = (4\pi x/L)\delta^{a3}$ and $\alpha^a = (4\pi y/L)\delta^{a3}$ which implement the shifts $c_{1,2} \to c_{1,2} + 4\pi/L$, the effective wave function transforms as in Eq. (10.15).

The phase factors thus obtained are nothing but the holonomies $\mathcal{E}_1 = \exp\left\{i \int_{\gamma_1} \mathcal{A}_1 \, dc_1\right\}$ and $\mathcal{E}_2 = \exp\left\{i \int_{\gamma_2} \mathcal{A}_2 \, dc_2\right\}$, with $\gamma_{1,2}$ being two cycles of the torus attached to the point (X, Y). The property

$$\mathcal{E}_1(X,Y)\mathcal{E}_2(X+1,Y)\mathcal{E}_1^{-1}(X,Y+1)\mathcal{E}_2^{-1}(X,Y) = e^{4\pi i k} = 1 \quad (10.18)$$

holds. The phase $4\pi k$ that one acquires by going around the sequence of two direct and two inverse cycles is nothing but the magnetic flux Φ on our dual torus. As we see, the condition (10.18) (a necessary condition for the wave functions to be uniquely defined) dictates the quantization of the flux.

The eigenfunctions of the Hamiltonian (10.12) satisfying the boundary conditions (10.15) are given by elliptic functions — a variety of theta functions. There are $\Phi/(2\pi) = 2k$ ground state wave functions. For $k > 0$, their explicit form is

$$\boxed{\Psi_{\text{tree}}^{\text{eff}}(X, Y) \propto e^{-\pi k z^* z} e^{\pi k (z^*)^2} Q_m^{2k}(z^*),} \qquad (10.19)$$

---

[4]The spatial coordinates $\mathbf{x}$ are not to be confused with the rescaled vector potentials X, Y.

where $z = X + iY$, $m = 0, \ldots, 2k - 1$, and the functions Q_m^q are defined in Eq. (B.5). The functions (10.19) are annihilated by the supercharge $\hat{Q}^{\text{eff}} \propto \hat{\bar{\eta}}$ and also by

$$\hat{Q}^{\text{eff}} \propto \hat{P}_- + \mathcal{A}_- \propto \frac{\partial}{\partial z} + \pi k z^* .$$

For negative k, the vacuum functions have the same form, except that z and z^* are interchanged and the functions acquire the extra fermionic factor η.

Not all $2|k|$ states satisfy the additional Weyl invariance condition and are admissible. For $SU(2)$, this condition amounts to[5] $\Psi^{\text{eff}}(-c_j) = \Psi^{\text{eff}}(c_j)$, which singles out $|k| + 1$ vacuum states, bosonic for $k > 0$ and fermionic for $k < 0$.

When $k = 0$, the effective Hamiltonian (10.12) describes free motion on the dual torus. There are two zero energy ground states: $\Psi^{\text{eff}} = \text{const}$ and $\Psi^{\text{eff}} = \text{const} \cdot \eta$ (we need not bother about the Weyl oddness of the factor η by the same reason as above). The index is zero.[6]

We thus derive

$$I_{SU(2)}^{\text{tree}} = (|k| + 1)\,\text{sgn}(k). \tag{10.20}$$

10.2.1.2 *Higher unitary and some other groups*

The effective Hamiltonian for the group $SU(N)$ involves $2r = 2(N - 1)$ slow bosonic variables c_j^a and $r = N - 1$ slow complex fermionic variables η^a belonging to the Cartan subalgebra of $su(N)$. In the orthogonal basis, it has the form

$$\hat{H} = \frac{g^2}{2L^2} \left[(\hat{P}_j^a + \mathcal{A}_j^a)^2 + \mathcal{B}(\eta^a \hat{\bar{\eta}}^a - \hat{\bar{\eta}}^a \eta^a) \right] , \tag{10.21}$$

---

[5]In contrast to what should be done in 4 dimensions, we did not include here the Weyl reflection of the fermion factor η entering the effective wave function for negative k. The reason is that, for negative k, the conveniently defined *fast* wave function (by which the effective wave function depending only on c_j and η should be multiplied) involves the Weyl-odd factor $c_1 + ic_2$. This oddness compensates the oddness of the factor η in the effective wave function [180].

[6]This *suggests* that supersymmetry is broken spontaneously at $k = 0$ and the ground states in the full Hamiltonian have positive energy but, as was discussed in detail in Chapter 4, the vanishing of the index is not a *sufficient* condition for the breaking. An additional study is required to find out whether the zero-energy ground states of the Hamiltonian (10.12) at $k = 0$ are shifted from zero or not when the corrections in the Born-Oppenheimer parameter to the effective Hamiltonian are taken into account.

Additional arguments in favor of the scenario with spontaneous breaking of supersymmetry at $k = 0$, based on Witten's approach when $g^2 L$ is assumed to be large and the theory becomes topological, were given in Ref. [182].

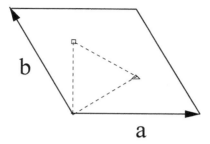

Fig. 10.1: Maximal torus and Weyl alcove for $SU(3)$. $\boldsymbol{a}$ and $\boldsymbol{b}$ are the simple coroots; we normalize here their length to one. The points $\triangle$ and $\square$ are the corresponding fundamental coweights.

where $a = 1, \ldots, r$;

$$\mathcal{A}_j^a = -\frac{\kappa L^2}{2}\epsilon_{jk}c_k^a \tag{10.22}$$

and $\mathcal{B} = \kappa L^2$ as before. By the same token as in the $SU(2)$ case, the motion is finite and extends over the manifold $\mathcal{M}_n = T_r \times T_r$, with T_r being the maximal torus of the $SU(N)$ group.

The Witten index of the effective Hamiltonian can be evaluated semi-classically by the CG method that reduces the functional integral for the index to an ordinary one. The latter represents a *generalized* magnetic flux (in mathematical language, it is the r-th Chern class of the $[U(1)]^r$ bundle over $\mathcal{M}_n$ with the connection $\mathcal{A}_j^a$):

$$I = \frac{\mathcal{B}^r}{(2\pi)^r}\int_{T_r \times T_r} \prod_{ja} dc_j^a . \tag{10.23}$$

The variables c_j^a [or rather $w_j^a = Lc_j^a/(4\pi)$ — the Cartesian coordinates of the rhombus in Fig. 10.1] belong to T_r. The area of T_2 depicted[7] in Fig. (10.1) is $\sqrt{3}/2$. Taking into account the factor $(4\pi/L)^4$, we derive

$$I^{SU(3)} = \frac{\mathcal{B}^2}{4\pi^2}\frac{3}{4}\left(\frac{4\pi}{L}\right)^4 = 3k^2 . \tag{10.24}$$

For $SU(N)$:

$$I^{SU(N)} = Nk^{N-1} . \tag{10.25}$$

But not all Nk^{N-1} vacuum states of (10.21) are Weyl-invariant.

---

[7]We have already drawn this figure in Chapter 6 (Fig. 6.1 there), but, not wishing to urge the reader to keep scrolling back and forth many pages, we are redrawing it here.

Let us show how to count Weyl-invariant states for $N = 3$. Look at Fig. 10.1. Each point (w^3, w^8) there is a *coweight* such that the corresponding group element in the maximal torus is $g^{T_2} = \exp\{4\pi i(w^3 t^3 + w^8 t^8)\}$. The dashed lines delimit the Weyl alcove — the maximal torus factorized over the Weyl group. The vertices of the alcove — the origin $(0,0)$ and two special points $\Box(0, \frac{1}{\sqrt{3}})$ and $\triangle(\frac{1}{2}, \frac{1}{2\sqrt{3}})$ (the fundamental coweights) — correspond to the center group elements $g = \mathbb{1}$ and $g = e^{\pm 2i\pi/3}\mathbb{1}$. The identities

$$\triangle \cdot \boldsymbol{a} = \Box \cdot \boldsymbol{b} = 1/2, \qquad \Box \cdot \boldsymbol{a} = \triangle \cdot \boldsymbol{b} = 0 \tag{10.26}$$

hold.

We will write now the explicit expressions for the $3k^2$ vacuum wave functions for the Hamiltonian (10.21) in the case of $SU(3)$. They are given by generalized theta functions defined on the coroot lattice of $SU(3)$. They satisfy the boundary conditions

$$\Psi(\boldsymbol{X} + \boldsymbol{a}, \boldsymbol{Y}) = e^{-2\pi i k\, \boldsymbol{a}\cdot \boldsymbol{Y}}\, \Psi(\boldsymbol{X}, \boldsymbol{Y}),$$
$$\Psi(\boldsymbol{X} + \boldsymbol{b}, \boldsymbol{Y}) = e^{-2\pi i k\, \boldsymbol{b}\cdot \boldsymbol{Y}}\, \Psi(\boldsymbol{X}, \boldsymbol{Y}),$$
$$\Psi(\boldsymbol{X}, \boldsymbol{Y} + \boldsymbol{a}) = e^{2\pi i k\, \boldsymbol{a}\cdot \boldsymbol{X}}\, \Psi(\boldsymbol{X}, \boldsymbol{Y}),$$
$$\Psi(\boldsymbol{X}, \boldsymbol{Y} + \boldsymbol{b}) = e^{2\pi i k\, \boldsymbol{b}\cdot \boldsymbol{X}}\, \Psi(\boldsymbol{X}, \boldsymbol{Y}), \tag{10.27}$$

where $\boldsymbol{X} = 4\pi \boldsymbol{c}_1/L$, $\boldsymbol{Y} = 4\pi \boldsymbol{c}_2/L$ and $\boldsymbol{a} = (1, 0)$, $\boldsymbol{b} = (-1/2, \sqrt{3}/2)$ are the simple coroots in the orthogonal basis. When $k = 1$, there are three such states:

$$\Psi_0 = \sum_{\boldsymbol{m}} \exp\left\{-2\pi \left[(\boldsymbol{m} + \boldsymbol{Y})^2 + i\boldsymbol{X} \cdot \boldsymbol{Y} + 2i\boldsymbol{X} \cdot \boldsymbol{m}\right]\right\},$$

$$\Psi_{\triangle} = \sum_{\boldsymbol{m}} \exp\left\{-2\pi \left[(\boldsymbol{m} + \boldsymbol{Y} + \triangle)^2 + i\boldsymbol{X} \cdot \boldsymbol{Y} + 2i\boldsymbol{X} \cdot (\boldsymbol{m} + \triangle)\right]\right\},$$

$$\Psi_{\Box} = \sum_{\boldsymbol{m}} \exp\left\{-2\pi \left[(\boldsymbol{m} + \boldsymbol{Y} + \Box)^2 + i\boldsymbol{X} \cdot \boldsymbol{Y} + 2i\boldsymbol{X} \cdot (\boldsymbol{m} + \Box)\right]\right\},$$

$$\tag{10.28}$$

where the sums run over the coroot lattice, $\boldsymbol{m} = m_a \boldsymbol{a} + m_b \boldsymbol{b}$ with integer $m_{a,b}$.

The functions (10.28) are Weyl-invariant because both the nodes of the coroot lattice and the points $\triangle$ and $\Box$ are Weyl-invariant.[8] But for $k > 1$,

---

[8]Geometrically, the elements of the Weyl group correspond to the reflections with respect to the dashed lines in Fig. 10.1 and compositions of such reflections.

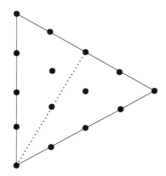

Fig. 10.2: $SU(3)$: 15 vacuum states for $k = 4$. The dotted line marks the boundary of the Weyl alcove for G_2.

the number of invariant states is less than $3k^2$. For an arbitrary k, the wave functions of all $3k^2$ eigenstates can be written in a similar way to (10.28):

$$\Psi_l = \sum_m \exp\{-2\pi k\left[(\boldsymbol{m} + \boldsymbol{Y} + \boldsymbol{w}_l)^2 + i\boldsymbol{X} \cdot \boldsymbol{Y} + 2i\boldsymbol{X} \cdot (\boldsymbol{m} + \boldsymbol{w}_l)\right]\},$$

(10.29)

where $\boldsymbol{w}_l$ are the coweights whose projections on the simple coroots $\boldsymbol{a}$, $\boldsymbol{b}$ are integer multiples of $1/(2k)$. Only the functions (10.29) with $\boldsymbol{w}_l$ lying on the borders of the Weyl alcove are Weyl invariant. For all other $\boldsymbol{w}_l$, one should construct Weyl invariant combinations

$$\Psi = \sum_{\hat{x} \in W} \hat{x}\Psi_{\boldsymbol{w}_l}.$$

(10.30)

As a result, the number of Weyl invariant states is equal to the number of the coweights $\boldsymbol{w}_l$ lying within the Weyl alcove (including the boundaries). For example, in the case $k = 4$, there are 15 such coweights shown in Fig. 10.2 and, accordingly, 15 vacuum states.

For a generic k, the number of the states is

$$\boxed{I^{\text{tree}}_{SU(3)}(k > 0) = \sum_{m=1}^{k+1} m = \frac{(k+1)(k+2)}{2}.}$$

(10.31)

The analysis for $SU(4)$ is similar. The Weyl alcove is the tetrahedron with the vertices corresponding to central elements of $SU(4)$. A pure geometric computation gives

$$I^{\text{tree}}_{SU(4)}(k > 0) = \sum_{m=1}^{k+1} \sum_{p=1}^{m} p = \frac{(k+1)(k+2)(k+3)}{6}.$$

(10.32)

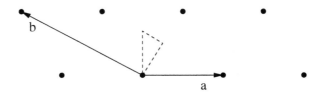

Fig. 10.3: Coroot lattice and Weyl alcove for G_2.

The generalization for an arbitrary N is obvious. It gives the result

$$I^{\text{tree}}(k, N) = \binom{k + N - 1}{N - 1} \tag{10.33}$$

We also performed a similar analysis for the symplectic groups and for G_2. Let us dwell on G_2. The simple coroots for G_2 are $\boldsymbol{a} = (1, 0)$ and $\boldsymbol{b} = (-3/2, \sqrt{3}/2)$. The lattice of coroots and the maximal torus look exactly in the same way as for $SU(3)$ (Fig. 10.3). Hence, before Weyl-invariance requirement is imposed, the index is equal to $3k^2$, as for $SU(3)$. The difference is that the Weyl group now involves 12 rather than 6 elements, and the Weyl alcove is two times smaller than for $SU(3)$. As a result, for $k = 4$, we have only 9 (rather than 15) Weyl-invariant states (see Fig.10.2). The general formula is

$$I^{\text{tree}}_{G_2}(k) = \begin{cases} \frac{(|k|+2)^2}{4} & \text{for even } k \\ \frac{(|k|+1)(|k|+3)}{4} & \text{for odd } k \end{cases} . \tag{10.34}$$

The result for $Sp(2r)$ is

$$I^{\text{tree}}(k, r) = [\text{sgn}(k)]^r \binom{|k| + r}{r} . \tag{10.35}$$

10.2.2 *Loops*

It has been known since [183] that the CS coupling κ in the pure YMCS theory is renormalized at the 1-loop level. For $\mathcal{N} = 1, 2, 3$ SYMCS theories, the corresponding calculations have been performed in [43]. The effect can be best understood by considering the fermion loop contribution to the renormalization of the structure $\propto A\partial A$ in the Chern-Simons term (see Fig.10.4). Assuming κ to be positive, we obtain

$$\Delta^{\text{ferm}}\kappa = -mc_V \int \frac{d^3 p_E}{(2\pi)^3} \frac{1}{(p_E^2 + m^2)^2} = -\frac{c_V}{8\pi}, \tag{10.36}$$

Fig. 10.4: Renormalization of the structure $\propto \epsilon^{\mu\nu\rho}\mathrm{Tr}\{A_\mu\partial_\nu A_\rho\}$ by a fermion loop.

where $m = \kappa g^2$ and $c_V \equiv h^\vee$ is the adjoint Casimir eigenvalue. That means that $\Delta k = -c_V/2$. For $SU(N)$, this coincides with (10.8).

A legitimate question is whether the second and higher loops also bring about a renormalization of the level k. The answer is negative. The proof is simple. Consider the case $k \gg c_V$. This is the perturbative regime where the loop corrections are ordered such that $\Delta k^{(1\ \mathrm{loop})} \sim O(1)$, $\Delta k^{(2\ \mathrm{loops})} \sim O(g^2/m) \sim O(1/k)$, etc. But corrections to k of order $\sim 1/k, \sim 1/k^2$, etc. are not allowed. To ensure gauge invariance, k_{ren} must be an integer. Hence, all higher loop contributions in k_{ren} must vanish, and they do.

However, k is renormalized also due to a gluon loop. The calculation gives [183] $\Delta^{\mathrm{gluon}} k = c_V$. In the large volume logic of Ref. [175], where we only have to take into account the effects due to the fermion determinant, the gluon loops play no role, but in the logic of the BO effective Hamiltonian at small box, they seem to be as relevant as the fermion ones. Taking them into account would mean that k should *increase* rather than decrease in the evaluation of the index in disagreement with [175].

To resolve this paradox, one has to accurately calculate the corrections to the BO Hamiltonian at small volume and look at their effects on the spectrum. Consider the simplest $SU(2)$ case. As was mentioned above, the coefficient κ (with the factor L^2) has the meaning of the magnetic field on the dual torus. The renormalization of κ translates into a renormalization of this magnetic field. At the tree level, the magnetic field was constant.

However, the renormalized field is not constant, but instead depends on the slow variables c_j. It can be found by evaluating the same Feynman graphs that determined renormalization of κ in the infinite volume, but now doing so in a finite torus and in the Abelian background $A_j^3 = c_j$. The integrals over the spatial momenta in Eq. (10.36) and in a similar expression for the gluon loop contribution should now be replaced by the sums:

$$\int \frac{d^2p}{(2\pi)^2} \to \frac{1}{L^2} \sum_{n_j},$$

while $p_{j=1,2}$ in the integrands should be replaced by

$$p_j \to \frac{2\pi n_j}{L} - c_j.$$

After integration over dp_0 in Eq. (10.36) and in the corresponding integral for $\Delta^{\text{bos}}\kappa$, we derive the following expressions for the induced magnetic fields:

$$\Delta\mathcal{B}^F(c_j) = -\frac{m}{2} \sum_{n_j} \frac{1}{\left[\left(c_j - \frac{2\pi n_j}{L}\right)^2 + m^2\right]^{3/2}}, \tag{10.37}$$

and

$$\Delta\mathcal{B}^B(c_j) = \frac{3m}{2} \sum_{n_j} \frac{\left(c_j - \frac{2\pi n_j}{L}\right)^2}{\left[\left(c_j - \frac{2\pi n_j}{L}\right)^2 + m^2\right]^{5/2}}. \tag{10.38}$$

For most values of c_j, the corrections (10.37), (10.38) are of order $\sim mL^3 = \kappa g^2 L^3$, which is small compared to $\mathcal{B}^{\text{tree}} \sim \kappa L^2$ if $g^2 L \ll 1$, which is our assumption. There are, however, four special points[9]

$$\boxed{c = 0, \quad c = (2\pi/L, 0), \quad c = (0, 2\pi/L), \quad c = (2\pi/L, 2\pi/L),} \tag{10.39}$$

at the vicinity of which the loop-induced magnetic field is *much larger* than the tree-level magnetic field. This actually means that the BO approximation, based on an assumption that the energy scale associated with the slow variables $\{c_j, \eta\}$ is small compared to the energy scale of the non-Abelian components and higher Fourier modes, breaks down in this region.

In fact, the loop corrections bring about effective flux lines similar to Abrikosov vortices located at the corners.[10] The width of these vortices

---

[9]We call them *corners*.

[10]Well, our problem is two-dimensional, and it is better to talk about "flux points" or imagine a flux line extending in an unphysical direction and piercing the plane of c_j.

is of the order of m. Gluon corrections generate the vortices of unite flux $\Phi^B = \frac{1}{2\pi} \int \Delta \mathcal{B}^B(c_j)\, d^2 c = 1$, while the fermion loops generate the vortices of flux $\Phi^F = -1/2$.

The crucial observation is that the vortices carrying an integer flux are *unobservable* and do not affect the spectrum! To understand this, consider a motion in the box (10.10) in the presence of a single infinitely narrow vortex of flux 1,

$$\mathcal{A}_j = -\frac{\varepsilon_{jk} c_k}{c^2}, \qquad (10.40)$$

and nothing else. Beyond the singularity at $c = 0$, the bosonic part of the Hamiltonian (10.12) acquires the form

$$\hat{H} = -\frac{g^2}{8\pi^2}\left(\frac{\partial}{\partial z^*} - \frac{1}{2z^*}\right)\left(\frac{\partial}{\partial z} + \frac{1}{2z}\right), \qquad (10.41)$$

[we remind the reader that $z = X + iY = (c_1 + ic_2)L/(4\pi)$]. It is easy to see that the spectrum of (10.41) coincides up to a common factor with the spectrum of the free Laplacian $\triangle = 4\partial\partial^*$: all the eigenstates of $\hat{H}$ are obtained from the eigenstates of $\triangle$ by multiplying the latter by $\sqrt{z^*/z} = e^{-i\phi}$. The situation is very similar to what happens for the Dirac string in the traditional description of the magnetic monopole: it is not observable [184].

A single vortex of a half-integer flux is not admissible: we would have to multiply the eigenstates of $\triangle$ by the factor $e^{-i\phi/2}$, which is not uniquely defined. But a quantum problem involving *four* such vortices is quite benign. If we disregard the integer vortices, the total flux including the tree one is $2k - 4/2 = 2(k-1)$ and that conforms to the renormalization (10.8) for $SU(2)$ theory.

This is also confirmed by a more accurate analysis. Let us *do* take into account the integer fluxes. The total flux is then $2k + 4 - 4/2 = 2(k+1)$. Now all we have to do is to solve the Schrödinger equation for the same effective Hamiltonian (10.12) as before, or just the equation

$$\hat{Q}^{\text{eff}}\Psi = 0 \qquad (10.42)$$

with the same effective supercharge (10.11) as before, but in the presence of the extra vortices in the corners, each of them carrying the flux $+1/2$. Consider the supercharge $\hat{Q}^{\text{eff}}$ at the vicinity of the vortex at the origin $c_j = 0$, but outside of its core,

$$m \ll c_j \ll 4\pi/L. \qquad (10.43)$$

The vector potential there is a half of the potential (10.40):

$$\mathcal{A}_j \approx -\frac{\epsilon_{jk} c_k}{2c^2} \qquad (10.44)$$

The equation (10.42) acquires the form

$$\left(\frac{\partial}{\partial z} + \frac{1}{4z}\right)\chi^{\text{eff}} = 0.$$ (10.45)

Its solution is

$$\Psi^{\text{eff}}(z, z^*) \sim \frac{F(z^*)}{z^{1/4}}.$$ (10.46)

The effective wave function on the entire torus can be restored from three conditions:

- it must behave as in (10.46) (with $z \to z$ or $z - \frac{1}{2}$ or $z - \frac{i}{2}$ or $z - \frac{1+i}{2}$) at the vicinity of each corner;
- it must satisfy the boundary conditions (10.18) with $k \to k+1$, which corresponds to the total flux $2k + 2$;
- the wave functions should *vanish* at the corners — then multiplication by $e^{-i\phi/2}$ does not make any harm.

The first two conditions give us the structure

$$\boxed{\Psi_m^{\text{eff}}(z, z^*) \propto \frac{Q_m^{2k+2}(z^*)}{[\Pi(z)\Pi(z^*)]^{1/4}},}$$ (10.47)

where

$$\Pi(z) = Q_3^4(z) - Q_1^4(z).$$ (10.48)

is a theta function of level 4 having zeros at the corners (10.39).[11] And the third condition dictates that not all functions $Q_m^{2k+2}(z^*)$ are admissible, but only those that have zeros right at the corners (10.39). Four out of $2k + 2$ eigenfunctions (10.47) are thereby excluded, and we are left with only $2k - 2$ vacuum states in agreement with (10.8).

In fact, the linear combinations of $Q_m^{2k+2}(z^*)$ that vanish at the points (10.39) are proportional to[12] $Q_m^{2k-2}(z^*)\Pi(z^*)$, $m = 0, \ldots, 2k - 3$. And the wave function (10.47) can be represented as

$$\Psi_m^{\text{eff}}(z, z^*) \propto Q_m^{2k-2}(z^*)\Pi^{-1/4}(z)\Pi^{3/4}(z^*).$$ (10.49)

Now, $2k - 2$ is a "pre-Weyl" index, which does not take into account the condition of Weyl invariance of the wave functions. The actual index is $I_W = (k - 1) + 1 = k$. This also holds for negative k. Indeed, the vacuum

---

[11] The function (10.48) is known from the studies of canonical quantization of pure CS theories [176, 177, 185]. It also enters the relation (B.8).

[12] This is a generalization of the relation (B.8) which may be derived in the same way — by comparing the positions of the zeroes on both sides.

states are in this case fermionic, the "pre-Weyl" index is $-(2|k|-2)$ and the actual index is $I_W = -(|k|-1+1) = -|k| = k$.

The analysis above can also be extended to the groups of higher ranks [186]. Similarly to what we saw for $SU(2)$, the gluon loops bring about certain singular gauge field configurations in the effective Hamiltonian — multidimensional analogs of the Dirac strings.[13] And they are not observable. The fermion loops bring about multidimensional strings of a *fractional* variety. The Schrödinger problem in the background of a single fractional string is ill-defined, but it is well defined for the background including several such strings [nine of them for $SU(3)$].

The fermion loops affect the index. In all the cases, it is obtained by substituting the value of k, renormalized according to $k \to k^{\text{tree}} - c_V/2$, in the tree-level expressions. We arrive at the result (10.5) for $SU(N)$.

For G_2, we obtain

$$I_{G_2} = \begin{cases} \frac{k^2}{4} & \text{for even } k \\ \frac{k^2-1}{4} & \text{for odd } k \end{cases}. \qquad (10.50)$$

And for $Sp(2r)$,

$$I_{Sp(2r)} = \begin{pmatrix} k + \frac{r-1}{2} \\ r \end{pmatrix} \qquad (10.51)$$

if k is positive (integer for odd r and half-integer for even r). For negative k, the index is restored via $I(-k) = (-1)^r I(k)$.

Let us go back to $SU(N)$ and to the result (10.5). As was already noted in Chapter 5, the formula (10.5) is *exactly the same* as the result (5.97) for the Witten index in the SQM system describing the Dirac operator living on $\mathbb{CP}^n$ with the gauge field (5.87) representing a generalization of the Dirac monopole. The expressions (5.97) and (10.5) coincide if one identifies the monopole charge q with the Chern-Simons level k and the complex dimension n of $\mathbb{CP}^n$ with the rank $N-1$ of $SU(N)$.

Such a coincidence cannot be accidental, though we cannot now pinpoint a *precise* reason for this *duality*.[14] There are, however, certain indications of the kinship of these two problems.

Consider the simplest case: the SQM including the monopole on $\mathbb{CP}^1 \simeq S^2$ vs. $SU(2)$ SYMCS theory. The former problem involves a

---

[13]The explicit expressions are presented and discussed in Ref. [186].

[14]Generically, duality can be defined as a situation when ostensively completely different theories give the coinciding answers for some relevant quantity (quantities).

homogeneous magnetic field on S^2. And the effective Hamiltonian (10.12) for the SYMCS theory involves a homogeneous magnetic field on 2-torus. In the $\mathbb{CP}^1$ case, the vacuum wave functions include the polynomials $P(z^*)$ or $P(z)$ normalized in a proper way, and to evaluate the index, one has to count such disctinct polynomials. For SYMCS, we have to count certain analytic elliptic functions — toroidal counterparts of the polynomials.

For $\mathbb{CP}^n$ with $n \geq 2$, we have to count polynomials depending on m complex arguments and to compare it with the number of generalized elliptic functions of the same complex dimension living on $T_n \times T_n$ for SYMCS.

But it is not clear by what underlying reason do these numbers exactly coincide. Especially, if one recalls that the index is directly determined by the topological invariant (5.105) for $\mathbb{CP}^n$, whereas, for SYMCS, the generalized magnetic flux (10.23) gives only the "pre-Weyl" index and we need then to impose the Weyl invariance requirement and to account for loops. The results (5.97) and (10.5) match only after that is done.

More reflections in this direction are desirable...

10.3   Theories with matter

10.3.1   *3D superspace, superfields and actions*

We did not need it before, but now let us explain how the Lagrangian (10.1) can be derived in superfield approach. Then we will introduce matter multiplets, add their contributions to the Lagrangian, go to components and calculate the indices.

We use a variant of the $\mathcal{N} = 1$ 3D superspace formalism outlined in [33].[15] The superspace (x^μ, θ^α) involves a real 2-component spinor θ^α. The spinor indices are lowered and raised with antisymmetric $\varepsilon_{\alpha\beta} = -\varepsilon^{\alpha\beta}, \varepsilon_{12} = 1$. Then the structure $\theta^2 \equiv \theta^\alpha \theta_\alpha = 2\theta^1\theta^2$ changes sign under conjugation. The 3D γ-matrices chosen as in (10.2) satisfy the identity (3.76)

The supersymmetric covariant derivatives are[16]

$$\mathcal{D}_\alpha = \frac{\partial}{\partial \theta^\alpha} - i(\gamma^\mu \theta)_\alpha \partial_\mu . \tag{10.52}$$

They satisfy the algebra $\{\mathcal{D}_\alpha, \mathcal{D}_\beta\} = -2i(\gamma^\mu)_{\alpha\beta}\partial_\mu$.

Gauge theories are described in terms of the real spinorial superfield Γ^a_α. As in 4D, one can choose the Wess-Zumino gauge reducing the number of

---

[15]There are some differences in conventions: for example our choice for the Minkowski signature is $(+ - -)$ rather than $(- + +)$.
[16]$(\gamma^\mu \theta)_\alpha \equiv (\gamma^\mu)_{\alpha\beta}\theta^\beta$.

components of Γ_α^a. In this gauge,

$$\Gamma_\alpha^a = (\gamma^\mu \theta)_\alpha A_\mu^a - i\theta^2 \lambda_\alpha^a \,. \tag{10.53}$$

The covariant superfield strength is then[17]

$$
\begin{aligned}
W_\alpha^a &= \frac{1}{2} \mathcal{D}^\beta \mathcal{D}_\alpha \Gamma_\beta^a - \frac{i}{2} f^{abc} \Gamma^{\beta b} \mathcal{D}_\beta \Gamma_\alpha^c \\
&= i\lambda_\alpha^a - \frac{i}{2} \epsilon^{\mu\nu\rho} F_{\mu\nu}^a (\gamma_\rho \theta)_\alpha + \frac{\theta^2}{2} (\gamma^\mu \nabla_\mu \lambda^a)_\alpha
\end{aligned}
\tag{10.54}
$$

with $\nabla_\mu \lambda_\alpha^a = \partial_\mu \lambda_\alpha^a + f^{abc} A_\mu^b \lambda_\alpha^c$, $F_{\mu\nu}^a = \partial_\mu A_\nu^a - \partial_\nu A_\mu^a + f^{abc} A_\mu^b A_\nu^c$.

In the superfield language, the Lagrangian (10.1) is represented as

$$
\mathcal{L} = \int d^2\theta \left[\frac{1}{4g^2} W^{\alpha a} W_\alpha^a + \frac{\kappa}{4} \left(i\Gamma^{a\alpha} W_\alpha^a + \frac{1}{6} f^{abc} \Gamma^{a\alpha} \Gamma^{b\beta} \mathcal{D}_\beta \Gamma_\alpha^c \right) \right]
\tag{10.55}
$$

with the convention $\int d^2\theta\, \theta^2 = 2$.

10.3.1.1 *Fundamental matter*

We add now matter multiplets. In the simplest case, the theory includes besides the gauge multiplet Γ_α^a a complex fundamental multiplet,

$$Q_j = q_j + \theta^\alpha \chi_{\alpha j} + \theta^2 F_j \,, \qquad \overline{Q}^j = q^{j*} + \bar{\chi}_\alpha^j \theta^\alpha - \theta^2 F^{j*} \,. \tag{10.56}$$

We add to the gauge Lagrangian (10.55) the terms

$$\mathcal{L}^{\text{fund}} = \frac{1}{2g^2} \int d^2\theta\, \overline{Q}^j \nabla^\alpha \nabla_\alpha Q_j + i\xi \int d^2\theta\, \overline{Q}^j Q_j \tag{10.57}$$

with $j = 1, 2$, $\nabla_\alpha = \mathcal{D}_\alpha - \hat{\Gamma}_\alpha$. After excluding the auxiliary fields F_j, this gives in components

$$\mathcal{L}^{\text{fund}} = -\frac{1}{g^2} (q^* \nabla^\mu \nabla_\mu q + m^2 q^* q) + \frac{i}{g^2} (\bar{\chi}^\alpha \not{\nabla}_{\alpha\beta} \chi^\beta + m \bar{\chi}^\alpha \chi_\alpha) \tag{10.58}$$

with $m = \xi g^2$, $\nabla_\mu = \partial_\mu - i\hat{A}_\mu$ and the doublet indices are not displayed.

The last term in (10.58) is a so-called *real* mass term which can be expressed in terms of $\mathcal{N} = 2$ $3D$ superfields only by adding an explicit θ-dependence in the integrand of the action [187]. To illustrate this, consider a free $3D$ theory with the Lagrangian

$$L = \frac{1}{4} \int d^2\Theta d^2\bar{\Theta}\, \overline{S} \exp\{2im\bar{\Theta}^\alpha \Theta_\alpha\} S \,, \tag{10.59}$$

---

[17]The full superfield expression quoted in Eq. (2.4.31b) in the book [33] includes also the term $\propto [\hat{\Gamma}^\beta, \{\hat{\Gamma}_\beta, \hat{\Gamma}_\alpha\}]$, but it vanishes in the WZ gauge.

where

$$S = s(x_L) + \sqrt{2}\,\Theta\psi(x_L) + \Theta^2 G(x), \quad \overline{S} = s^*(x_R) + \sqrt{2}\,\overline{\Theta}\overline{\psi}(x_R) + \overline{\Theta}^2 G^*(x)$$

with $x_{L,R}^\mu = x^\mu \mp i\Theta\gamma^\mu\overline{\Theta}$ are the $\mathcal{N} = 2$ superfields having canonical dimension $m^{1/2}$. The component form of this Lagrangian is

$$\mathcal{L} = \partial^\mu s^* \partial_\mu s + G^* G - m^2 s^* s + i\psi\,\slashed{\partial}\overline{\psi} + im\overline{\psi}^\alpha\psi_\alpha, \qquad (10.60)$$

where we may set $G = 0$. The corresponding classical Hamiltonian reads

$$H = \int d^2x \left[\pi\pi^* + (\partial_j s^*)(\partial_j s) + m^2 s^* s - im\,\varepsilon_{\alpha\beta}\psi_\alpha\overline{\psi}_\beta\right]$$

(we have put down here all the spinor indices).

The extra exponential factor in the integrand in (10.59) seems to break the $\mathcal{N} = 2$ supersymmetry of the massless theory. It turns out, however, that the symmetry is still there, though the algebra is now modified [187] and includes central charges as in (2.69).

To understand it, one can observe that the Lagrangian (10.60) (with $G = 0$) can be derived by dimensional reduction from the *massless* free $4D$ Lagrangian,

$$\mathcal{L} = \partial^\mu s^* \partial_\mu s - \frac{\partial s^*}{\partial z}\frac{\partial s}{\partial z} + i\psi\left(\slashed{\partial} - \gamma^3\frac{\partial}{\partial z}\right)\overline{\psi}, \qquad (10.61)$$

where the derivatives over the third spatial coordinate z are singled out and γ^3 is the component of the four-dimensional σ^μ that was not used up to now in our 3-dimensional description, i.e. σ^2: $(\gamma^3)_{\alpha\beta} = (\sigma^2)_{\alpha\beta} = -i\varepsilon_{\alpha\beta}$.

But it is not the ordinary dimentional reduction where the fields are assumed not to depend on z. We assume instead that the dependence is there, but it has a special form:

$$s(t, x, y; z) = s_1(t, x, y)e^{imz}, \quad \psi(t, x, y; z) = \psi_1(t, x, y)e^{imz}. \qquad (10.62)$$

Then m acquires the meaning of the *momentum* P_z of a one-particle state in the reduced direction — an eigenvalue of the operator

$$\hat{P}_z = \int dx\,dy\,dz \left[\hat{\pi}\partial_z s + \hat{\pi}^\dagger\partial_z s^* + i\psi_\alpha\partial_z\hat{\overline{\psi}}_\alpha\right]. \qquad (10.63)$$

In the $3D$ theory with the fields (10.62), it is reduced to

$$\hat{P}_z \to m\int dx\,dy\left[i(\hat{\pi}_1 s_1 - \hat{\pi}_1^\dagger s_1^*) + \psi_{1\alpha}\hat{\overline{\psi}}_{1\alpha}\right] \equiv m\hat{Z} \qquad (10.64)$$

with the operator $\hat{Z}$ having only the integer eigenvalues.

The original $4D$ supersymmetric theory had the algebra (3.2). And the reduced theory enjoys the same algebra with the operator $\hat{P}_z$ being replaced by $m\hat{Z}$. We now have

$$\{\hat{Q}_\alpha, \hat{\overline{Q}}_\beta\} = 2[\delta_{\alpha\beta}\hat{H} - (\sigma^1)_{\alpha\beta}\hat{P}_x - (\sigma^3)_{\alpha\beta}\hat{P}_y + im\varepsilon_{\alpha\beta}\hat{Z}]. \qquad (10.65)$$

The extra term in the RHS of Eq. (10.65) is a central charge. It commutes with the Hamiltonian.

The procedure we just described is akin to the Dirac quantization procedure (alias *Hamiltonian reduction*, alias *moment map*), where the Hilbert state is restricted to involve only the states satisfying some constraints. The difference is that the RHS of the constraint to be imposed on the wave function is not zero as in (10.3) and is not just a constant, but instead represents a nontrivial operator[18] $m\hat{Z}$.

The algebra (10.65) includes, as a subalgebra, an ordinary $\mathcal{N} = 1$ superalgebra with the supercharges $\hat{\mathcal{Q}} = \hat{Q}_1 + \hat{Q}_2$ and $\hat{\bar{\mathcal{Q}}} = \hat{\bar{Q}}_1 + \hat{\bar{Q}}_2$. One can also see this by looking at (10.60) and restricting ψ_α to be real there. After excluding G, G^*, it acquires the same structure as in Eq. (10.58).

Note that one can also write in $\mathcal{N} = 2$ 3D theories *complex* mass terms like $\frac{m}{4} \int d^2\theta\, S^2$ + c.c. and evaluate the indices in such theories. But we will focus on the real masses.

10.3.1.2 *Adjoint matter*

Now consider the theory with a single real adjoint multiplet,

$$\Phi^a = \phi^a + i\psi^a_\alpha \theta^\alpha + i\theta^2 D^a . \tag{10.70}$$

The gauge invariant kinetic term has the form

$$\mathcal{L}^{\text{kin}} = -\frac{1}{4g^2} \int d^2\theta\, \nabla^\alpha \Phi^a \nabla_\alpha \Phi^a , \tag{10.71}$$

---

[18]It is instructive to see what happens in the quantum mechanical limit. The Hamiltonian of the QM model reads

$$\hat{H} = \hat{\pi}\hat{\pi}^\dagger + m^2 ss^* - im\,\varepsilon_{\alpha\beta}\psi_\alpha \hat{\bar{\psi}}_\beta . \tag{10.66}$$

The supercharges are

$$\hat{Q}_\alpha = \hat{\pi}\psi_\alpha + m\,\varepsilon_{\alpha\gamma}\psi_\gamma s^*, \qquad \hat{\bar{Q}}_\beta = \pi^\dagger \hat{\bar{\psi}}_\beta + m\,\varepsilon_{\beta\delta}\hat{\bar{\psi}}_\delta s . \tag{10.67}$$

It is straigtforward to see that the anticommutators $\{\hat{Q}_\alpha, \hat{Q}_\beta\}$ and $\{\hat{\bar{Q}}_\alpha, \hat{\bar{Q}}_\beta\}$ vanish, whereas the anticommutator $\{\hat{Q}_\alpha, \hat{\bar{Q}}_\beta\}$ is

$$\{\hat{Q}_\alpha, \hat{\bar{Q}}_\beta\} = \delta_{\alpha\beta}\hat{H} + im\varepsilon_{\alpha\beta}\hat{Z} \tag{10.68}$$

with

$$\hat{Z} = i(\hat{\pi}s - \hat{\pi}^\dagger s^*) + \psi_\gamma \hat{\bar{\psi}}_\gamma . \tag{10.69}$$

where $\nabla_\alpha \Phi^a = \mathcal{D}_\alpha \Phi^a - i f^{abc} \Gamma^b_\alpha \Phi^c$ and the coefficient $1/(4g^2)$ is chosen for further convenience. One can also add the mass term,

$$\mathcal{L}_M \;=\; i\frac{\zeta}{2} \int d^2\theta\, \Phi^a \Phi^a \,. \tag{10.72}$$

Adding together (10.55), (10.71), (10.72), expressing the Lagrangian in components, and excluding the auxiliary fields D^a, we obtain

$$\mathcal{L}^{\mathrm{adj}} \;=\; \frac{1}{2g^2}\left\{ -\frac{1}{2} F^a_{\mu\nu} F^{a\mu\nu} + \nabla_\mu \phi^a \nabla^\mu \phi^a + i\lambda^a \bar{\nabla}\lambda^a + i\psi^a \bar{\nabla}\psi^a \right\}$$
$$+ \frac{\kappa}{2}\left[\varepsilon^{\mu\nu\rho}\left(A^a_\mu \partial_\nu A^a_\rho + \frac{1}{3} f^{abc} A^a_\mu A^b_\nu A^c_\rho \right) + i\lambda^{a\alpha} \lambda^a_\alpha \right]$$
$$+ \frac{i\zeta}{2}\,\psi^{a\alpha}\psi^a_\alpha \;-\; \frac{1}{2}\zeta^2 g^2 (\phi^a)^2 \,. \tag{10.73}$$

This Lagrangian involves, besides the gauge field, the adjoint fermion λ with the mass $m_\lambda = \kappa g^2$, the adjoint fermion ψ with the mass $m_\psi = \zeta g^2$ and the adjoint scalar with the same mass. The point $\zeta = \kappa$ is special. In this case, the Lagrangian (10.73) enjoys the $\mathcal{N} = 2$ supersymmetry [43].

Consider finally the model including a complex fundamental *and* an adjoint matter multiplet. The Lagrangian of the model reads

$$\mathcal{L} = \int d^2\theta \left[\frac{1}{4g^2} W^{\alpha a} W^a_\alpha + \frac{\kappa}{4}\left(i\Gamma^{a\alpha} W^a_\alpha + \frac{1}{6} f^{abc} \Gamma^{a\alpha} \Gamma^{b\beta} \mathcal{D}_\beta \Gamma^c_\alpha \right) \right]$$
$$- \frac{1}{4g^2} \int d^2\theta\, \nabla^\alpha \Phi^a \nabla_\alpha \Phi^a + i\frac{\zeta}{2} \int d^2\theta\, \Phi^a \Phi^a - \frac{i}{g^2} \int d^2\theta\, \overline{Q}^j (t^a)_j{}^k Q_k \Phi^a$$
$$+ \frac{1}{2g^2} \int d^2\theta\, \overline{Q}^j \nabla^\alpha \nabla_\alpha Q_j + i\xi \int d^2\theta\, \overline{Q}^j Q_j \,. \tag{10.74}$$

We also included here the Yukawa term $\propto \int d^2\theta\, \overline{Q}Q\Phi$. We have chosen the particular coefficient $-i/g^2$ in front of this term so that the theory enjoys the extended $\mathcal{N} = 2$ supersymmetry when $\kappa = \zeta$ and $\xi = 0$. In this case, the masses of all massive fields are equal to the gluino mass, and this degeneracy is not violated in any order of perturbation theory.

10.3.1.3   *From* $\mathcal{N} = 2$ *to* $\mathcal{N} = 1$

It is straightforward to see that the component Lagrangian (10.73) is $\mathcal{N} = 2$ supersymmetric at the point $\kappa = \zeta$: it is symmetric under the exchange $\lambda \leftrightarrow \psi$ and the extra supersymmetry transformations represent the compositions of the manifest $\mathcal{N} = 1$ transformations and this exchange.

To be convinced that the Lagrangian (10.74) with $\xi = 0$ enjoys $\mathcal{N} = 2$ supersymmetry, one also could go down to components and observe it in a similar way, but it is instructive to derive it using the superfield language [188]. We illustrate it here for the Abelian variant of the theory, which is simpler.

Consider the $\mathcal{N} = 2$ Lagrangian

$$\mathcal{L} = \frac{i\kappa}{32} \int d^2\Theta d^2\bar{\Theta} \, \bar{\mathcal{D}}^\alpha V \mathcal{D}_\alpha V \,, \tag{10.75}$$

where Θ^α are complex Grassmann coordinates,

$$V = -2\Theta\gamma^\mu\bar{\Theta}A_\mu + 2i\Theta^2(\bar{\Theta}\bar{\Lambda}) - 2i\bar{\Theta}^2(\Theta\Lambda) + D\Theta^2\bar{\Theta}^2 \tag{10.76}$$

with complex Λ and

$$\mathcal{D}_\alpha = \frac{\partial}{\partial\Theta^\alpha} - i(\gamma^\mu\bar{\Theta})_\alpha\partial_\mu, \qquad \bar{\mathcal{D}}_\alpha = -\frac{\partial}{\partial\bar{\Theta}^\alpha} + i(\Theta\gamma^\mu)_\alpha\partial_\mu \,. \tag{10.77}$$

The superfield (10.76) and the covariant derivatives (10.77) are the familiar $4D$ vector superfield (3.26) and the $4D$ covariant derivatives (3.5) reduced to three dimensions.

Now we split the real and imaginary part of Θ:

$$\Theta^\alpha = \frac{1}{\sqrt{2}}(\theta^\alpha + i\eta^\alpha), \qquad \bar{\Theta}^\alpha = \frac{1}{\sqrt{2}}(\theta^\alpha - i\eta^\alpha) \tag{10.78}$$

with real θ, η. In these terms,

$$\mathcal{D}_\alpha = \frac{1}{\sqrt{2}}\mathcal{D}_\alpha - \frac{i}{\sqrt{2}}\left[\frac{\partial}{\partial\eta^\alpha} - i(\gamma^\mu\eta)_\alpha\partial_\mu\right],$$

$$\bar{\mathcal{D}}_\alpha = -\frac{1}{\sqrt{2}}\mathcal{D}_\alpha - \frac{i}{\sqrt{2}}\left[\frac{\partial}{\partial\eta^\alpha} - i(\gamma^\mu\eta)_\alpha\partial_\mu\right], \tag{10.79}$$

where $\mathcal{D}_\alpha$ is the $\mathcal{N} = 1$ covariant derivative defined in (10.52).

The $\mathcal{N} = 2$ superfield (10.76) can be traded for two $\mathcal{N} = 1$ superfields:[19]

$$V(x^\mu; \theta, \eta) = -2i\eta^\alpha\Gamma_\alpha(x^\mu, \theta) - 2i\eta^2\Phi(x^\mu, \theta) \,, \tag{10.80}$$

where Γ_α and Φ are the Abelian counterparts of (10.53) and (10.70). Then

$$\mathcal{D}_\alpha V = i\sqrt{2}\,\eta^\beta\mathcal{D}_\alpha\Gamma_\beta - \sqrt{2}\,\Gamma_\alpha - \frac{i\eta^2}{\sqrt{2}}(\gamma^\mu\partial_\mu\Gamma)_\alpha - 2\sqrt{2}\,\eta_\alpha\Phi - i\sqrt{2}\,\eta^2\mathcal{D}_\alpha\Phi,$$

$$\bar{\mathcal{D}}_\alpha V = -i\sqrt{2}\,\eta^\beta\mathcal{D}_\alpha\Gamma_\beta - \sqrt{2}\,\Gamma_\alpha - \frac{i\eta^2}{\sqrt{2}}(\gamma^\mu\partial_\mu\Gamma)_\alpha - 2\sqrt{2}\,\eta_\alpha\Phi + i\sqrt{2}\,\eta^2\mathcal{D}_\alpha\Phi.$$

$$\tag{10.81}$$

---

[19]The sign of the second term is a convention.

Substituting this in (10.75) and integrating over $d^2\eta$, we derive[20]

$$\mathcal{L} = \frac{i\kappa}{8} \int d^2\theta\, \Gamma^\alpha \mathcal{D}^\beta \mathcal{D}_\alpha \Gamma_\beta + \frac{i\kappa}{2} \int d^2\theta\, \Phi^2 \tag{10.82}$$

in agreement with (10.55) and (10.72).

To fix the coefficient of the Yukawa term in (10.74), we introduce the charged $\mathcal{N} = 2$ chiral superfield $\mathcal{Q}$ and write the Lagrangian

$$\mathcal{L}_\mathcal{Q} = \frac{1}{4g^2} \int d^2\Theta d^2\bar{\Theta}\, \overline{\mathcal{Q}}(x_R^\mu) \mathcal{Q}(x_L^\mu) e^V \,, \tag{10.83}$$

where

$$x_{L,R}^\mu = x^\mu \mp i\Theta\gamma^\mu\bar{\Theta}\,,$$

so that $\mathcal{D}_\alpha x_R^\mu = \bar{\mathcal{D}}_\alpha x_L^\mu = 0$. The Lagrangian (10.83) represents the $3D$ counterpart of the corresponding term in (3.32).[21]

The chiral superfields $\mathcal{Q}$ and $\overline{\mathcal{Q}}$ satisfying $\bar{\mathcal{D}}_\alpha \mathcal{Q} = \mathcal{D}_\alpha \overline{\mathcal{Q}} = 0$ may be expressed via the $\mathcal{N} = 1$ superfields $Q, \overline{Q}$ as follows:

$$\mathcal{Q} = Q + i\eta^\alpha \mathcal{D}_\alpha Q + \frac{\eta^2}{4} \mathcal{D}^\alpha \mathcal{D}_\alpha Q\,,$$

$$\overline{\mathcal{Q}} = \overline{Q} - i\eta^\alpha \mathcal{D}_\alpha \overline{Q} + \frac{\eta^2}{4} \mathcal{D}^\alpha \mathcal{D}_\alpha \overline{Q}\,. \tag{10.84}$$

Substituting (10.84) and (10.80) into (10.83) and integrating over $d^2\eta$, we derive

$$\mathcal{L} = \frac{1}{2g^2} \int d^2\theta\, \overline{Q}\nabla^\alpha \nabla_\alpha Q - \frac{i}{g^2} \int d^2\theta\, \overline{Q}Q\Phi\,. \tag{10.85}$$

in agreement with (10.74).

10.3.2   *Index calculations*

i) We start our discussion with the $SU(2)$ model including an extra adjoint multiplet. In the $\mathcal{N} = 2$ theory, the mass of two fermions and of the scalar coincide. One can, however, consider a $\mathcal{N} = 1$ theory (10.73) including an adjoint matter multiplet whose mass does not coincide with κg^2 [189].

---

[20]In fact, the supergauge term following from Eq. (10.75) has the form

$$\frac{i\kappa}{16} \int d^2\theta [2i\Gamma^\alpha(\gamma^\mu \partial_\mu \Gamma)_\alpha + \mathcal{D}^\alpha \Gamma^\beta \mathcal{D}_\alpha \Gamma_\beta]\,.$$

But this expression coincides with the corresponding term in the Abelian version of (10.55), as can be seen from comparing their component representations.

[21]In four dimensions, we had to introduce a couple of chiral multiplets to cancel the anomaly, but we need not do so in $3D$.

Consider first the case $\zeta > 0$. Then the mass of the matter fermion is positive. The matter loop leads to an extra renormalization of k.

We note that the status of this renormalization is different from the one due to gluino. We have seen that, for the latter, the induced magnetic field on the dual torus is concentrated at the corners (10.39), which follows from the equality $m_\lambda L \ll 1$. On the other hand, the mass of the matter fields $m_\psi = \zeta g^2$ is an independent parameter. It is convenient to make it *large*, $m_\psi L \gg 1$. In this case, the corresponding induced flux density becomes roughly constant, as the tree flux density is. This brings about the renormalization $k \to k - 1$.

The index of such a theory coincides with the index of the $\mathcal{N} = 1$ SYMCS theory with a renormalized k. We derive

$$I_{\zeta>0} = k - 1 \,. \tag{10.86}$$

For $k = 1$, the index is zero and supersymmetry may be spontaneously broken.

We calculated the index for large m_ψ, but the value of the index does not change under smooth deformations, and we can be sure that the index of the theory (10.73) is equal to $k-1$ for all positive values of ζ. In particular, for $\mathcal{N} = 2$ theory with $\kappa = \zeta > 0$. It need not be and is not so for negative ζ because the deformation from positive to negative matter masses is not smooth: we have to pass the point $m_\psi = 0$, where the spectrum in the matter sector becomes continuous, and the index is ill-defined. The result (10.86) coincides with the result of Refs. [190], which was derived using the D-brane magic.

For negative ζ, two things happen:

- To begin with, the fermion matter mass has the opposite sign and so does the renormalization of k due to the matter loop. We seem to obtain $I_{\zeta<0} = k + 1$.
- However, this is wrong, due to another effect. For a positive ζ, the ground state of the fast Hamiltonian in the matter sector is bosonic. But for a negative ζ, it is fermionic, $\Psi \propto \prod_{a=1}^{3} \psi^a$, changing the sign of the index.

We obtain

$$I_{\zeta<0} = -k - 1 \,, \tag{10.87}$$

and this is also true for the $\mathcal{N} = 2$ theory with $\kappa = \zeta < 0$. Supersymmetry breaking is expected here for $k = -1$.

Combining (10.86) and (10.87), we derive

$$\boxed{I_W\,(\mathcal{N}=2) \;=\; |k| - 1\,.}$$ (10.88)

In contrast to (10.86) and (10.87), this expression is not analytic at $k = 0$.

As was mentioned, formula (10.88) does not work strictly speaking for $k = \zeta = 0$ due to the presence of continuous spectrum. We saw, however, in Chapter 9 that, for *some* theories with continuous spectrum, the index that disregards continuum and counts only normalized vacuum states can still be well defined. And bearing in mind that the expressions (10.86) and (10.87) give the value $I_W = -1$ for $k = 0$ both for positive and negative ζ, we may tentatively attribute this value to I_W also for $\mathcal{N} = 2$ theory with $k = \zeta = 0$.

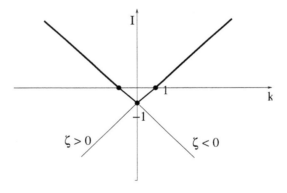

Fig. 10.5: The indices in the theory (10.73) with $\zeta > 0$, $\zeta < 0$, and $\zeta = \kappa$ (bold lines).

The three index formulas (10.86), (10.87), and (10.88) are represented together in Fig. 10.5.

ii) Consider now the theory including *three* adjoint multiplets such that the masses of two of them have the same sign, while the sign of the third mass is opposite. The interest of this theory lies in the fact that, at a special point,

$$m_1 \;=\; m_2 \;=\; -m_3 \;=\; \kappa g^2\,,$$ (10.89)

the theory enjoys $\mathcal{N} = 3$ supersymmetry [43] ! The Lagrangian of $\mathcal{N} = 3$

theory reads

$$
\mathcal{L} = \frac{1}{g^2} \int d^3x \left[-\frac{1}{4} F^a_{\mu\nu} F^{a\mu\nu} + \frac{1}{2} \left(\nabla_\mu \phi^a_{1,2,3} \nabla^\mu \phi^a_{1,2,3} - m^2 \phi^a_{1,2,3} \phi^a_{1,2,3} \right) \right.
$$
$$
\left. + \frac{i}{2} \left(\lambda^a \slashed{\nabla} \lambda^a + \psi^a_{1,2,3} \slashed{\nabla} \psi^a_{1,2,3} \right) \right]
$$
$$
+ \frac{\kappa}{2} \int d^3x \left[\epsilon^{\mu\nu\rho} \left(A^a_\mu \partial_\nu A^a_\rho + \frac{1}{3} f^{abc} A^a_\mu A^b_\nu A^c_\rho \right) + i\lambda^a \lambda^a + i\psi^a_{1,2} \psi^a_{1,2} - i\psi^a_3 \psi^a_3 \right]
$$

$$(10.90)$$

with $\lambda^a \lambda^a \equiv \lambda^{a\alpha} \lambda^a_\alpha$, etc. In the $\mathcal{N} = 1$ theory with large $|m_{1,2,3}|$, the index can be determined as in the previous case. It should be noted that the shift of k due to the loops of three fermion matter flavors is *the same* as before: the shifts due to fermions of opposite mass signs cancel.

Let $k > 0$. Then $k^{\text{ren}} = k - 1$, but we have to take into account that the fast vacuum wave function for the fermion with negative sign of mass has odd fermion number $F = 3$ changing the sign of the index. On the other hand, for negative k, we have $m_{1,2} < 0$ and $m_3 > 0$. The renormalized level is equal to $k^{\text{ren}} = k + 1$ and this is also the value of the index because the fast vacuum wave function has in this case the even fermion charge $F = 3 + 3 = 6$ and $(-1)^F$ is positive. Combining the results for positive and negative k, we obtain the value $1 - |k|$ for the index, which is also valid for $\mathcal{N} = 3$ theory:

$$
\boxed{I_W(\mathcal{N} = 3) = 1 - |k|.}
$$

$$(10.91)$$

Thus, the Witten index in the $\mathcal{N} = 3$ theory is equal to the index in the $\mathcal{N} = 2$ theory, but with the opposite sign.

iii) The simplest $\mathcal{N} = 1$ SYMCS + matter theory includes a single fundamental matter multiplet, and its Lagrangian was written in Eqs. (10.57), (10.58). Again, the matter fermion loops affect k. The shift of k is half as much as in the adjoint case.[22] There are two fermion components χ_1, χ_2 and the fast ground state wave function in the matter sector involves the fermionic factor $\chi_1 \chi_2$ for negative ξ or not at all for positive ξ. In both cases, the state is bosonic. We obtain,

$$
\boxed{I = k - \frac{1}{2} \mathrm{sgn}(\xi).}
$$

$$(10.92)$$

Note that, for consistency, k should be half-integer here. The reason is the same as for pure $\mathcal{N} = 1$ SYMCS theories with higher groups [see the

---

[22]When calculating the adjoint real fermion loop, the color and reality give the factor $c_V \delta^{ab}/2 = \delta^{ab}$. For the complex fundamental loop, the factor is $\mathrm{Tr}\{t^a t^b\} = (1/2)\delta^{ab}$.

remark after Eq. (10.7)]: the large gauge transformations, besides adding $2\pi i k$ to the Minkowski action, change the sign of the fermion determinant in the functional integral.

iv) Finally, consider the model (10.74) [the gauge group is again chosen to be $SU(2)$]. The dynamics here is more rich. Especially interesting is the special point $\zeta = \kappa,\ \xi = 0$ when the theory posseses $\mathcal{N} = 2$ supersymmetry. The index of such a theory was calculated by Intriligator and Seiberg in a $\mathcal{N} = 2$ language [191]. And we will do so for the $\mathcal{N} = 1$ theory (10.74) and then capitalize on the invariance of the index under deformations [189].

The mass of the multiplet Φ in the deformed theory is $M = \zeta g^2$. The mass of the fundamental multiplet is $m = \xi g^2$. We assume both of them to be large by absolute value: $|M|L, |m|L \gg 1$. (As we always keep $g^2 L$ small, this also means that $|M|, |m| \gg g^2$ and $|\zeta|, |\xi| \gg 1$.) Then k is renormalized by fermion loops with quasi-homogeneous flux densities.

There are four different cases:[23]

(1) All the constants κ, ζ, ξ are positive. Then,

$$k \to k - 1_{\text{adj. matter}} - \left(\frac{1}{2}\right)_{\text{fund. matter}} = k - \frac{3}{2}, \qquad (10.93)$$

giving the value $k - \frac{3}{2}$ of the index.

(2) $\xi > 0,\ \kappa, \zeta < 0$. In this case,

$$k \to k + 1_{\text{adj. matter}} - \left(\frac{1}{2}\right)_{\text{fund. matter}} = k + \frac{1}{2}. \qquad (10.94)$$

Multiplying it by -1 due to the fermionic nature of the fast ground state wave function in the adjoint matter sector [see the discussion before Eq.(10.87)], we obtain $I = -k - 1/2$.

(3) $\xi < 0,\ \kappa, \zeta > 0$. Then,

$$k \to k - 1_{\text{adj. matter}} + \left(\frac{1}{2}\right)_{\text{fund. matter}} = k - \frac{1}{2}, \qquad (10.95)$$

and this is also the value of the index.

(4) All the constants are negative. Then,

$$k \to k + 1_{\text{adj. matter}} + \left(\frac{1}{2}\right)_{\text{fund. matter}} = k + \frac{3}{2}. \qquad (10.96)$$

The index involves the extra minus factor: $I = -k - 3/2$.

---

[23]When comparing with [191], note that the mass sign convention for the matter fermions is *opposite* there compared to our convention. We call the mass positive if it has the same sign as the masses of fermions in the gauge multiplet for positive k (and hence positive ζ). In other words, for positive k, ξ, the shifts of k due to both adjoint and fundamental fermion loops are negative.

Note that negative ξ does not bring about an extra change of sign because of the presence of two components in the fundamental doublet [see the remark before Eq.(10.92)].

For the time being, we have

$$m > 0: \quad I = \begin{cases} k - \frac{3}{2}, & k, M > 0 \\ -k - \frac{1}{2}, & k, M < 0 \end{cases}$$

$$m < 0: \quad I = \begin{cases} k - \frac{1}{2}, & k, M > 0 \\ -k - \frac{3}{2}, & k, M < 0 \end{cases}. \tag{10.97}$$

These results look surprising because they exhibit dependence on the sign of the masses. Especially troublesome is the dependence on $\text{sgn}(m)$ such that we cannot go to the $\mathcal{N} = 2$ point where the fundamental fields are massless.

But it is not the end of the story yet! The presence of the Yukawa term in (10.74) may lead to appearance of *extra* vacuum states on the Higgs branch by the same mechanism as it does in the G_2 model and other similar models considered at the end of Chapter 7 and in Chapter 9. In the half of the cases listed above, an extra vacuum state is there and it affects the index.

The component bosonic potential following from (10.74) reads

$$V = -\frac{2}{g^2}(D^a)^2 + 2\zeta \phi^a D^a - \frac{4}{g^2} F^{j*} F_j + 2i\xi(F^{j*} q_j - q^{j*} F_j)$$
$$+ \frac{2}{g^2} \left[i\phi^a \left(q^* t^a F - F^* t^a q \right) - D^a q^* t^a q \right]. \tag{10.98}$$

Excluding the auxiliary fields, we obtain

$$g^2 V = (mq^* - \phi^a q^* t^a)^j (mq - \phi^a t^a q)_j + \frac{1}{2} (M\phi^a - q^* t^a q)^2 . \tag{10.99}$$

When $M \sim m \gg g^2$, this is not renormalized by loops. The potential vanishes when

$$\begin{cases} mq = \phi^a t^a q, \\ M\phi^a = q^* t^a q. \end{cases} \tag{10.100}$$

The equations (10.100) have a trivial solution $\phi^a = q_j = 0$, but there is also a nontrivial one. By a gauge rotation, one can always bring ϕ^a in the form $\phi^a = \phi \delta^{3a}$ with positive ϕ. Let $m > 0$. Then the first equation in (10.100) implies $q_2 = 0$ and $\phi = 2m$, while the second gives $2M\phi = |q_1|^2$. This system has a unique solution (the phase of q can, of course, be unwinded by a gauge transformation) when $M > 0$, i.e. $k > 0$. Similarly, when $m < 0$, it is q_1 that vanishes and the solution exists for negative k and M.

Note that the $SU(2)$ gauge symmetry is broken completely at this minimum. No light fields are left, there is no BO dynamics and a classical vacuum corresponds to a single quantum state.

Adding when appropriate this extra (bosonic) state to the index (10.97), we obtain the final universal result [189, 191]:

$$\boxed{I^{\mathcal{N}=2 \text{ and massive fund. mult.}} = |k| - \frac{1}{2}.} \qquad (10.101)$$

Supersymmetry is broken for $|k| = 1/2$. One can observe that the modification compared to (10.88) is minimal here: $|k| - 1$ is replaced by $|k| - 1/2$, reflecting the fact that k now has to be half-integer rather than integer. The result (10.101) does not depend on the sign of m, and, by the same reasoning as in the paragraph after Eq. (10.88), we can go to the $\mathcal{N} = 2$ point with $m = 0$, closing our eyes to the presence of the continuous spectrum in the theory with massless matter. Presumably, the index (10.101) counts the normalized vacuum states in this limit. It would be interesting to perform something similar to the analysis in Sect. 9.2 and demonstrate their presence explicitly.

Many other $\mathcal{N} = 1$ and $\mathcal{N} = 2$ $3D$ models with matter can be written and the Witten index there can be evaluated. Certain nontrivialities arise for the theories based on higher gauge groups. We address the reader to Refs. [191–193] (where $\mathcal{N} = 2$ theories were studied) and to Ref. [189] (where also $\mathcal{N} = 1$ theories were discussed) for more details.

Chapter 11

Related supersymmetric indices

11.1 S^3 index □ Weak supersymmetry

When we discussed supersymmetric field theories in the previous chapters, we regularized them in the infrared by putting them in a finite toroidal spatial box. The original supersymmetry algebra is kept intact under such regularisation, and the Witten index was a β-independent integer.

But there is an interesting alternative way of regularization when the theory is placed not on T^3, but on S^3. In this case, the algebra is necessarily modified. Indeed, the momenta $\hat{P}_a$, which commute on $\mathbb{R}^3$ and on T^3, do not commute anymore: they realize now the isometries of S^3 and satisfy the $su(2)$ algebra.

For an explicit construction, consider the embedding $x^2 + y^2 + z^2 + t^2 = \rho^2$ of S^3 of radius ρ into $\mathbb{R}^4$. Consider then the operators $\hat{J}^a = -\frac{i}{2}\eta^a_{\mu\nu}x_\mu\partial_\nu$, where $\eta^a_{\mu\nu}$ are the 't Hooft symbols (6.60),

$$\hat{J}^1 = \frac{i}{2}(t\partial_x - x\partial_t + z\partial_y - y\partial_z),$$

$$\hat{J}^2 = \frac{i}{2}(t\partial_y - y\partial_t + x\partial_z - z\partial_x),$$

$$\hat{J}^3 = \frac{i}{2}(t\partial_z - z\partial_t + y\partial_x - x\partial_y). \tag{11.1}$$

Their commutators are $[\hat{J}^a, \hat{J}^b] = i\varepsilon^{abc}\hat{J}^c$. At the vicinity of the north pole of S^3, $x_\mu = (\mathbf{0}, \rho)$, $\hat{J}^a$ generate tangent space translations: $\hat{J}^a \approx \frac{i\rho}{2}\partial_a = -\frac{\rho}{2}\hat{P}_a$.

The supersymmetry may be broken completely (if, for example, one naïvely replaces the flat Minkowski metric by a curved $S^3 \times \mathbb{R}$ metric in the Lagrangian), but, proceeding in a clever way and adding certain extra terms in the Lagrangian, one may keep a *part* of the original supersymmetry [194].

The new algebra has the form

$$
\begin{aligned}
&[\hat{P}_a, \hat{P}_b] = \frac{-2i}{\rho}\varepsilon_{abc}\hat{P}_c, \\[4pt]
&[\hat{Q}_\alpha, \hat{P}_a] = -\frac{1}{\rho}(\sigma_a\hat{Q})_\alpha, \qquad [\hat{\bar{Q}}^\alpha, \hat{P}_a] = \frac{1}{\rho}(\hat{\bar{Q}}\sigma_a)^\alpha, \\[4pt]
&\{\hat{Q}_\alpha, \hat{\bar{Q}}^\beta\} = \left(2\hat{H} - \frac{\hat{R}}{\rho}\right)\delta_\alpha{}^\beta + (\sigma_a)_\alpha{}^\beta \hat{P}_a, \\[4pt]
&[\hat{Q}_\alpha, \hat{R}] = -\hat{Q}_\alpha, \qquad [\hat{\bar{Q}}^\alpha, \hat{R}] = \hat{\bar{Q}}^\alpha, \\[4pt]
&[\hat{H}, \hat{P}_a] = [\hat{H}, \hat{R}] = [\hat{R}, \hat{P}_a] = 0, \\[4pt]
&\{\hat{Q}_\alpha, \hat{Q}_\beta\} = \{\hat{\bar{Q}}^\alpha, \hat{\bar{Q}}^\beta\} = [\hat{Q}_\alpha, \hat{H}] = [\hat{\bar{Q}}^\alpha, \hat{H}] = 0,
\end{aligned}
\tag{11.2}
$$

where $\alpha = 1, 2$ (bearing in mind further quantum-mechanical applications, we do not distiguish between the dotted and undotted indices; the convention $\bar{Q}^\alpha = (Q_\alpha)^\dagger$ is used) and σ_a are the Pauli matrices.

This algebra involves, besides $\hat{H}, \hat{P}_a, \hat{Q}_\alpha, \hat{\bar{Q}}^\alpha$, an extra $U(1)$ generator $\hat{R}$. The presence of the latter is necessary: if one keeps the requirement $[\hat{Q}_\alpha, \hat{H}] = 0$, the Jacobi identities would not hold without $\hat{R}$ and the algebra would not be consistent.

The mathematical notation for the algebra (11.2) is $su(2|1)$. To be more precise, the algebra $su(2|1)$ includes only four bosonic generators: $2\hat{H} - \hat{R}/\rho$ and $\hat{P}_a$. Here we are dealing with a *central extension* of this algebra, with the Hamiltonian $\hat{H}$, which commutes with all the other operators, playing the role of the central charge. In the limit $\rho \to \infty$, the standard 4-dimensional $\mathcal{N} = 1$ supersymmetry algebra is reproduced.

An important remark is that the algebra (11.2) *is* not specific for the $4D$ supersymmetric field theories placed on $S^3 \times \mathbb{R}$. This algebra also shows up in SQM systems not related to any field theory [28]. Following [28], we will call *weak* this variety of supersymmetry.

The presence of the combination $2\hat{H} - \hat{R}/\rho$ rather than $\hat{H}$ (which is related to the Lagrangian by Legendre transformation and has a dynamical meaning), in the anticommutator $\{\hat{Q}_\alpha, \hat{\bar{Q}}^\beta\}$ invalidates the usual claim that all the positive-energy states of the Hamiltonian are paired, each pair including a bosonic and a fermionic state. As a result, the excited states may contribute to the supertrace

$$
I(\beta) = \mathrm{Tr}\left[(-1)^{\hat{F}}e^{-\beta\hat{H}}\right]
\tag{11.3}
$$

and the index is not an integer number anymore, but instead represents a nontrivial function of temperature. If the theory involves extra conserved charges $\hat{\mathcal{M}}_j$ that commute with both the Hamiltonian and the supercharges, one may introduce the associated chemical potentials μ_j and consider the supertraces

$$\tilde{I}(\beta, \mu_j) \;=\; \mathrm{Tr}\left[(-1)^{\hat{F}} e^{-\beta \hat{H}} e^{\mu_j \hat{\mathcal{M}}_j}\right]. \qquad (11.4)$$

These supertraces may represent complicated functions of β and μ_j [195], but the point is that they represent topological invariants in the same sense as the ordinary Witten index does — they stay invariant under the deformations of the theory that keep the algebra intact!

In the literature, this functional index is usually referred to as *superconformal index* because its discoverers [196] were primarily interested in its applications to superconformal theories. Indeed, the index (11.4) is very useful for studying their dynamics (see e.g. the review [197]). But this index has a more general scope. First of all, a $4D$ theory to be placed on $S^3 \times \mathbb{R}$ need not necessarily be conformal. Second, as we mentioned, there exist supersymmetric systems enjoying the algebra isomorphic to (11.2) and not related to any field theory. That is why we will not call this index superconformal but, still keeping trace of its $4D$ implications, prefer to refer to it as S^3 supersymmetric index.

Following [198], we will now give a simple proof of the invariance of this index under deformations. Then we will illustrate this general theorem by two examples: *(i)* the simplest weak supersymmetry model of Ref. [28] and *(ii)* Römelsberger's model — a weak supersymmetric quantum mechanical model involving a complex bosonic dynamical variable and arising when the massless Wess-Zumino $4D$ model is put on $S^3 \times \mathbb{R}$ and the higher spherical harmonics of the fields are suppressed.

Theorem 11.1. *Let* $\hat{Q}_\alpha, \hat{\bar{Q}}^\alpha$ *and* $\hat{H} \equiv \hat{P}_0$ *but not* $\hat{P}_a$ *or* $\hat{R}$ *be functions of parameter* γ *such that the algebra (11.2) stays intact. Then*

$$\frac{d}{d\gamma} \mathrm{Tr}\left[(-1)^{\hat{F}} e^{-\beta \hat{H}}\right] \;=\; 0. \qquad (11.5)$$

Proof. By expanding $e^{-\beta \hat{H}}$ into the series and using the cyclic property of the supertrace and the fact that $[(-1)^{\hat{F}}, \hat{H}] = 0$, we deduce

$$\frac{d}{d\gamma} \mathrm{Tr}\left[(-1)^{\hat{F}} e^{-\beta \hat{H}}\right] \;=\; -\beta\, \mathrm{Tr}\left[(-1)^{\hat{F}} \frac{d\hat{H}}{d\gamma} e^{-\beta \hat{H}}\right]. \qquad (11.6)$$

The third line in (11.2) reads

$$\{\hat{Q}_\alpha, \hat{\bar{Q}}^\beta\} = 2\hat{H}\delta_\alpha{}^\beta + \text{the terms} \propto \hat{R}, \hat{P}_a.$$

Capitalizing on the assumed γ-independence of the extra terms[1] and the property $[(-1)^{\hat{F}}, \hat{H}] = 0$, we deduce

$$\frac{d\hat{H}}{d\gamma} = \frac{1}{4}\left\{\hat{Q}_\alpha, \frac{d\hat{\bar{Q}}^\alpha}{d\gamma}\right\} + \frac{1}{4}\left\{\frac{d\hat{Q}_\alpha}{d\gamma}, \hat{\bar{Q}}^\alpha\right\}. \tag{11.7}$$

Lemma 1.

$$\text{Tr}\left[(-1)^{\hat{F}}\{\hat{Q}_\alpha, \hat{V}\}e^{-\beta\hat{H}}\right] = 0 \tag{11.8}$$

for any (not too wild) V.

Proof. Take for definiteness $\alpha = 1$. By definition,

$$\text{Tr}\left[(-1)^{\hat{F}}\hat{O}\right] = \sum_B \langle B|\hat{O}|B\rangle - \sum_F \langle F|\hat{O}|F\rangle, \tag{11.9}$$

where $|B\rangle$ and $|F\rangle$ are the bosonic and fermionic states.

The weak SUSY algebra (11.2) includes the ordinary $\mathcal{N} = 2$ SQM sub-algebra $\mathcal{A}_1$ with the generators $\hat{Q}_1, \hat{\bar{Q}}^1$ and

$$\hat{H}_1 = \frac{1}{2}\{\hat{Q}_1, \hat{\bar{Q}}^1\} = \hat{H} + \frac{\hat{P}_3}{2} - \frac{\hat{R}}{2\rho}. \tag{11.10}$$

We choose the eigenstates of $\hat{H}_1$ (which are also eigenstates of $\hat{H}$ due to $[\hat{H}_1, \hat{H}] = 0$) as the basis in Hilbert space. From the viewpoint of $\hat{H}_1$, the spectrum includes:

i) The singlet states annihilated by the action of both $\hat{Q}_1$ and $\hat{\bar{Q}}^1$. If the symmetry $\mathcal{A}_1$ is not broken spontaneously, they are the ground states of $\hat{H}_1$,

ii) The degenerate doublets (B, F) satisfying

$$\hat{Q}_1|B\rangle = |F\rangle, \quad \hat{Q}_1|F\rangle = 0, \quad \langle B|\hat{Q}_1 = 0, \quad \langle F|\hat{Q}_1 = \langle B|. \tag{11.11}$$

---

[1] Using a loose physical terminology, one may call these extra terms central charges, though, as was mentioned above, the only central charge in the strict mathematical sense of this word — the operator that commutes with all other operators in the algebra — is the Hamiltonian.

Using $[\hat{Q}_1, \hat{H}] = 0$, we may rewrite (11.8) as

$$\text{Tr}\left[(-1)^{\hat{F}}\{\hat{Q}_1, \hat{V}\}e^{-\beta\hat{H}}\right] = \text{Tr}\left[(-1)^{\hat{F}}\hat{Q}_1\hat{V}e^{-\beta\hat{H}}\right] + \text{Tr}\left[(-1)^{\hat{F}}\hat{V}e^{-\beta\hat{H}}\hat{Q}_1\right]$$
$$= \sum_B \langle B|\hat{Q}_1\hat{V}e^{-\beta\hat{H}} + \hat{V}e^{-\beta\hat{H}}\hat{Q}_1|B\rangle - \sum_F \langle F|\hat{Q}_1\hat{V}e^{-\beta\hat{H}} + \hat{V}e^{-\beta\hat{H}}\hat{Q}_1|F\rangle.$$

$$(11.12)$$

It is immediately seen that the singlet states do not contribute. Next, using (11.11), we obtain

$$\text{Tr}\left[(-1)^{\hat{F}}\{\hat{Q}_1, \hat{V}\}e^{-\beta\hat{H}}\right]$$
$$= \sum_{\text{doublets}} \langle B|\hat{V}e^{-\beta\hat{H}}|F\rangle - \sum_{\text{doublets}} \langle B|\hat{V}e^{-\beta\hat{H}}|F\rangle = 0.$$

$$(11.13)$$

$\square$

By the same token we can prove

$$\text{Tr}\left[(-1)^{\hat{F}}\{\hat{\bar{Q}}^\alpha, \hat{V}\}e^{-\beta\hat{H}}\right] = 0 \qquad (11.14)$$

for any $\hat{V}$.

By combining (11.6), (11.7), (11.8), (11.14), we arrive at (11.5). $\square$

Remarks

(1) The whole reasoning above also works for *generalized* indices (11.4) where $\hat{\mathcal{M}}_j$ are operators that commute with the Hamiltonian, with at least one pair of the supercharges and do not depend on γ. These functions may represent complicated functions of β and μ_j, but they are invariant under the deformations described above.

(2) The theorem just proven represents a particular case of the so-called *equivariant index theorem* known to mathematicians [1].

11.1.1 *Weak supersymmetric harmonic oscillator and its deformation*

Being confronted with a complicated physical problem, Enrico Fermi used to ask his collaborators and himself — what is the *hydrogen atom* for this problem? In other words — can one simplify the problem to make it easily tractable, while keeping its essential features? In our case, it is possible. The simplest weak supersymmetric system [28] is even much simpler than the hydrogen atom: it includes only one real degree of freedom and the oscillator-like Hamiltonian,

$$\hat{H} = \frac{\hat{p}^2 + x^2}{2} + \hat{F} - 1,$$ (11.15)

where $\hat{p} = -i\partial/\partial x$ and

$$\hat{F} = \chi_\alpha \frac{\partial}{\partial \chi_\alpha} \equiv \chi_\alpha \hat{\bar{\chi}}^\alpha$$ (11.16)

is the fermion charge operator.

This Hamiltonian commutes with the supercharges

$$\hat{Q}_\alpha = (\hat{p} - ix)\chi_\alpha, \qquad \hat{\bar{Q}}^\beta = (\hat{p} + ix)\hat{\bar{\chi}}^\beta,$$ (11.17)

the anticommutators $\{\hat{Q}_\alpha, \hat{Q}_\beta\}$ and $\{\hat{\bar{Q}}^\alpha, \hat{\bar{Q}}^\beta\}$ vanish, but the anticommutator $\{\hat{Q}_\alpha, \hat{\bar{Q}}^\beta\}$ is not reduced to the Hamiltonian, involving four other operators:

$$\{\hat{Q}_\alpha, \hat{\bar{Q}}^\beta\} = (2\hat{H} - \hat{Y})\delta_\alpha{}^\beta + \hat{Z}_\alpha{}^\beta,$$ (11.18)

where

$$\hat{Y} = \hat{F} - 1, \qquad \hat{Z}_\alpha{}^\beta = 2\chi_\alpha\hat{\bar{\chi}}^\beta - \delta_\alpha{}^\beta\hat{F}$$ (11.19)

($\hat{Z}_\alpha{}^\alpha = 0$). The other nonzero commutators of the algebra are

$$[\hat{Q}_\alpha, \hat{Z}_\beta{}^\gamma] = \delta_\beta{}^\gamma\hat{Q}_\alpha - 2\delta_\alpha{}^\gamma\hat{Q}_\beta, \qquad [\hat{\bar{Q}}^\alpha, \hat{Z}_\beta{}^\gamma] = 2\delta_\beta{}^\alpha\hat{\bar{Q}}^\gamma - \delta_\beta{}^\gamma\hat{\bar{Q}}^\alpha,$$

$$[\hat{Q}_\alpha, \hat{Y}] = -\hat{Q}_\alpha, \qquad [\hat{\bar{Q}}^\alpha, \hat{Y}] = \hat{\bar{Q}}^\alpha,$$

$$[\hat{Z}_\alpha{}^\beta, \hat{Z}_\gamma{}^\delta] = 2\left(\delta_\gamma{}^\beta\hat{Z}_\alpha{}^\delta - \delta_\alpha{}^\delta\hat{Z}_\gamma{}^\beta\right).$$ (11.20)

As we see, the full algebra, (11.18), (11.20), has the $su(2)$ subalgebra including $\hat{Z}_\alpha{}^\beta$, two doublets $\hat{Q}_\alpha, \hat{\bar{Q}}^\alpha$ and two singlets $\hat{H}, \hat{Y}$. It represents a central extension of the $su(2|1)$ algebra[2] and is isomorphic to (11.2), as one can be explicitly convinced by replacing in (11.18), (11.20)

$$\hat{H} \to \rho\hat{H}, \qquad \hat{Q}_\alpha \to \sqrt{\rho}\hat{Q}_\alpha, \qquad \hat{Y} \to \hat{R}, \qquad \hat{Z}_\alpha{}^\beta \to \rho\hat{P}_a(\sigma_a)_\alpha{}^\beta \quad (11.21)$$

---

[2]In Refs. [199], a quantum system enjoying the *non-extended* $su(2|1)$ algebra with only one singlet operator having been associated with the Hamiltonian was considered. But it such a system, the commutators $[\hat{Q}_\alpha, \hat{H}]$ and $[\hat{\bar{Q}}_\alpha, \hat{H}]$ do not vanish so that $Q_\alpha, \bar{Q}^\alpha$ are not integrals of motion and are not associated with a symmetry of the Lagrangian.

A different *nonlinear* modification of the supersymmetry algebra was suggested in [200].

and using the identities

$$(\sigma_a)_\alpha{}^\gamma(\sigma_a)_\beta{}^\delta = 2\delta_\alpha{}^\delta\delta_\beta{}^\gamma - \delta_\alpha{}^\gamma\delta_\beta{}^\delta,$$

$$\varepsilon_{abc}(\sigma_a)_\alpha{}^\beta(\sigma_b)_\gamma{}^\delta = i\left[\delta_\gamma{}^\beta(\sigma_c)_\alpha{}^\delta - \delta_\alpha{}^\delta(\sigma_c)_\gamma{}^\beta\right]. \qquad (11.22)$$

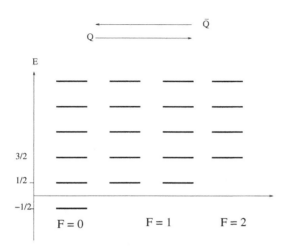

Fig. 11.1: Weak supersymmetric spectrum

The spectrum of the Hamiltonian (11.15) consists of four towers. The left tower involves the ordinary bosonic oscillator states $\Psi_n^{F=0} = \Phi_n(x)$. The lowest state has the energy $E = -1/2$. Then there are two middle towers with fermion charge 1 and the wave functions $\Psi_n^{F=1} = \Phi_n(x)\chi_\alpha$. The right tower has the states of fermion charge 2 with the wave functions $\Psi_n^{F=2} = \Phi_n(x)\chi_1\chi_2$.

In contrast to the ordinary supersymmetic system where the ground state has a zero or maybe positive energy (if supersymmetry is spontaneously broken), the ground state $\Phi_0(x)$ of (11.15) annihilated by the action of $\hat{Q}_\alpha$ and $\hat{\bar{Q}}^\alpha$ has the energy $E = -1/2$. The violation of the familiar supersymmetry pattern is, of course, related to the presence of the extra terms in $\{\hat{Q}_\alpha, \hat{\bar{Q}}^\beta\}$. If one wishes, one can add a constant and bring the ground state energy to zero, but there is no impelling reason to do so.

At the first excited level, we have *three* states of energy $E = 1/2$: $\Phi_1(x)$ and $\hat{Q}_\alpha\Phi_1(x) = \Phi_0(x)\chi_\alpha$. The boson and the fermion states are not paired! This happens because $\hat{Q}_\beta(\Phi_0\chi_\alpha) = 0$ and the fourth component of the multiplet is missing. For the energies $E = 3/2, 5/2, \ldots$, we have $\hat{Q}_1 \hat{Q}_2 \Phi_{n>1}(x) \neq 0$, giving rise to habitual degenerate quartets.

The index (11.3) has two contributions: from the ground state and from the triplet of first excitations. It reads

$$\boxed{I \; = \; e^{\beta/2} - e^{-\beta/2} \; = \; 2\sinh(\beta/2).} \tag{11.23}$$

An alternative interpretation of the result (11.23) is the following. Represent the index as

$$I \; = \; \mathrm{Tr}\left[(-1)^{\hat{F}} e^{-\beta \hat{H}_1} e^{\beta(\hat{Z}_1^1 - \hat{Y})/2}\right], \tag{11.24}$$

where

$$\hat{H}_1 \; = \; \frac{\{\hat{Q}_1, \hat{\bar{Q}}^1\}}{2} \; = \; \frac{1}{2}\left(\hat{p}^2 + x^2 + \chi_1 \hat{\chi}^1 - \hat{\chi}^1 \chi_1\right). \tag{11.25}$$

From the viewpoint of $\hat{H}_1$, we are calculating the generalized Witten index including the fugacities $e^{\beta \hat{Z}_1^1/2}$ and $e^{-\beta \hat{Y}/2}$ associated with the conserved charges $\hat{Z}_1^1$ and $\hat{Y}$. Only the vacuum zero energy states of $\hat{H}_1$ — the state $\Phi_0(x)$ and the state $\Phi_0(x)\chi_2$ contribute in (11.24). The different contributions of the two states is now attributed not to their different energies, but to different fugacities.

11.1.1.1 *Deformation*

The system (11.15), (11.17) is the simplest weak supersymmetric system, but it can be deformed to include nontrivial interactions, while keeping the algebra (11.18), (11.20) intact. According to Theorem 11.1, the index stays invariant under such deformation.

It will be more convenient for us to first study a *classical* counterpart of this system. The supercharges, Hamiltonian, Y and Z_α^β depend now on commuting (p, x) and anticommuting $(\chi_\alpha, \bar{\chi}^\alpha)$ phase space variables. As was spelled out in Sect. 2.3, the algebra of commutators and anticommutators is replaced by the algebra of Poisson brackets.[3] The basic Poisson brackets are

$$\{p, x\}_P = 1, \qquad \{\chi_\alpha, \bar{\chi}^\beta\}_P = i\delta_\alpha{}^\beta. \tag{11.26}$$

To keep the same functional form of the algebra when both the commutators $[\hat{A}, \hat{B}]$ (if one of the operators is even) and anticommutators $\{\hat{A}, \hat{B}\}$ (if both

---

[3] And when the expressions for observables are not so simple, the calculation of Poisson brackets is technically easier than the calculation of (anti)commutators — that is the convenience of using the classical language.

$\hat{A}$ and $\hat{B}$ are odd) are mapped to $-i\{A,B\}_P$, we have to pose

$$Q_\alpha = (p-ix)\chi_\alpha, \quad \bar{Q}^\alpha = (p+ix)\bar{\chi}^\alpha, \quad H = \frac{p^2+x^2}{2} + \chi_\alpha\bar{\chi}^\alpha,$$

$$Y = \chi_\alpha\bar{\chi}^\alpha, \quad Z_\alpha{}^\beta = 2\chi_\alpha\bar{\chi}^\beta - \delta_\alpha{}^\beta\chi_\gamma\bar{\chi}^\gamma. \tag{11.27}$$

We seek the deformed classical supercharges in the form

$$Q_\alpha = [p - iV(x)]\chi_\alpha - ib(x)\,\chi_\alpha\chi_\beta\bar{\chi}^\beta,$$
$$\bar{Q}^\alpha = [p + iV(x)]\bar{\chi}^\alpha + ib(x)\,\bar{\chi}^\alpha\chi_\beta\bar{\chi}^\beta. \tag{11.28}$$

It is easy to see that $\{Q_\alpha, Q_\beta\}_P$ and $\{\bar{Q}^\alpha, \bar{Q}^\beta\}_P$ still vanish.

Now we require for the Poisson bracket $-i\{Q_\alpha, \bar{Q}^\beta\}_P$ to be still represented in the form (11.18) with $Y, Z_\alpha{}^\beta$ given in (11.27) but with a deformed Hamiltonian. The pure bosonic term and the 4-fermionic term in the bracket are proportional to $\delta_\alpha{}^\beta$. They contribute to the deformed Hamiltonian. A nontrivial constraint follows from considering bifermionic terms. They include along with the structure $\propto \delta_\alpha{}^\beta\chi_\gamma\bar{\chi}^\gamma$ also the structure $\propto \chi_\alpha\bar{\chi}^\beta$. We require that the coefficient of the latter in $\{Q_\alpha, \bar{Q}^\beta\}_P$ is equal to $2i$, as in the undeformed case. This gives the condition

$$V' - bV = 1. \tag{11.29}$$

The deformed classical Hamiltonian reads[4]

$$H = \frac{p^2 + V^2}{2} + V'\chi_\alpha\bar{\chi}^\alpha + \frac{b'}{2}(\chi_\alpha\bar{\chi}^\alpha)^2. \tag{11.30}$$

The property

$$\{Q_\alpha, H\}_P = \{\bar{Q}^\alpha, H\}_P = 0 \tag{11.31}$$

holds, as it should. Also the Poisson brackets of the supercharges with the phase space functions Y and $Z_\alpha{}^\beta$ have the same form as in (11.20).

The classical deformed supercharges and the Hamiltonian have their quantum counterparts. As was observed in [24] and spelled out in Chapter 5, to keep supersymmetry at the quantum level, it suffices to order the operators in the quantum supercharges following the Weyl symmetric prescription. We derive:

$$\hat{Q}_\alpha = (\hat{p} - iV)\chi_\alpha - ib\chi_\alpha\left(\chi_\beta\hat{\bar{\chi}}^\beta - \frac{1}{2}\right),$$
$$\hat{\bar{Q}}^\alpha = (\hat{p} + iV)\hat{\bar{\chi}}^\alpha + ib\left(\chi_\beta\hat{\bar{\chi}}^\beta - \frac{1}{2}\right)\hat{\bar{\chi}}^\alpha, \tag{11.32}$$

[4]This system and also the weak supersymmetric system with complex field to be considered later admit a superfield description worked out in [201].

The commutators of these supercharges with the operators (11.19) have the same form (11.20) as in the undeformed case. Then the quantum Hamiltonian is obtained by subtracting from the anticommutator $\{\hat{Q}_\alpha, \hat{\bar{Q}}^\beta\}$ [which should be calculated with implementing the condition (11.29)] the extra terms present in the RHS of Eq. (11.18).[5] The result is:

$$\hat{H} = \frac{\hat{p}^2 + V^2}{2} + V'(\hat{F} - 1) + \frac{b'}{2}(\hat{F}^2 - \hat{F}) + \frac{b^2}{8} \qquad (11.33)$$

with $\hat{F}$ defined in (11.16). This Hamiltonian was in fact discussed in a different language back in [202].

As was argued above, the index (11.23) stays invariant under such deformation. It follows that the ground state and the triplet of first excitations are not shifted. Explicit calculations confirm this [28]. In fact, the system (11.33) is related to certain *quasi-exactly solvable* models, where the energies of the ground state and the first excitation are fixed [203].

The system (11.15), (11.17) can be generalized [204] to the case when the index α takes $N > 2$ diffefent values and the system involves $\mathcal{N} = 2N$ real supercharges. In this case, the algebra reads:

$$\{\hat{Q}_\alpha, \hat{\bar{Q}}^\beta\} = 2\left(\hat{H} - \frac{N-1}{N}\hat{Y}\right)\delta_\alpha{}^\beta + \hat{Z}_\alpha{}^\beta,$$

$$[\hat{Q}_\alpha, \hat{Z}_\beta{}^\gamma] = \frac{2}{N}\delta_\beta{}^\gamma \hat{Q}_\alpha - 2\delta_\alpha{}^\gamma \hat{Q}_\beta,$$

$$[\hat{\bar{Q}}^\alpha, \hat{Z}_\beta{}^\gamma] = 2\delta_\beta{}^\alpha \hat{\bar{Q}}^\gamma - \frac{2}{N}\delta_\beta{}^\gamma \hat{\bar{Q}}^\alpha, \qquad (11.34)$$

$$[\hat{Q}_\alpha, \hat{Y}] = -\hat{Q}_\alpha, \qquad [\hat{\bar{Q}}^\alpha, \hat{Y}] = \hat{\bar{Q}}^\alpha,$$

$$[\hat{Z}_\alpha{}^\beta, \hat{Z}_\gamma{}^\delta] = 2\left(\delta_\gamma{}^\beta \hat{Z}_\alpha{}^\delta - \delta_\alpha{}^\delta \hat{Z}_\gamma{}^\beta\right),$$

where

$$\hat{Q}_\alpha = \chi_\alpha(\hat{p} - ix), \qquad \hat{H} = \frac{\hat{p}^2 + x^2}{2} + \hat{Y},$$

$$\hat{Y} = \hat{F} - N/2, \qquad \hat{Z}_\alpha{}^\beta = 2\chi_\alpha\hat{\bar{\chi}}^\beta - \frac{2}{N}\delta_\alpha{}^\beta\hat{F}. \qquad (11.35)$$

The spectrum of the Hamiltonian now involves:

(1) The ground bosonic oscillator state $\Phi_0(x)$ with zero fermion charge and the energy $E = (1 - N)/2$. It gives the contribution $\exp\{\beta(N-1)/2\}$ to the index.

---

[5]Alternatively, one can calculate the *Grönewold-Moyal bracket* (2.40) of the classical supercharges (11.28) and order the result [which does not quite coincide with (11.30)] following the Weyl prescription.

(2) A state $\Phi_1(x)$ of zero fermion charge with energy $E = (3 - N)/2$ and N fermion states with $F = 1$ and the same energy that are obtained from Φ_1 by the action of the supercharges $\hat{Q}_\alpha$. There are no more states at this energy level because $\hat{Q}_\alpha \hat{Q}_\beta \Phi_1 = 0$. The multiplet is not complete and these states give the contribution $(1 - N) \exp\{\beta(N - 3)/2\}$ to the index.

(3) At the level $E = (5 - N)/2$, we have a state $\Phi_2(x)$ with $F = 0$, N states $\hat{Q}_\alpha \Phi_2$ and also $N(N - 1)/2$ states $\hat{Q}_\alpha \hat{Q}_\beta \Phi_2$. The action of the product $\hat{Q}_\alpha \hat{Q}_\beta \hat{Q}_\gamma$ on Φ_2 gives zero. If $N > 2$, this multiplet is not complete and gives a nonzero contribution to the index.

(4) The last incomplete multiplet has the energy $E = (N - 1)/2$. The multiplets with still larger energies are complete including 2^N states. Their contribution to the index vanishes.

Using the identity

$$\sum_{j=0}^{k}(-1)^j C_N^j = (-1)^k C_{N-1}^k \tag{11.36}$$

for the binomial coefficients, we derive

$$I_W = \sum_{k=0}^{N-1}(-1)^k C_{N-1}^k \exp\left\{\beta\left(\frac{N-1}{2} - k\right)\right\}$$
$$= [2\sinh(\beta/2)]^{N-1}. \tag{11.37}$$

As was the case for $N = 2$, the oscillator Hamiltonian (11.35) may be deformed to include nontrivial nonlinear interactions. But the Ansatz (11.28) does not work anymore. Imposing the condition (11.29), we can get rid of extra unkosher bi-fermionic terms in $\{Q_\alpha, \bar{Q}^\beta\}_P$, but the 4-fermionic terms $\propto \chi_\alpha \bar{\chi}^\beta \chi_\gamma \bar{\chi}^\gamma$ do not easily cancel. For $N = 2$, such terms were in fact proportional to $\delta_\alpha{}^\beta$ and gave rise to 4-fermion contribution in the deformed Hamiltonian, but it is generically not so anymore when we have three or more different fermion flavours. To cancel these terms for a generic $V(x)$, we need to include higher nonlinear fermionic terms in the supercharges.

Consider the case $N = 3$. We seek the classical supercharges in the form

$$Q_\alpha = [p - iV(x)]\chi_\alpha - ib(x)\,\chi_\alpha \chi_\gamma \bar{\chi}^\gamma - ic(x)\,\chi_\alpha(\chi_\gamma\bar{\chi}^\gamma)^2 \tag{11.38}$$

The terms $\propto \chi_\alpha \bar{\chi}^\beta \chi_\gamma \bar{\chi}^\gamma$ cancel out in the Poisson bracket $\{Q_\alpha, \bar{Q}^\beta\}_P$ if the condition

$$b' - b^2 - 2Vc = 0 \tag{11.39}$$

is fulfilled.

Note that (11.39) together with (11.29) are the conditions for the *classical* Poisson algebra to be left intact after the deformation. But, in contrast to what we have seen in the $\mathcal{N} = 2$ case, the necessary conditions for the *quantum* algebra involving Weyl-ordered supercharges $\hat{Q}_\alpha$ and $\hat{\bar{Q}}^\beta$ to be the same as in the undeformed case are modified acquiring the form

$$V' - bV - \frac{bc}{2} = 1,$$

$$b' - b^2 - 2cV - \frac{c^2}{2} = 0. \tag{11.40}$$

The origin of the extra terms $\propto bc$ and $\propto c^2$ are the extra contributions in the GM bracket (2.40) of the classical supercharges involving six derivatives over the fermion variables. In the $N = 2$ case, there was also such contribution in the GM bracket, but it only gave rise to the extra scalar contribution $b^2/8$ in the quantum Hamiltonian and did not involve the terms $\propto \chi_\alpha \bar{\chi}^\beta$ to be get rid of. And for $N \geq 3$, the situation is more complicated.

For $N = 4$, we also have to add the term $-id(x)\, \chi_\alpha (\chi_\gamma \bar{\chi}^\gamma)^3$ in Q_α and, in order to cancel the unkosher 6-fermion terms in the classical Poisson algebra, we need to impose the extra condition,

$$c' - 3bc - 3Vd = 0, \tag{11.41}$$

on top of (11.29) and (11.39). The author was too lazy to work out the corresponding conditions for the quantum algebra (11.34) to be left intact. A dedicated reader can do so!

Obviously, such a procedure, allowing the algebra to be kept intact after deformation, can be generalized to an arbitrary N. The Witten index is invariant under this condition, and that means that the energies of the states belonging to uncomplete supermultiplets at the first N levels are not shifted in the deformed theory. The energies of the states from complete 2^N-plets at the upper levels may well shift.

11.1.2 *Römelsberger's model*

One of the ways to put the theory of free massless chiral superfield $\Phi = \phi(x_L^\mu) + \sqrt{2}\,\theta^\alpha \chi_\alpha(x_L^\mu)$ on S^3 of radius ρ while keeping as much of the original supersymmetries as possible is to write the Hamiltonian

$$\hat{H} = \frac{1}{\rho^3}\hat{\pi}^\dagger \hat{\pi} + \rho\, \phi^* \phi + \frac{\rho^2}{2}(\chi_\alpha \hat{\bar{\chi}}^\alpha - \hat{\bar{\chi}}^\alpha \chi_\alpha) + \cdots, \tag{11.42}$$

where ϕ etc. are the *constant* field modes. The dots stand for the contribution of all other modes, which we will disregard. The canonical (anti)commutators are

$$[\hat{\pi}, \phi] = [\hat{\pi}^\dagger, \phi^*] = -i, \qquad \{\chi_\alpha, \hat{\bar{\chi}}^\beta\} = \frac{1}{\rho^3}\delta_\alpha^\beta.$$

The Hamiltonian (11.42) commutes with the supercharges

$$\hat{Q}_\alpha = \sqrt{2}\chi_\alpha(\hat{\pi} - i\rho^2\phi^*), \qquad \hat{\bar{Q}}^\alpha = \sqrt{2}\hat{\bar{\chi}}^\alpha(\hat{\pi}^\dagger + i\rho^2\phi). \quad (11.43)$$

Now, $\{\hat{Q}_\alpha, \hat{Q}_\beta\} = 0$, but the anticommutator $\{\hat{Q}_\alpha, \hat{\bar{Q}}^\beta\}$ involves the familiar extra terms:

$$\{\hat{Q}_\alpha, \hat{\bar{Q}}^\beta\} = 2\left(\hat{H} + \frac{\hat{L}}{\rho}\right)\delta_\alpha^\beta + 2\rho^2\hat{Z}_\alpha^\beta, \quad (11.44)$$

where

$$\hat{L} = i(\phi\hat{\pi} - \phi^*\hat{\pi}^\dagger) \quad (11.45)$$

has the meaning of the angular momentum in the ϕ plane and $\hat{Z}_\alpha^\beta$ was defined in (11.19). The commutators $[\hat{Q}_\alpha, \hat{Z}_\beta{}^\gamma], [\hat{\bar{Q}}^\alpha, \hat{Z}_\beta{}^\gamma]$ and $[\hat{Z}_\alpha{}^\beta, \hat{Z}_\gamma{}^\delta]$ are the same as (11.20). In addition,

$$[\hat{Q}_\alpha, \hat{L}] = \hat{Q}_\alpha, \qquad [\hat{\bar{Q}}^\alpha, \hat{L}] = -\hat{\bar{Q}}^\alpha. \quad (11.46)$$

This algebra is isomorphic to (11.2) and (11.20).

In the following, we set for simplicity $\rho = 1$, which gives

$$\hat{H} = \hat{\pi}^\dagger\pi + \phi^*\phi + \hat{F} - 1, \quad \hat{Q}_\alpha = \sqrt{2}\chi_\alpha(\hat{\pi} - i\phi^*), \quad (11.47)$$

where $\hat{F}$ is the fermion charge (11.16), and

$$\{\hat{Q}_\alpha, \hat{\bar{Q}}^\beta\} = 2\left(\hat{H} + \hat{L}\right)\delta_\alpha^\beta + 2\hat{Z}_\alpha^\beta. \quad (11.48)$$

The spectrum of $\hat{H}$ is the spectrum of the 2-dimensional oscillator shifted by $F - 1$. We have

$$E_{ml}^F = 2m + |l| + F, \quad (11.49)$$

where n is the number of the radial excitation. Note now that the supercharges also commute with the operator $\hat{K} = \hat{L} + \hat{F}$. We can thus consider a modified Hamiltonian

$$\boxed{\hat{H}_\lambda = \hat{H} + \lambda\hat{K}.} \quad (11.50)$$

The bosonic part of this Hamiltonian describes a 2-dimensional oscillator supplemented by a magnetic field. The anticommutator $\{\hat{Q}_\alpha, \hat{\bar{Q}}^\beta\}$ is expressed as

$$\{\hat{Q}_\alpha, \hat{\bar{Q}}^\beta\} = 2\delta_\alpha{}^\beta(\hat{H}_\lambda - \lambda\hat{K} + \hat{L}) + 2\hat{Z}_\alpha{}^\beta. \tag{11.51}$$

Of course, that does not mean an essential modification of the algebra, it is still a central extention of $su(2|1)$. But the anticommutator $\{\hat{Q}_\alpha, \hat{\bar{Q}}^\beta\}$, being expressed in terms of $\hat{H}_\lambda$, does not have the same functional form as (11.48) and includes an extra operator $\hat{K}$. Theorem 11.1 does not apply in this case, the index of $\hat{H}_\lambda$ need not be the same as the index of $\hat{H}$, and it is not.

The spectrum of $\hat{H}_\lambda$ reads

$$E^F_{ml} = 2m + |l| + F + \lambda(l + F). \tag{11.52}$$

If we want the spectrum to be bounded from below, the parameter λ should not exceed unity.

Well, in principle, there is nothing wrong with the *free* Hamiltonian with $\lambda > 1$. It would represent an example of the system with *benign* ghosts [15]. But we will be interested in *nonlinear* deformations keeping the weak supersymmetry algebra intact. If $\lambda > 1$, such a deformation would make the ghosts *malignant* and unitarity of the theory would be destroyed.

As was mentioned, the Hamiltonians in (11.47) and (11.50) are related to the theory of free massless chiral multiplet placed on $S^3 \times \mathbb{R}$. Suppose that we want to put there a theory involving a superpotential $\mathcal{W}(\Phi)$ in such a way that the weak SUSY algebra is kept intact. One can then derive that [201, 205]

(i) It is only possible for a superpotential $\mathcal{W}(\Phi) \propto \Phi^n$.

(ii) The parameter λ must take a particular value:

$$\lambda = 1 - 2/n. \tag{11.53}$$

As far as algebraic properties are concerned, n need not be integer. But we will be interested only in the models with integer n, where the potential does not involve ugly branchings at the origin. In the first place, in the renormalizable massless Wess-Zumino model with $n = 3$.

For a generic λ, the spectrum (11.52) has a complicated structure. It involves an infinite number of "castles", each consisting of four towers, as in Fig. 11.1. The first such castle grows from the "basement" $\Psi^F_{nl} = \Psi^0_{00}$ with zero energy. At its ground floor, we find a bosonic state Ψ^0_{01} with energy $E = 1 + \lambda$ and a couple of fermionic states Ψ^1_{00} with the same energy.

Higher floors of this castle represent degenerate quartets. At the first floor with energy $E = 2(1 + \lambda)$, we find the bosonic states Ψ_{02}^0 and Ψ_{00}^2 and a couple of fermionic states Ψ_{01}^1.

The basement of the second castle is the state $\Psi_{0,-1}^0$ with energy $E = 1 - \lambda$. The ground floor has the energy $E = 2$ and includes the bosonic state Ψ_{10}^0 and two fermionic states $\Psi_{0,-1}^1$. At the first floor with energy $E = 3 + \lambda$, we have the state Ψ_{11}^0, two states Ψ_{10}^1 and the state $\Psi_{0,-1}^2$. And so on. The higher castles grow from the states $\Psi_{0,-m}^0$ with energy $E = m(1 - \lambda)$. The gap between the floors in all the castles is $\Delta E = 1 + \lambda$.

The index (11.3) acquires the contributions from the basements and ground floors in each castle. The calculation gives

$$
\begin{aligned}
I(\lambda) &= \left[1 - e^{-\beta(1+\lambda)}\right]\left[1 + e^{-\beta(1-\lambda)} + e^{-2\beta(1-\lambda)} + \cdots\right] \\
&= \frac{1 - e^{-\beta(1+\lambda)}}{1 - e^{-\beta(1-\lambda)}},
\end{aligned}
\tag{11.54}
$$

which is in agreement with Eq.(31) in the last Römelsberger's paper [205].[6]

However, the spectrum simplifies a lot for the special values of λ in (11.53). Take $n = 3$ giving $\lambda = 1/3$. Then the triplet on the ground floor of the m-th castle becomes degenerate with the basement of the castle $m + 2$. As a result, the spectrum now involves two bosonic singlets with energies $E = 0$ and $E = 2/3$, while all the other states belong to the supersymmetric quartets — the "old" ones that exist for any value of λ and the "new" ones that appear as a result of such degeneracy. The index is

$$
I(3) = 1 + e^{-2\beta/3}.
\tag{11.55}
$$

The extra degeneracies also appear for higher n when the triplet on the ground floor of the m-th castle becomes degenerate with the basement of the castle $m + n - 1$. We are left with $n - 1$ singlet bosonic states, the other states are in the quartets, and the index is

$$
\boxed{I(n) = \sum_{m=0}^{n-2} e^{-2\beta m/n}.}
\tag{11.56}
$$

The results (11.55) and (11.56) [which also follow from (11.54)] are quite natural. If the Wess-Zumino model with superpotential $W(\Phi) \propto \Phi^n$ is placed on $T^3 \times \mathbb{R}$, the Witten index is known to be equal to $n - 1$. But on S^3 the degenerate vacuum states are equidistantly split with the gap

---

[6] Note the difference in notations. Römelsberger's Ξ is the same as our H.

$\Delta E = 2/n$ [actually, $\Delta E = 2/(n\rho)$], and the index becomes a function of the ratio β/ρ.

We go back to the simplest case $n = 3$. The deformed supercharges and the Hamiltonian read[7]

$$\hat{Q}_\alpha = \sqrt{2}\left[(\hat{\pi} - i\phi^*)\chi_\alpha + i\gamma\hat{\bar{\chi}}_\alpha(\phi^*)^2\right],$$
$$\hat{\bar{Q}}^\alpha = \sqrt{2}\left[(\hat{\pi}^\dagger + i\phi)\hat{\bar{\chi}}^\alpha + i\gamma\chi^\alpha\phi^2\right], \tag{11.57}$$

$$\hat{H}_\lambda = \hat{\pi}^\dagger\pi + \phi^*\phi + \hat{F} - 1 + \frac{1}{3}[\hat{L} + \hat{F}]$$
$$-\gamma(\chi^\alpha\chi_\alpha\,\phi + \hat{\bar{\chi}}_\alpha\hat{\bar{\chi}}^\alpha\phi^*) + \gamma^2(\phi^*\phi)^2, \tag{11.58}$$

where γ is the deformation parameter.

They satisfy the same algebra as in the undeformed case:

$$\{\hat{Q}_\alpha, \hat{\bar{Q}}^\beta\} = 2\delta_\alpha{}^\beta\left(\hat{H}_\lambda + \frac{2\hat{L}}{3} - \frac{\hat{F}}{3}\right) + 2\hat{Z}_\alpha{}^\beta. \tag{11.59}$$

By Theorem 11.1, the index stays invariant under such deformation.

This implies the following properties of the deformed spectrum:

- The energies of the singlet bosonic states are still $E_0 = 0$ and $E_1 = 2/3$.
- The quartets stay degenerate, but their position may be shifted.

These conclusions were confirmed in Ref. [198] by an explicit perturbative calculation to the order γ^2. The singlets are not shifted, but the quartets are. For example a "new" quartet [that appears in the spectrum for a special value $\lambda = 1/3$, has the energy $E = 2m/3$ in the undeformed case $(m \geq 2)$ and includes the states $\Psi^0_{0,-m}$, $2\Psi^1_{0,2-m}$, $\Psi^0_{1,3-m}$] is shifted up by

$$\delta E_\gamma = \frac{m(m-1)}{4}\gamma^2. \tag{11.60}$$

Note that the undeformed model has additional degeneracies which *are* lifted under deformation. For example, the "new" quartet with the states $\{\Psi^0_{0,-4}, 2\Psi^1_{0,-2}, \Psi^0_{1,-1}\}$ has the same energy $E = 8/3$ as the "old" quartet (the first floor of the first castle) with the states $\{\Psi^0_{02}, 2\Psi^1_{01}, \Psi^2_{00}\}$ before the deformation. After the deformation, the new quartet is shifted up by $3\gamma^2$, while the old one only by $3\gamma^2/2$.

---

[7]Cf. Eqs. (2.67), (2.68) describing the supercharges and Hamiltonian of the massless WZ model on $\mathbb{R}^3$ reduced to (0+1) dimensions.

11.2 CFIV index

This index was suggested by Cecotti, Fendley, Intriligator and Vafa [206]. It is defined for the theories in (1+1) and also (0+1) dimensions including at least two different complex supercharges and reads

$$\boxed{I_{CFIV} = \text{Tr}\left[(-1)^{\hat{F}}\hat{F}e^{-\beta\hat{H}}\right].}$$ (11.61)

Note first that, for extended SQM models with discrete spectrum enjoying the algebra (2.65), this index plays the same role as the ordinary Witten index and acquires contributions only from vacuum states.

To be more precise, one can prove it under the condition that the Hamiltonian commutes not only with $(-1)^{\hat{F}}$ (which is the case for all supersymmetric Hamiltonians), but also with $\hat{F}$ so that the eigenstates of the Hamiltonian can be chosen to have a definite fermion number. Then the supercharges $\hat{Q}_\alpha$ (which change the fermion parity of the states) can be chosen in such a way that $[\hat{Q}_\alpha, \hat{F}] = -\hat{Q}_\alpha$, i.e. the states $\hat{Q}_\alpha\Psi_0$ have the fermion number $F_0 + 1$ if Ψ_0 has fermion number F_0. The conjugated supercharges $\hat{\bar{Q}}^\alpha$ decrease the fermion number of a state by 1.

All the Hamiltonians discussed in this book have this property except the Hamiltonian (3.64) of the maximally supersymmetric SQM system. However, this system has a continuous spectrum, and the index is not well defined there anyway.

The proof is simple. We know from Theorem 2.6 that the excited states of an $\mathcal{N} = 4$ SQM Hamiltonian form the quartets $(\Psi_0, \hat{Q}_\alpha\Psi_0, \hat{Q}_1\hat{Q}_2\Psi_0)$. Under the condition specified above, these states have fermion charges $(F_0, F_0 + 1, F_0 + 2)$. The contribution of any such quartet in the CFIV index (11.61) is

$$F_0 - 2(F_0 + 1) + (F_0 + 2) = 0.$$ (11.62)

The CFIV index can also be defined for weak SQM systems. As we have seen, these systems include incomplete supermultiplets, which contribute to the ordinary Witten index and may contribute also to the CFIV index so that the latter becomes a nontrivial function of β and of chemical potentials.

Consider e.g. the system (11.33). Its ground state has zero fermion charge and does not contribute in (11.61). But at the next level with the energy $\beta/2$, we have not a quartet, but a triplet for which the cancellation (11.62) does not work. Only two fermion states contribute there, and we derive

$$I_{CFIV}^{N=2} = -2e^{-\beta/2}.$$ (11.63)

The CFIV index may also be defined for a system with the Hamiltonian in Eq. (11.35) and its deformations keeping the algebra (11.34). The calculation gives

$$I_{CFIV}[\text{model (11.35)}] = -Ne^{-\beta/2}\left(2\sinh\frac{\beta}{2}\right)^{N-2}. \qquad (11.64)$$

As we mentioned, for ordinary (not weak) extended SQM systems, the CFIV index acquires contributions only from the vacuum states and does not depend on β. But the reasoning above based on Theorem 2.6 does not work, however, for certain supersymmetric *field* systems including solitons and associated *short* supersymmetric multiplets.

As was first noticed in Ref. [115], the existence of solitons may modify the ordinary supersymmetry algebra and bring about central charges. In this case, Theorem 2.6 does not apply and, in addition to the quartets, the excited spectrum may include *short multiplets* consisting of a soliton (a bosonic state) and a soliton equiped with a zero fermion mode. These short multiplets contribute in the CFIV index.

As the simplest example, take the Wess-Zumino model. In Chapter 3, we considered this model in (3+1) dimensions. In Eqs. (2.67), (2.68), we wrote the supercharges and the Hamiltonian of the reduced QM version of this model for a particular choice of the superpotential. And now we will discuss the 2-dimensional version.

The Lagrangian of the model reads

$$\mathcal{L}_{WZ} = \partial_\mu\phi^*\partial^\mu\phi + i\psi\sigma^\mu\partial_\mu\bar{\psi} - \bar{\mathcal{W}}'(\phi^*)\mathcal{W}'(\phi)$$
$$-\frac{1}{2}\mathcal{W}''(\phi)\psi^\alpha\psi_\alpha - \frac{1}{2}\bar{\mathcal{W}}''(\phi^*)\bar{\psi}_\alpha\bar{\psi}^\alpha. \qquad (11.65)$$

It is the same as in Eq. (3.19), only now $\mu = 0, 3$ takes only two values and we suppressed the dots over the spinor indices (they make no sense in two dimensions). The field $\bar{\psi}_\alpha$ is conjugate to[8] ψ_α. The indices are still raised and lowered as in (0.2). Following (0.3) and (0.4), we define

$$(\sigma^\mu)_{\alpha\beta} = (\mathbb{1}, \sigma_3)_{\alpha\beta} \quad \text{and} \quad (\bar{\sigma}^\mu)^{\alpha\beta} = \varepsilon^{\alpha\gamma}\varepsilon^{\beta\delta}(\sigma^\mu)_{\gamma\delta} = (\mathbb{1}, -\sigma_3)^{\alpha\beta}.$$

The Lagrangian (11.65) is invariant up to a total derivative under the transformations [cf. (3.20)]

$$\delta\phi = \epsilon^\alpha\psi_\alpha \implies \delta\phi^* = \bar{\epsilon}_\alpha\bar{\psi}^\alpha,$$
$$\delta\psi_\alpha = -i(\sigma^\mu)_{\alpha\beta}\bar{\epsilon}^\beta\partial_\mu\phi - \epsilon_\alpha\bar{\mathcal{W}}'(\phi^*) \implies \delta\psi^\alpha = i\bar{\epsilon}_\beta(\bar{\sigma}^\mu)^{\beta\alpha}\partial_\mu\phi - \epsilon^\alpha\bar{\mathcal{W}}'(\phi^*),$$
$$\delta\bar{\psi}_\alpha = i\epsilon^\beta(\sigma^\mu)_{\beta\alpha}\partial_\mu\phi^* - \bar{\epsilon}_\alpha\mathcal{W}'(\phi) \implies \delta\bar{\psi}^\alpha = -i(\bar{\sigma}^\mu)^{\alpha\beta}\epsilon_\beta\partial_\mu\phi^* - \bar{\epsilon}^\alpha\mathcal{W}'(\phi).$$

$$(11.66)$$

---

[8]This comes from field theory conventions. Note the difference with the QM convention $(\psi_\alpha)^\dagger = \bar{\psi}^\alpha$ used in Chapter 2 and in the first section of this chapter.

The standard Noether procedure gives us the conserved supercurrents:

$$S_\alpha^\mu = \psi_\alpha \partial^\mu \phi^* + i(\sigma^\mu)_{\alpha\beta}\bar{\psi}^\beta \bar{\mathcal{W}}'(\phi^*) + \varepsilon^{\mu\nu}(\sigma_3)_\alpha{}^\beta \psi_\beta \partial_\nu \phi^*,$$
$$\bar{S}_{\alpha\mu} = \bar{\psi}_\alpha \partial_\mu \phi - i\psi^\beta (\sigma^\mu)_{\beta\alpha} \mathcal{W}'(\phi) + \varepsilon^{\mu\nu}\bar{\psi}_\beta (\sigma_3)^\beta{}_\alpha \partial_\nu \phi \qquad (11.67)$$

with $\varepsilon^{03} = 1$. The quantum supercharges are[9]

$$\hat{Q}_\alpha = \int_{-\infty}^{\infty} dz\, \hat{S}_\alpha^0 = \int_{-\infty}^{\infty} dz \left[\psi_\alpha \hat{\pi} - i\varepsilon_{\alpha\beta}\hat{\bar{\psi}}_\beta \bar{\mathcal{W}}'(\phi^*) - (\sigma_3)_{\alpha\beta}\psi_\beta \partial_z \phi^* \right],$$
$$\hat{\bar{Q}}_\alpha = \int_{-\infty}^{\infty} dz\, \hat{\bar{S}}_\alpha^0 = \int_{-\infty}^{\infty} dz \left[\hat{\bar{\psi}}_\alpha \hat{\pi}^\dagger + i\varepsilon_{\alpha\beta}\psi_\beta \mathcal{W}'(\phi) - (\sigma_3)_{\alpha\beta}\hat{\bar{\psi}}_\beta \partial_z \phi \right].$$

$$(11.68)$$

The anticommutator $\{\hat{Q}_\alpha, \hat{\bar{Q}}_\beta\}$ has the standard form:

$$\{\hat{Q}_\alpha, \hat{\bar{Q}}_\beta\} = \delta_{\alpha\beta}\hat{H} - (\sigma_3)_{\alpha\beta}\hat{P}, \qquad (11.69)$$

where

$$\hat{H} = \int_{-\infty}^{\infty} dz \left[\hat{\pi}\hat{\pi}^\dagger + |\partial_z \phi|^2 + i(\sigma_3)_{\alpha\beta}\psi_\alpha \partial_z \hat{\bar{\psi}}_\beta \right.$$
$$\left. + \bar{\mathcal{W}}'(\phi^*)\mathcal{W}'(\phi) + \frac{1}{2}\mathcal{W}''(\phi)\varepsilon_{\alpha\gamma}\psi_\alpha \psi_\gamma - \frac{1}{2}\bar{\mathcal{W}}''(\phi^*)\varepsilon_{\alpha\gamma}\hat{\bar{\psi}}_\alpha \hat{\bar{\psi}}_\gamma \right] \quad (11.70)$$

is the Hamiltonian and

$$\hat{P} = \int_{-\infty}^{\infty} dz\, (\hat{\pi}\partial_z \phi + \hat{\pi}^\dagger \partial_z \phi^* + i\psi_\alpha \partial_z \hat{\bar{\psi}}_\alpha) \qquad (11.71)$$

is the momentum operator.

Consider, however the anticommutator $\{\hat{Q}_\alpha, \hat{Q}_\beta\}$, which vanishes in the ordinary SUSY algebra. But in our case we obtain

$$\boxed{\{\hat{Q}_\alpha, \hat{Q}_\beta\} = 2i(\sigma_1)_{\alpha\beta}\int_{-\infty}^{\infty} \partial_z \bar{\mathcal{W}}(\phi^*)\, dz} \qquad (11.72)$$

and similarly

$$\boxed{\{\hat{\bar{Q}}_\alpha, \hat{\bar{Q}}_\beta\} = -2i(\sigma_1)_{\alpha\beta}\int_{-\infty}^{\infty} \partial_z \mathcal{W}(\phi)\, dz.} \qquad (11.73)$$

We see the integrals of the total derivatives in the right-hand sides. If we put our system in a finite box and impose the periodic boundary conditions on the fields, the right hand sides in Eqs. (11.72), (11.73) vanish, the standard SUSY algebra holds and the CFIV index vanishes too. But these integrals

---

[9]All the indices are put down; $\varepsilon_{12} = 1$.

do not vanish on the infinite line for certain choices of the superpotential. In particular, they do not vanish for[10]

$$W(\phi) = \mu^2\phi - \frac{\alpha\phi^3}{3}. \tag{11.74}$$

with real μ, α. Such a system has a soliton[11] (7.37) satisfying the equation $\partial_z\phi = W'(\phi^*) = W'(\phi)$:

$$\phi(z) = \frac{\mu}{\sqrt{\alpha}}\tanh[\mu\sqrt{\alpha}(z - z_0)]. \tag{11.75}$$

The values of $\phi(z)$ and $W[\phi(z)]$ at $z = \infty$ and $z = -\infty$ are different, and the standard SUSY algebra is modified. It now includes central charges. In this case, Theorem 2.6 does not apply and the spectrum possesses a short multiplet including the soliton (11.75) and this soliton equipped by the fermion zero mode. The latter is derived from (11.75) by a supertransformation (11.66). We have

$$\delta\psi_\alpha = -i(\sigma_3)_{\alpha\beta}\varepsilon^{\beta\gamma}\bar{\epsilon}_\gamma\partial_z\phi - \epsilon_\alpha W'(\phi^*) \propto \frac{i(\sigma_1)_{\alpha\gamma}\bar{\epsilon}_\gamma - \epsilon_\alpha}{\cosh^2[\mu\sqrt{\alpha}(z - z_0)]}. \tag{11.76}$$

If $\epsilon_\alpha = i(\sigma_1)_{\alpha\gamma}\bar{\epsilon}_\gamma$, i.e. $\epsilon_1 = i\bar{\epsilon}_2$, this vanishes. But if $\epsilon_1 = -i\bar{\epsilon}_2$, it does not, giving the zero mode.

The classical energy of the soliton (11.75) is

$$M = \int_{-\infty}^{\infty} dz\,\{(\partial_z\phi)^2 + [W'(\phi)]^2\} = 2\int_{-\infty}^{\infty} dz\,\phi'(z)W(\phi)$$

$$= 2[W(\infty) - W(-\infty)] = \frac{8\mu^3}{3\sqrt{\alpha}}. \tag{11.77}$$

There are no quantum corrections to this value due to supersymmetry.

Let Ψ_- be the pure soliton state and Ψ_+ be the state describing the soliton equipped by the fermion zero mode. It is natural to attribute to Ψ_- the fermion charge $F_- = F_0 = -1/2$ and to the state Ψ_+ the charge $F_+ = F_0 + 1 = 1/2$ [207].

To understand this, consider first the free massive fermion theory.[12] Regularize it in the infrared by putting it on a finite spatial interval. The momenta p are then quantized. The spectrum involves the excitations with positive energies (their creation operators are $a_p^\dagger$) and excitations with

---

[10]We have already met this superpotential in Eq. (7.33), but now ϕ is dimensionless, $[\psi] = [\mu] = m^{1/2}$ and $[W] = [\alpha] = m$.

[11]It was a domain wall in $4D$ theory.

[12]Or the theory (11.70) in the sector involving the classical vacuum state, $\phi_{\text{vac}} = \mu/\sqrt{\alpha}$, and its perturbative excitations. Then ϕ_{vac} plays the role of mass.

negative energies (their creation operators are $\tilde{a}_p^\dagger$). The physical vacuum state $|vac\rangle$ is the state where all the negative energy levels (the Dirac sea levels) are occupied. The states $a_p^\dagger|vac\rangle$ are particles with momentum p and the states $\tilde{a}_p|vac\rangle$ are the holes in the sea — the antiparticles with momentum $-p$. The fermion charge operator is

$$\hat{F} = \sum_p [a_p^\dagger a_p + (\tilde{a}_p^\dagger \tilde{a}_p - 1)]$$

so that the particles have positive, the antiparticles negative and the vacuum zero fermion charge.

Consider now the soliton sector. In this case, besides $a_p^\dagger$ and $\tilde{a}_p^\dagger$, there is also the operator $c^\dagger$ creating the zero mode. $c^\dagger$ neither creates a particle nor does it annihilate an antiparticle, and the associated fermion charge is neither $c^\dagger c$, nor $c^\dagger c - 1$, but something in between:

$$\hat{F}_c = c^\dagger c - \frac{1}{2} = \frac{1}{2}(c^\dagger c - cc^\dagger).$$

Then the state $|sol\rangle$ has $F = -1/2$ and the state $c^\dagger|sol\rangle$ has $F = 1/2$, as indicated above. These two states (as well as the ordinary particle and antiparticle states) are related to each other by charge conjugation.

To calculate actually the functional trace (11.61) in our theory placed in a spatial box $-\frac{L}{2} \le z \le \frac{L}{2}$, we have to specify boundary conditions. To assure that the allowed field configurations have finite energy also in the limit $L \to \infty$, the fields have to acquire in this limit vacuum values at both ends of the interval. But in our case we have *two* vacua, and we have two options:

- If the vacua at $z \to \pm\infty$ are chosen to be the same, there is no contribution to the index.
- But a different choice is

$$\phi\left(\pm\frac{L}{2}\right) = \frac{\mu}{\sqrt{\alpha}} \tanh\left(\pm\mu\sqrt{\alpha}\,\frac{L}{2}\right) \xrightarrow{\text{for large } L} \pm\frac{\mu}{\sqrt{\alpha}} \quad (11.78)$$

A soliton centered at the origin satisfies these conditions and contributes in the index!

The quantum states in the soliton sector are distinguished by their momenta. The energy of the state with momentum p is

$$E_p = \sqrt{p^2 + M^2}. \quad (11.79)$$

The contribution of the corresponding doublet in the index is

$$I_{\text{CFIV}}(p) = \left[\frac{1}{2}(-1)^{1/2} - \frac{1}{2}(-1)^{-1/2}\right] \exp\{-\beta\sqrt{p^2 + M^2}\}$$

$$= i \exp\{-\beta\sqrt{p^2 + M^2}\}. \quad (11.80)$$

In a finite box, the momenta are quantized, $p_n = 2\pi n/L$ with integer n. All together, we derive[13]

$$I_{\text{CFIV}} = i \sum_n \exp\{-\beta\sqrt{p_n^2 + M^2}\} = \frac{iL}{2\pi}\int_{-\infty}^{\infty} dp\, e^{-\beta\sqrt{p^2+M^2}}$$

$$\boxed{= \frac{iLM}{\pi}K_1(\beta M)\,.} \tag{11.81}$$

Does the CFIV index have index nature? In other words, is it invariant under smooth deformations of the theory as Witten index is? Not completely — it is clear from the expression (11.81)! The index depends on the mass of the kink, and the latter depends on the parameters of the Lagrangian according to (11.77). However, it is invariant under *some* deformations of the Lagrangian, the so-called *D-term* deformations.

The component Lagrangian (11.65) can be expressed via a chiral superfield Φ as in Eqs. (3.14), (3.17):

$$\mathcal{L} = \frac{1}{4}\int d^2\theta d^2\bar\theta\,\overline{\Phi}\Phi + \frac{1}{2}\int d^2\theta\,\mathcal{W}(\Phi) + \frac{1}{2}\int d^2\bar\theta\,\overline{\mathcal{W}}(\overline{\Phi})\,. \tag{11.82}$$

The first term represents an integral over full superspace, it is a *D-term*. The second and the third terms are integrals over the chiral subspaces of the full superspace, they are *F-terms*. The statement is that the CFIV index is not invariant under deformations of the superpotential, but it *is* invariant under *D*-term deformations. We address the reader to Ref. [206] for the general proof, and just illustrate it by inspecting the one-kink contribution (11.81).

Consider the deformed Lagrangian

$$\mathcal{L}_\gamma = \frac{\gamma}{4}\int d^2\theta d^2\bar\theta\,\overline{\Phi}\Phi + \frac{1}{2}\int d^2\theta\,\mathcal{W}(\Phi) + \frac{1}{2}\int d^2\bar\theta\,\overline{\mathcal{W}}(\overline{\Phi})\,. \tag{11.83}$$

with the superpotential (11.74) and an arbitrary real parameter γ. Integrating over $d^2\theta$ and $d^2\bar\theta$ and excluding the auxiliary variables, we derive:

$$\mathcal{L}_\gamma^{\text{bos}} = \gamma[\partial_0\phi\partial_0\phi^* - \partial_z\phi\partial_z\phi^*] - \frac{\mathcal{W}'(\phi)\mathcal{W}'(\phi^*)}{\gamma}\,. \tag{11.84}$$

The static solution satisfies the BPS equation

$$\gamma\partial_z\phi = \mathcal{W}'(\phi) \tag{11.85}$$

and has the form (11.75) with $z \to z/\gamma$. The mass of the deformed soliton is calculated in the same way as of the undeformed one and reads

---

[13]This is the contribution of the one-soliton sector. There are also 3-soliton contributions and more [206].

[cf. Eq. (7.35)]

$$M = \int_{-\infty}^{\infty} dz \left\{ \gamma (\partial_z \phi)^2 + \frac{[\mathcal{W}'(\phi)]^2}{\gamma} \right\} = \gamma \int_{-\infty}^{\infty} dz \left[\partial_z \phi - \frac{\mathcal{W}'(\phi)}{\gamma} \right]^2$$
$$+ 2 \int_{-\infty}^{\infty} dz \, \partial_z \phi \, \mathcal{W}'(\phi) = 2 \int_{-\infty}^{\infty} dz \, \partial_z \phi \, \mathcal{W}'(\phi), \tag{11.86}$$

which does *not* depend on γ.

The energy of the moving soliton is still (11.79). To see it, consider a soliton moving with a small velocity v :

$$\phi^{\text{moving}}(z, t) = \phi^{\text{stat}}(z - vt). \tag{11.87}$$

This is a good solution to the equations of motion as long as $v \ll 1$. We obtain an extra contribution to the energy:

$$E^{\text{kin}} = \gamma \int_{-\infty}^{\infty} dz \, (\partial_0 \phi^{\text{moving}})^2 = v^2 \gamma \int_{-\infty}^{\infty} dz \, (\partial_z \phi^{\text{stat}})^2 = v^2 \frac{M}{2}. \tag{11.88}$$

This is a nonrelativistic kinetic energy. The Lorentz invariance dictates the form (11.79) for the relativistic one.

The result (11.81) for the CFIV index is thereby reproduced.

11.2.1 *Four dimensions?*

Can one define an object that would play the same role for 4-dimensional supersymmetric theories, being sensitive to the presence of short soliton multiplets, as the CFIV index plays in two dimensions? At first sight, it seems to be feasible.

Consider the pure $\mathcal{N} = 2$ SYM theory. As was mentioned in Chapter 3 and at the end of Chapter 7, this theory has a scalar vacuum valley (called *Coulomb branch*) and its spectrum includes a massless chiral multiplet associated with the motion along this valley. Besides, the spectrum includes many massive states — the *dyons* characterized by a nonzero electric and magnetic charges, n_e and n_m, and their superpartners. The mass of these multiplets is [32]

$$M_{n_e, n_m} = \sqrt{2} |a n_e - a_D n_m|, \tag{11.89}$$

where a is the vacuum average of the massless scalar field and a_D is a nontrivial function of a. In the perturbative regime,

$$a_D \approx i a \left(\frac{4\pi}{g_0^2} - \frac{2}{\pi} \ln \frac{M_0}{a} \right), \tag{11.90}$$

where g_0 is the gauge coupling constant at the scale M_0.

Each multiplet has four degenerate states [115] — two bosonic states with fermion charges $F = 0$ and $F = 2$ and two fermionic states with $F = 1$. As we discussed above, such multiplet does not contribute to the CFIV index (11.61). Neither the vacuum states contribute — they carry zero fermion charge. As a result, the index (11.61) vanishes.

One may try to consider the object

$$I? \;=\; \mathrm{Tr}\{\hat{F}^2(-1)^{\hat{F}}e^{-\beta\hat{H}}\}\,, \qquad\qquad (11.91)$$

which acquires contributions from all the short multiplets. The problem is, however, that $I?$ has no index nature — it is *not* invariant under any deformation of a theory [206] and is not a useful instrument to study its properties.

Mathematical appendices

A. Some facts about Lie algebras and Lie groups

Mathematicians distinguish between the Lie algebras and their envelopping algebras and prefer to give to the former an abstract definition. But we will not do that. For us, a Lie algebra is an algebra of some special matrices g, which can be multiplied one by another, but we will concentrate on three relevant algebraic operations: mutiplication by a number, addition and commutation.

The exponentials e^{ig} form Lie groups. We are interested only in the compact groups, in which case g are Hermitian. They may be all the Hermitian matrices of a given order N forming the algebra $u(N)$, in which case e^{ig} form the group $U(N)$. They may also belong to a subalgebra $\mathfrak{g}$ of $u(N)$, in which case e^{ig} form a subgroup $\mathcal{G}$ of $U(N)$.

Let $\mathfrak{g}$ be *simple*, i.e. it cannot be represented as a direct sum of smaller algebras.[14] It is convenient to choose there a *complex* basis called *Cartan-Weyl basis*. It includes commuting Hermitian elements generating the Cartan subalgebra $\mathfrak{h}$ (their number r is called the *rank* of the group), and a set of *root vectors* E_{α_j}. Then for any $h \in \mathfrak{h}$

$$\boxed{[h, E_{\alpha_j}] = \alpha_j(h) E_{\alpha_j}.} \qquad \text{(A.1)}$$

The functions $\alpha_j(h)$ are linear forms on the Cartan subalgebra, which are called the *roots*. If we choose in $\mathfrak{h}$ an orthonormal basis e_k (the canonical choice being $\text{Tr}\{e_k e_l\} = \frac{1}{2}\delta_{kl}$), the roots are represented as r-dimensional real vectors.

---

[14]A *semi-simple* algebra may be represented as such sum, but all the summands are non-Abelian.

The root system has the following properties:

• For any root α, $-\alpha$ is also a root. The corresponding root vectors are Hermitially conjugated to each other: $E_{-\alpha} = E_\alpha^\dagger$. The elements of $\mathfrak{g}$ are linear combinations $g = \sum_{k=1}^r b_k e_k + \sum_\alpha (c_\alpha E_\alpha + c_\alpha^* E_{-\alpha})$ with real b_k.

• Let α, β be two arbitrary roots. Then, if $\alpha + \beta$ is also a root,

$$[E_\alpha, E_\beta] \propto E_{\alpha+\beta}. \tag{A.2}$$

If $\alpha + \beta$ is not a root, the commutator vanishes.

• In the set of all roots, one can distinguish a subset of r *simple* roots. Then all other roots represent linear combinations of simple roots with integer coefficients, which either are all positive (and in this case, we are dealing with the *positive* roots) or all negative for *negative* roots.

• In the following, the notation α will be mostly reserved for positive roots and the negative roots will be denoted by $-\alpha$. For the positive roots, the sum of the integer coefficients of their expansion into simple roots is called the *level* of the root. A root with the maximal level is called the *highest* root. The coefficients of the expansion of the highest root are called *Dynkin labels* of the algebra.

• The commutator $[E_\alpha, E_{-\alpha}]$ belongs to $\mathfrak{h}$. A *coroot* corresponding to the root α and denoted as $\alpha^\vee$ is proportional to this commutator, $\alpha^\vee = c[E_\alpha, E_{-\alpha}]$, with the coefficient c assumed to be positive and determined from the condition

$$e^{2\pi i \alpha^\vee} = \mathbb{1}, \quad \text{but} \quad e^{i\phi\alpha^\vee} \neq \mathbb{1} \ \text{ if } \phi < 2\pi. \tag{A.3}$$

In the convenient *Chevalley normalization* of the root vectors, $c = 1$.

• In this normalization, one can fix [208] the proportionality coefficient in (A.2):

$$[E_\alpha, E_\beta] = \pm(q+1)E_{\alpha+\beta}, \tag{A.4}$$

where the sign depends on convention and q is the greatest positive integer such that $\alpha - q\beta$ is a root.

• Let E_α be a simple root vector. A *fundamental coweight* ω_α is an element of the Cartan subalgebra that commutes with all other simple root vectors, while $[\omega_\alpha, E_\alpha] = E_\alpha$.

Consider the algebra $su(3)$ as the simplest nontrivial example. A canonical orthonormal basis in its Cartan subalgebra is

$$t^3 = \frac{1}{2}\text{diag}(1, -1, 0), \qquad t^8 = \frac{1}{2\sqrt{3}}\text{diag}(1, 1, -2). \tag{A.5}$$

The root vectors are

$$E_\alpha = \begin{pmatrix} 0 & 1 & 0 \\ 0 & 0 & 0 \\ 0 & 0 & 0 \end{pmatrix}, \qquad E_\beta = \begin{pmatrix} 0 & 0 & 0 \\ 0 & 0 & 1 \\ 0 & 0 & 0 \end{pmatrix}, \qquad E_{\alpha+\beta} = \begin{pmatrix} 0 & 0 & 1 \\ 0 & 0 & 0 \\ 0 & 0 & 0 \end{pmatrix},$$

$$E_{-\alpha} = \begin{pmatrix} 0 & 0 & 0 \\ 1 & 0 & 0 \\ 0 & 0 & 0 \end{pmatrix}, \qquad E_{-\beta} = \begin{pmatrix} 0 & 0 & 0 \\ 0 & 0 & 0 \\ 0 & 1 & 0 \end{pmatrix}, \qquad E_{-(\alpha+\beta)} = \begin{pmatrix} 0 & 0 & 0 \\ 0 & 0 & 0 \\ 1 & 0 & 0 \end{pmatrix}. \quad (A.6)$$

The simple roots in the basis (A.5) are

$$\boldsymbol{\alpha} = (1,0), \qquad \boldsymbol{\beta} = \left(-\frac{1}{2}, \frac{\sqrt{3}}{2}\right). \qquad (A.7)$$

Note that $\alpha^2 = \beta^2 = 1$ and $\boldsymbol{\alpha} \cdot \boldsymbol{\beta} = -1/2$, i.e. the angle between these vectors is equal to $2\pi/3$.

The simple coroots are

$$\alpha^\vee = \text{diag}(1,-1,0), \qquad \beta^\vee = \text{diag}(0,1,-1). \qquad (A.8)$$

The fundamental coweights are

$$\omega_\alpha = \frac{1}{3}\text{diag}(2,-1,-1), \qquad \omega_\beta = \frac{1}{3}\text{diag}(1,1,-2). \qquad (A.9)$$

Note that

$$e^{2\pi i \omega_\alpha} = e^{-2\pi i/3}\mathbb{1}, \quad \text{and} \quad e^{2\pi i \omega_\beta} = e^{2\pi i/3}\mathbb{1}, \qquad (A.10)$$

which are elements of the group center. One can show that $\exp\{2\pi i \omega_{\alpha_j}\}$ belong to the center also for all other groups.

Nonzero commutators in the basis (A.6), (A.8) are[15]

$$[E_\alpha, E_{-\alpha}] = \alpha^\vee, \quad [E_\beta, E_{-\beta}] = \beta^\vee, \quad [\alpha^\vee, E_{\pm\alpha}] = \pm 2E_{\pm\alpha},$$
$$[\beta^\vee, E_{\pm\beta}] = \pm 2E_{\pm\beta}, \quad [\alpha^\vee, E_{\pm\beta}] = \mp E_{\pm\beta}, \quad [\beta^\vee, E_{\pm\alpha}] = \mp E_{\pm\alpha},$$
$$[E_{\pm\alpha}, E_{\pm\beta}] = \pm E_{\pm(\alpha+\beta)},$$
$$[\alpha^\vee, E_{\pm(\alpha+\beta)}] = [\beta^\vee, E_{\pm(\alpha+\beta)}] = \pm E_{\pm(\alpha+\beta)},$$
$$[E_{\pm(\alpha+\beta)}, E_{\mp\beta}] = \pm E_{\pm\alpha}, \quad [E_{\pm(\alpha+\beta)}, E_{\mp\alpha}] = \mp E_{\pm\beta}. \qquad (A.11)$$

- For a generic Lie algebra, the roots (the real r-dimensional vectors that represent them) may have different length. If the length of the longer roots is normalized to 1, the lengths of the shorter roots may be $1/\sqrt{2}$ or

---

[15]One can observe that the commutators in the last two lines of Eq. (A.11) follow from the commutators in the first three lines, from the fact $[E_\alpha, E_{-\beta}] = [E_\beta, E_{-\alpha}] = 0$ and from the Jacobi identity.

Group	Algebra	extended Dynkin diagrams

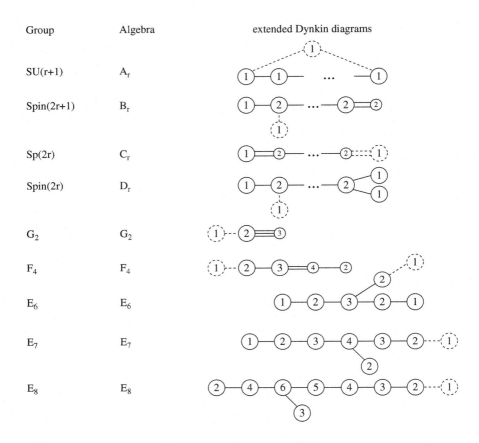

Fig. A.1: Extended Dynkin diagrams for all classical Lie groups.

$1/\sqrt{3}$ (for g_2). Two different long simple roots may be orthogonal or form the angle $2\pi/3$ between them. The same concerns two short simple roots. The angle between a long simple root and a short simple root of length $1/\sqrt{2}$ may be $\pi/2$ or $3\pi/4$. The algebra g_2 involves a long simple root and a short simple root of length $1/\sqrt{3}$. They form the angle $5\pi/6$.

• Given a set of simple roots, one can define a *Cartan matrix*,

$$A_{jk} = \frac{2\boldsymbol{\alpha}_j \cdot \boldsymbol{\alpha}_k}{\boldsymbol{\alpha}_k \cdot \boldsymbol{\alpha}_k} = \alpha_j(\alpha_k^\vee). \tag{A.12}$$

Its elements are integer numbers. Its determinant never vanishes, this follows from the fact that α_j form a basis in $\mathbb{R}^r$. For the algebras of rank 2,

$$A^{su(3)} = \begin{pmatrix} 2 & -1 \\ -1 & 2 \end{pmatrix}, \quad A^{so(5)} = \begin{pmatrix} 2 & -2 \\ -1 & 2 \end{pmatrix}, \quad A^{g_2} = \begin{pmatrix} 2 & -3 \\ -1 & 2 \end{pmatrix}. \text{ (A.13)}$$

• A convenient way to represent a generic simple algebra or a generic simple group $\mathcal{G} = \exp\{i\mathfrak{g}\}$ is to draw a *Dynkin diagram*. All of them are represented in Fig. A.1. The large circles there are the long simple roots, smaller circles represent short simple roots. If the roots are orthogonal, they are not related by a line. A single line means the angle $2\pi/3$, a double line the angle $3\pi/4$ and a triple line the angle $5\pi/6$. The dashed circles stand for $-\theta$, where θ is the highest root (that is why the diagrams drawn in Fig. A.1 are *extended*; on the ordinary Dynkin diagrams, the highest root is not shown). The numbers in the circles are the Dynkin labels.

• As is shown in Fig. A.1, a simple Lie group (or algebra) can belong to one of four infinite series A_r, B_r, C_r, D_r, and there are also five exceptional groups $G_2, F_4, E_{6,7,8}$.

A symplectic group $Sp(2r)$ may be defined as a subgroup of $SU(2r)$ with the elements g satisfying the condition $g^T \Omega g = \Omega$, where Ω is a skew-symmetric matrix with eigenvalues ± 1. It can be chosen in the form

$$\Omega = \begin{pmatrix} 0 & \mathbb{1}_r \\ -\mathbb{1}_r & 0 \end{pmatrix}. \tag{A.14}$$

The groups $Spin(N)$ represent double covers of $SO(N)$, which rotate spinors rather than real vectors. $Spin(3) \simeq SU(2)$ and $Spin(5) \simeq Sp(4)$.

• The simplest exceptional group is G_2 and, bearing in mind the applications in Chapters 6,7,9,10 we will describe here its properties in some more detail.

• Before doing so, we note that, for many purposes (like vacuum counting discussed in Sect. 6.4), it is convenient to consider on top of a_α also the *modified* Dynkin labels

$$a_\alpha^\vee = a_\alpha |\alpha|^2. \tag{A.15}$$

For long roots, $a_\alpha^\vee = a_\alpha$, while, for short roots, $a_\alpha^\vee = a_\alpha/2$ for the algebras B_r and C_r and $a_\alpha^\vee = a_\alpha/3$ for the short root in g_2. As we see from the values of a_α shown in Fig. A.1, $a_\alpha^\vee = 1$ for all but one short roots.

• The sum of all the modified Dynkin labels on an extended Dynkin diagram (with $a_\theta = a_\theta^\vee = 1$) is called the *dual Coxeter number* $h^\vee$ (see Sect. 6.4 for more details).

G_2 and octonions

As is well known, the group $SO(3)$ can be interpreted as the group of automorphisms of quaternion algebra. Given three imaginary units $e_{\alpha=1,2,3}$ satisfying the algebra

$$e_\alpha e_\beta = -\delta_{\alpha\beta} + \varepsilon_{\alpha\beta\gamma} e_\gamma, \qquad (A.16)$$

one can rotate them, $e'_a = O_{ab} e_b$ with $O \in SO(3)$, and be convinced that the algebra (A.16) holds also for e'_a.

In a similar way, G_2 is the group of automorphisms of *octonion* algebra.

An octonion is an object $a_0 + a_\alpha e_\alpha$, where a_0 and $a_{1,\ldots,7}$ are real numbers and e_α satisfy the algebra

$$\boxed{e_\alpha e_\beta = -\delta_{\alpha\beta} + f_{\alpha\beta\gamma} e_\gamma.} \qquad (A.17)$$

One of the possible choices for the antisymmetric tensor $f_{\alpha\beta\gamma}$ is

$$f_{165} = f_{341} = f_{523} = f_{271} = f_{673} = f_{475} = f_{246} = 1 \qquad (A.18)$$

and all other nonzero components are restored by antisymmetry. The values (A.18) can be mnemonized by looking at the *Fano plane* depicted in Fig. A.2. $f_{\alpha\beta\gamma}$ is nonzero only for the indices lying on the same line, with the arrows indicating the order of indices when $f_{\alpha\beta\gamma}$ is positive.

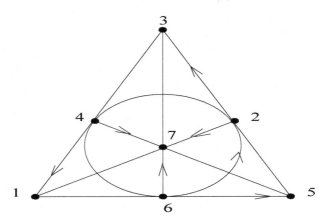

Fig. A.2: Fano plane

One need not necessarily invoke octonions and define G_2 as a subgroup of $SO(7)$ that leaves invariant the form $f_{\alpha\beta\gamma} A_\alpha B_\beta C_\gamma$ for any triple of 7-dimensional vectors A, B, C.

Thirdly, one can define G_2 as a subgroup of $Spin(7)$ leaving invariant a given 7-dimensional spinor. Such a spinor has 8 components, the requirement of its invariance imposes 7 conditions, and it follows that G_2 has dimension $21 - 7 = 14$.

At the end of this section, we give three more definitions.

(i) The *maximal torus* is an Abelian subgroup of $\mathcal{G}$ generated by its Cartan subalgebra.

(ii) The *centralizer* of an element $g \in \mathcal{G}$ or of a set of elements $\{g_i \in \mathcal{G}\}$ is the subgroup of $\mathcal{G}$ that commutes with g (resp. with all the elements g_i in the set).

(iii) We define the *Dynkin index* $I(R)$ of a representation R in group $\mathcal{G}$ as[16]

$$\mathrm{Tr}\{T^a T^b\} = I(R)\delta^{ab}, \tag{A.19}$$

where T^a are the generators in the representation R normalized in a standard way so that $[T^a, T^b] = i f^{abc} T^c$, where f_{abc} are the structure constants of the group.

Speaking of f_{abc}, they are normalized so that

$$f^{acd} f^{bcd} = h^\vee \delta^{ab}. \tag{A.20}$$

It follows that the Dynkin index of the adjoint representation is equal to $h^\vee$ for any simple Lie group.

- $I = 1/2$ for the fundamental representation in $SU(N)$ and for the spinor representation in $Spin(N)$
- $I = 1$ for the vector representation in $Spin(N \geq 5)$ [$Spin(3) \simeq SU(2)$ is a special case and $Spin(4) \simeq SU(2) \times SU(2)$ is not simple] and for the fundamental 7-plet in G_2.

B. Theta functions

We recall here certain mathematical facts concerning the properties of analytical functions on a torus. They are mostly taken from the textbook [94], but we are using a different notation, which we find clearer and more appropriate for our purposes.

Theta functions play the same role for the torus as ordinary polynomials for the Riemann sphere. They are analytic, but satisfy certain nontrivial quasiperiodic boundary conditions with respect to shifts along the cycles of

---

[16]Sometimes it is defined with an extra factor 2 [209].

the torus. A generic torus is characterized by a complex modular parameter τ, but we will stick to the simplest choice $\tau = i$ so that the torus represents a square $x, y \in [0, 1]$ ($z = x + iy$) glued around. The simplest θ-function satisfies the boundary conditions

$$\theta(z + 1) = \theta(z),$$
$$\theta(z + i) = e^{\pi(1-2iz)}\theta(z). \tag{B.1}$$

This defines a *unique* (up to a constant complex factor) analytic function. Its explicit form is

$$\theta(z) = \sum_{n=-\infty}^{\infty} \exp\{-\pi n^2 + 2\pi i n z\}. \tag{B.2}$$

This function [call it theta function of level 1 and introduce an alternative notation $\theta(z) \equiv Q^1(z)$] has only one zero in the square $x, y \in [0, 1]$, exactly in its middle, $\theta(\frac{1+i}{2}) = 0$.

For any integer $q > 0$, theta functions of level q can be defined such that

$$Q^q(z + 1) = Q^q(z),$$
$$Q^q(z + i) = e^{q\pi(1-2iz)}Q^q(z). \tag{B.3}$$

Like the conditions (B.1), the boundary conditions (B.3) are consistent in the sense that two ways to reach the point $z + 1 + i$ from the point z give the same result:

$$Q^q(z + 1 + i) = Q^q(z + i + 1) = e^{q\pi(1-2iz)}Q^q(z). \tag{B.4}$$

The functions satisfying (B.3) lie in a vector space of dimension q. The basis in this vector space can be chosen as

$$\boxed{Q_m^q(z) = \sum_{n=-\infty}^{\infty} \exp\left\{-\pi q\left(n + \frac{m}{q}\right)^2 + 2\pi i q z\left(n + \frac{m}{q}\right)\right\},}$$
$$m = 0, \ldots, q - 1. \tag{B.5}$$

In Mumford's notation [94], $Q_m^q(z)$ can be expressed as

$$Q_m^q(z) = \theta_{m/q,0}(qz, iq), \tag{B.6}$$

where $\theta_{a,b}(z, \tau)$ are theta functions of rational characteristics.

$Q_m^q(z)$ can be called "elliptic polynomials" of degree q. Indeed, each $Q_m^q(z)$ has q simple zeros at

$$z_s^{(m)} = \frac{2s + 1}{2q} + i\left(\frac{1}{2} - \frac{m}{q}\right), \qquad s = 0, \ldots, q - 1 \tag{B.7}$$

(add i to bring it onto the fundamental domain $x, y \in [0, 1]$ when necessary). A product $Q^q(z)Q^{q'}(z)$ of two such "polynomials" of degrees q, q' gives a "polynomial" of degree $q + q'$. There are many relations between the theta functions of different level and their products, which follow. We can amuse the reader with a relation [94]

$$Q_5^6(z) - Q_1^6(z) = \frac{2\pi^{3/4}}{\Gamma(1/4)}[Q_3^4(z) - Q_1^4(z)]Q_0^2(z). \tag{B.8}$$

The function $\Pi(z) = Q_3^4(z) - Q_1^4(z)$ was defined in Eq. (10.48). The fact that the theta function of level 6 in LHS is proportional to the product $\Pi(z)Q_0^2(z)$ follows from the fact that it has the same zeros as $\Pi(z)Q_0^2(z)$. As follows from (B.7), the function $Q_0^2(z)$ has zeros at

$$z = \frac{1}{4} + \frac{i}{2} \quad \text{and} \quad z = \frac{3}{4} + \frac{i}{2}.$$

One can verify using the definition (B.5) that $\Pi(z)$ has 4 zeros at the points $z = 0, \frac{1}{2}, \frac{i}{2}, \frac{i+1}{2}$, which correspond to Eq. (10.39). And the function $Q_5^6(z) - Q_1^6(z)$ has 6 zeros:

$$z_1 = 0, \qquad z_2 = \frac{1}{2}, \qquad z_3 = \frac{i}{2},$$

$$z_4 = \frac{1}{2} + \frac{i}{2}, \qquad z_5 = \frac{1}{4} + \frac{i}{2}, \qquad z_6 = \frac{3}{4} + \frac{i}{2}. \tag{B.9}$$

The ratios of different elliptic functions of the same level give double periodic meromorphic elliptic functions. For example, the ratio of a properly chosen linear combination $\alpha Q_0^2(z) + \beta Q_1^2(z)$ and $[Q^1(z)]^2$ is the Weierstrass function.

Bibliography

[1] H. Cartan, *La transgression dans un groupe de Lie et dans un fibré principal*, Colloque de topologie (espaces fibrés), Centre belge de recherches mathématiques, Masson et Cie, Paris, 1951, p.57.

[2] M.F. Atiyah and I.M. Singer, *The index of elliptic operators*, Annals Math. **87** (1968) 484, 546; **93** (1971) 119, 139.

[3] E. Witten, *Supersymmetry and Morse theory*, J. Diff. Geom. **17** (1982) 661.

[4] A.V. Smilga, *Differential Geometry through Supersymmetric Glasses*, World Scientific, 2020.

[5] A.V. Smilga, *Digestible Quantum Field Theory*, Springer, 2018.

[6] A. Zee, *Quantum Field Theory in a Nutshell*, Princeton University Press, 2010.

[7] M.E. Peskin and D.V. Schroeder, *Introduction to Quantum Field Theory*, Perseus Books Publishing, 1995.

[8] A.V. Smilga, *Lectures on Quantum Chromodynamics*, World Scientific, 2001.

[9] E. Cartan, *Les groupes d'holonomie des espaces généralisés*, Acta Math. **48** (1926) 1.

[10] C. Ehresmann, *Les connexions infinitésimales dans un espace fibré différentiable*, Seminaire N. Bourbaki **1** (1952) 153 [exposé 24, mars 1950].

[11] A.M. Jaffe, *The millennium grand challenge in mathematics*, Notices of AMS **53** (2006) 652.

[12] M. Fierz, *Über die relativistische Theorie kräftefreier Teilchen mit beliebigen Spin*, Helv. Phys. Acta **12** (1939) 3;
W. Pauli, *The connection between spin and statistics*, Phys. Rev. **58** (1940) 716.

[13] R. Casalbuoni, *The classical mechanics for Bose-Fermi systems*, Nuovo Cim. **A33** (1976) 389;
F.A. Berezin and M.S. Marinov, *Particle spin dynamics as the Grassmann variant of classical mechanics*, Ann. Phys. **104** (1977) 336.

[14] F.A. Berezin, *The Method of Second Quantization*, Academic Press, NY, 1966.

[15] A.V. Smilga, *Classical and quantum dynamics of higher-derivative systems*, Int. J. Mod. Phys. **A32** (2017) 1730025, `arXiv:1710.11538 [hep-th]`.

[16] Yu.A. Golfand and E.P. Likhtman, *Extension of the algebra of Poincare group generators by bispinor generators*, JETP Lett. **13** (1971) 323.

[17] There is a vast literature on this subject. We refer the reader to the recently published book — C.M. Bender et al, *PT Symmetry in Quantum and Classical Physics*, World Scientific, 2019.

[18] D. Robert and A.V. Smilga, *Supersymmetry vs. ghosts*, J. Math. Phys. **49** (2008) 042104, arXiv:math-ph/0611023.

[19] E. Witten, *Dynamical breaking of supersymmetry*, Nucl. Phys. **B188** (1981) 513.

[20] L.D. Landau, *Diamagnetismus der Metalle*, Z. Phys. **64** (1930) 629.

[21] S.P. Novikov, *Hamiltonian formalism and a multivalued analog of Morse theory*, Russ. Math. Surveys **37** (1982) 1.

[22] A.V. Smilga, *SUSY anomaly in quantum mechanical systems*, Phys. Lett. **B199** (1987) 516.

[23] J.W. van Holten and R.H. Rietdijk, *Symmetries and motions in manifolds*, J. Geom. Phys. **11** (1993) 559, arXiv:hep-th/9205074.

[24] A.V. Smilga, *How to quantize supersymmetric theories*, Nucl. Phys. **B292** (1987) 363.

[25] H.J. Grönewold, *On the principles of elementary quantum mechanics*, Physica **12** (1946) 405;
J.E. Moyal, *Quantum mechanics as a statistical theory*, Proc. Cambr. Phil. Soc. **45** (1949) 99.

[26] A. Salam and J. Strathdee, *Supergauge transformations*, Nucl. Phys. **B76** (1974) 477.

[27] A.V. Smilga, *Comments on the Newlander-Nirenberg theorem*, in [Varna 2019, Proceedings, Lie Theory and its Application in Physics, ed. V. Dobrev, Springer, 2020, p.167], arXiv:1902.08549 [math-ph].

[28] A.V. Smilga, *Weak supersymmetry*, Phys. Lett. B **585** (2004) 173, arXiv:hep-th/0311023.

[29] J. Wess and J. Bagger, *Supersymmetry and Supergravity*, Princeton University Press, 1992.

[30] J. Wess and B. Zumino, *A Lagrangian model invariant under supergauge transformations*, Phys. Lett. **B49** (1974) 52.

[31] S. Ferrara and B. Zumino, *Supergauge invariant Yang-Mills theories*, Nucl. Phys. **B79** (1974) 413.

[32] N. Seiberg and E. Witten, *Electric-magnetic duality, monopole condensation and confinement in $\mathcal{N} = 2$ supersymmetric Yang-Mills theory*, Nucl. Phys. **B426** (1994) 19, arXiv:hep-th/9407087.

[33] S.J. Gates, Jr., M.T. Grisaru, M. Roček and W. Siegel, *Superspace: or One Thousand and One Lessons in Supersymmetry*, Frontiers in physics: v. 58 (Benjamin/Cummings, 1983), arXiv:hep-th/0108200.

[34] S.G. Matinyan, G.K. Savvidy and N.G. Ter-Arutyunyan-Savvidy, *Classical Yang-Mills mechanics: nonlinear color oscillations*, JETP **53** (1981) 421;
T.S. Biro, S.G. Matinyan and B. Muller, *Chaos and gauge field theory*, World Scientific, 1994.

[35] ATLAS collaboration, *Observation of a new particle in the search for the Standard Model Higgs boson with the ATLAS detector at the LHS*, Phys. Lett. **716** (2012) 1.

[36] CMS collaboration, *Search for supersymmetry in events with a photon, jets, b-jets, and missing transverse momentum in proton-proton collisions at 13 TeV*, Eur. Phys. J. **C79** (2019) 444, arXiv:1901.06726.

[37] M. Claudson and M.B. Halpern, *Supersymmetric ground state wave functions*, Nucl. Phys. **B250** (1985) 689.

[38] B.M. Zupnik, *Six-dimensional supergauge theories in the harmonic superspace*, Sov. J. Nucl. Phys. **44** (1986) 512.

[39] E.A. Ivanov, A.V. Smilga and B.M. Zupnik, *Renormalizable supersymmetric gauge theory in six dimensions*, Nucl. Phys. **B726** (2005) 131, arXiv:hep-th/0505082.

[40] G. Bossard, E. Ivanov and A. Smilga, *Ultraviolet behavior of 6D supersymmetric Yang-Mills theories and harmonic superspace*, JHEP **12** (2015) 085, 1509.08027 [hep-th].

[41] A.S. Galperin, E.A. Ivanov, V.I. Ogievetsky and E.S. Sokatchev, *Harmonic Superspace*, Cambridge Univ. Press., 2001.

[42] M.B. Green, J.H. Schwarz and E. Witten, *Superstring Theory*, Cambridge Univ. Press, 2012.

[43] H.-C. Kao, K.-M. Lee and T. Lee, *The Chern-Simons coefficient in supersymmetric Yang-Mills Chern-Simons theories*, Phys. Lett. **B373** (1996) 94., arXiv:hep-th/9506170.

[44] G. Dunne, *Aspects of Chern-Simons theory*, arXiv:hep-th/9902115.

[45] A.J. Niemi and G.W. Semenoff, *Axial-anomaly-induced fermion fractionization and effective gauge-theory actions in odd-dimensional space-times*, Phys. Rev. Lett. **51** (1983) 2077.

[46] N. Redlich, *Parity violation and gauge noninvariance of the effective gauge field action in three dimensions*, Phys. Rev. **D29** (1984) 2366.

[47] E. Witten, *Constraints on supersymmetry breaking*, Nucl. Phys. **B202** (1982) 253.

[48] L. Alvarez-Gaumé, *Supersymmetry and the Atiyah-Singer index theorem*, Commun. Math. Phys. **90** (1983) 161.

[49] D. Friedan and O. Windey, *Supersymmetric derivation of the Atiyah-Singer index and the chiral anomaly*, Nucl. Phys. **B235** (1984) 395.

[50] I.M. Gelfand, *On elliptic equations*, Russ. Math. Surv. **15** (1960) 113.

[51] M.F. Atiyah, V.K. Patodi and I.M. Singer, *Spectral asymmetry and Riemannian geometry*, Math. Proc. Camb. Phil. Soc. **77** (1975) 43; **78** (1976) 405; **79** (1976) 71.

[52] R.P Feynman and A.R. Hibbs, *Quantum Mechanics and Path Integrals*, McGraw-Hill, 1965.

[53] F.A. Berezin, *Feynman path integrals in a phase space*, Sov. Phys. Usp. **23** (1981) 763.

[54] T. Matsubara, *A new approach to quantum-statistical mechanics*, Prog. Theor. Phys. **14** (1955) 351.

[55] D. Boyanovsky, *Supersymmetry breaking at finite temperature: the Goldstone fermion*, Phys. Rev. **D29** (1984) 743;
V.L. Lebedev and A.V. Smilga, *Supersymmetric sound*, Nucl. Phys **B318** (1989) 669.

[56] S. Cecotti and L. Girardello, *Functional measure, topology and dynamical supersymmetry breaking*, Phys. Lett. **B110** (1982) 39;
L. Girardello, C. Imbimbo and S. Mukhi, *On constant configurations and the evaluation of the Witten index*, Phys. Lett. **B132** (1983) 69.

[57] P. Yi, *Witten index and threshold bound states of D-branes*, Nucl. Phys. **B505** (1997) 307, arXiv:hep-th/9704098;
S. Sethi and M. Stern, *D-brane bound states redux*, Comm. Math. Phys. **194** (1998) 675, arXiv:hep-th/9705046.

[58] M.B. Halpern and C. Schwartz, *Asymptotic search for ground states of SU(2) matrix theory*, Int. J. Mod. Phys. **A13** (1998) 4367, arXiv:hep-th/9712133.

[59] G.M. Graf and J. Hoppe, *Asymptotic ground state for 10-dimensional reduced supersymmetric SU(2) Yang-Mills theory*, arXiv:hep-th/9805080.

[60] J. Frohlich, G.M. Graf, D. Hasler, J. Hoppe and S.-T. Yau, *Asymptotic form of zero energy wave functions in supersymmetric matrix models*, Nucl.Phys. **B567** (2000) 231, arXiv:hep-th/9904182.

[61] V.G. Kac and A.V. Smilga, *Normalized vacuum states in $N = 4$ supersymmetric Yang–Mills quantum mechanics with any gauge group*, Nucl. Phys. **B571** (2000) 515, arXiv:hep-th/9908096.

[62] S. Fubini and E. Rabinovici, *Superconformal quantum mechanics*, Nucl. Phys. **B245** (1984) 17.

[63] A.V. Smilga, *Non-integer flux — why it does not work*, J.Math.Phys. **53** (2012) 042103, arXiv:1104.3986 [math-ph].

[64] D.Z. Freedman and P.K. Townsend, *Antisymmetric tensor gauge theories and nonlinear sigma models*, Nucl. Phys. **B177** (1981) 282.

[65] A.C. Davis, A.J. Macfarlane, P. Popat and J.W. van Holten, *The quantum mechanics of the supersymmetric nonlinear sigma model*, J. Phys. **A17** (1984) 2945.

[66] E.A. Ivanov and A.V. Smilga, *Dirac operator on complex manifolds and supersymmetric quantum mechanics*, Int. J. Mod. Phys. **A27** (2012) 1230024, arXiv:1012.2069 [hep-th].

[67] N. Hitchin, *Harmonic spinors*, Adv. Math. **14** (1974) 1, p.15.

[68] T.T. Wu and C.N. Yang, *Dirac monopole without strings: monopole harmonics*, Nucl. Phys. **B107** (1976) 365.

[69] A. Kirchberg, J.D. Lange and A. Wipf, *Extended supersymmetries and the Dirac operator*, Annals Phys. **315** (2005) 467, arXiv:hep-th/0401134.

[70] D.V. Bykov and A.V. Smilga, *Monopole harmonics on $\mathbb{CP}^{n-1}$*, arXiv:2302.11691 [hep-th].

[71] P. Windey, *Supersymmetric quantum mechanics and the Atiyah-Singer index theorem*, Acta Phys.Polon. **B15** (1984) 435;
A. Mostafazadeh, *Supersymmetry and the Atiyah-Singer index theorem. 1: Peierls brackets, Green's functions, and a supersymmetric proof of the index theorem*, J. Math. Phys. **35** (1994) 1095 arXiv: hep-th/9309060.

[72] A.A. Belavin, A.M. Polyakov, A.S. Schwartz and Yu.S. Tyupkin, *Pseudoparticle solutions of the Yang-Mills equations*, Phys. Lett. **B59** (1975).

[73] M.F. Atiyah, V.G. Drinfeld, N.J. Hitchin and Yu.I. Manin, *Construction of instantons*, Phys.Lett. **A65** (1978) 185;
E. Corrigan, D.B. Fairlie, S. Templeton and P. Goddard, *A Green's function for the general selfdual gauge field*, Nucl.Phys. **B140** (1978) 31.

[74] G. 't Hooft, *Computation of the quantum effects due to a four-dimensional pseudoparticle*, Phys. Rev. **D14** (1976) 3432.

[75] F. Hirzebruch, *Arithmetic genera and the theorem of Riemann-Roch for algebraic varieties*, Proc. Nat. Acad. Sci. USA **40** (1954) 110;
Topological Methods in Algebraic Geometry, Spinger-Verlag, Berlin, 1978.

[76] J.-M. Bismut, *A local index theorem for non Kähler manifolds*, Math. Ann. **284** (1989) 681.

[77] N.E. Mavromatos, *A note on the Atiyah-Singer index theorem for manifolds with totally antisymmetric H torsion*, J. Phys. **A21** (1988) 2279.

[78] A.V. Smilga, *Supersymmetric proof of the Hirzebruch-Riemann-Roch theorem for non-Kähler manifolds*, SIGMA **8** (2012) 003.

[79] A.V. Smilga, *Dolbeault complex on $S^4\backslash\{\cdot\}$ and $S^6\backslash\{\cdot\}$ through supersymmetric glasses*, SIGMA **7** (2011) 105.

[80] A. Lichnerowicz, *Spineurs harmoniques*, C. R. Acad. Sci. Paris **257** (1963) 7.

[81] L. O'Raifeartaigh, *Spontaneous symmetry breaking for chiral scalar superfields*, Nucl. Phys. **B96** (1975) 331.

[82] M. Huq, *Spontaneous breakdown of fermion number conservation and supersymmetry*, Phys. Rev. **D14** (1976) 3548.

[83] D.V. Volkov and V.P. Akulov, *Is the neutrino a Goldstone particle?*, Phys. Lett. **B46** (1973) 109;
A. Salam, *On Goldstone fermion*, Phys. Lett. **B49** (1974) 465.

[84] A.V. Smilga, **a)** *Witten index calculation in supersymmetric gauge theory*, Nucl. Phys. **B250** (1985) 689; **b)** *Calculation of the Witten index in extended supersymmetric Yang-Mills theory*, Yadernaya Fizika **43** (1986) 215 (in Russian).

[85] M. Born and J.R. Oppenheimer, *Zur Quantentheorie der Molekeln*, Annalen der Physik **389** (1927) 457.

[86] A.V. Smilga, *Perturbative corrections to effective zero-mode Hamiltonian in supersymmetric QED*, Nucl. Phys. **B291** (1987) 241;
Born-Oppenheimer corrections to the effective zero-mode Hamiltonian in SYM theory, JHEP **04** (2002) 054, arXiv:hep-th/0201048.

[87] Z. Komargodski, K. Ohmori, K. Roumpedakis and S. Seifnashri, *Symmetries and strings of adjoint QCD_2*, J. High Energ. Phys. **03** (2021) 103, arXiv: 2008.07567 [hep-th].

[88] P. Fayet and J. Illiopoulos, *Spontaneously broken supergauge symmetries and goldstone spinors*, Phys. Lett. **B51** (1974) 461.

[89] A. Keurentjes, A.A. Rosly and A.V. Smilga, *Isolated vacua in supersymmetric Yang-Mills theories*, Phys. Rev. **D58** (1998) 081701, arXiv:hep-th/9805183.

[90] P. van Baal, *Some results for SU(N) gauge fields on the hypertorus*, Comm. Math. Phys. **85** (1982) 529.

[91] G. 't Hooft, *A property of electric and magnetic flux in non-Abelian gauge theories*, Nucl. Phys. **B153** (1979) 141;
Some twisted selfdual solutions for the Yang-Mills equations on a hypertorus, Commun. Math. Phys. **81** (1981) 267.

[92] R. Jackiw and C. Rebbi, *Vacuum periodicity in a Yang-Mills quantum theory*, Phys. Rev. Lett. **37** (1976) 172.

[93] K.G. Selivanov and A.V. Smilga, *Classical Yang-Mills vacua on T^3: explicit constructions*, Phys. Rev. **D63** (2001) 125020, arXiv:hep-th/0010243.

[94] D. Mumford, *Tata Lectures on Theta*, Birkhauser, Boston, 1983.

[95] C.G. Callan, R. Dashen and D.J. Gross, *Mechanism for quark confinement*, Phys. Lett. **66B** (1977) 375.

[96] V. Novikov, M. Shifman, A. Vainshtein and V. Zakharov, *Instanton effects in supersymmetric theories*, Nucl. Phys. **B229** (1983) 407;
D. Amati, G.C. Rossi and G. Veneziano, *Instanton effects in supersymmetric gauge theories*, Nucl. Phys. **B249** (1985) 1.

[97] T. Bhattacharya, A. Gocksh, C. Korthals Altes and R.D. Pisarski, *Z(N) interface tension in a hot SU(N) gauge theory*, Nucl. Phys. **B383** (1992) 497, arXiv:hep-ph/9205231.

[98] A.V. Smilga, *Are Z(N) bubbles really there?*, Ann. Phys. **234** (1994)1;
Physics of thermal QCD, Phys. Repts. **291** (1997) 1, arXiv:hep-ph/9612347, Chap. 3.

[99] J.E. Kiskis, *Phase of the Wilson line*, Phys. Rev. **D49** (1995) 3781, arXiv:hep-lat/9407001.

[100] G. Veneziano and S. Yankielowicz, *An effective Lagrangian for the pure $\mathcal{N} = 1$ supersymmetric Yang-Mills theory*, Phys. Lett. **113B** (1982) 231.

[101] A. Kovner and M. Shifman, *Chirally symmetric phase of supersymmetric gluodynamics*, Phys. Rev. **D56** (1997) 2396, arXiv:hep-th/9702174.

[102] A. Kovner, M.A. Shifman and A. Smilga, *Domain walls in supersymmetric Yang-Mills theories*, Phys. Rev. **D56** (1997) 7978, arXiv:hep-th/9706089.

[103] E. Witten, *Toroidal compactification without vector structure*, JHEP **02** (1998) 006, arXiv:hep-th/9712028.

[104] V.G. Kac and A.V. Smilga, *Vacuum structure in supersymmetric Yang–Mills theories with any gauge group*, in: [Many Faces of the Superworld: Yuri Golfand memorial volume], World Scientific, 2000, p.185, arXiv:hep-th/9902029.

[105] A. Keurentjes, *Nontrivial flat connections on the 3 torus I: G(2) and the orthogonal groups*, JHEP **05** (1999) 001, arXiv:hep-th/9901154; *Nontrivial flat connections on the three torus. 2. The exceptional groups F4 and E6, E7, E8*, JHEP **05** (1999) 014.

[106] A. Borel, R. Friedman and J. Morgan, *Almost commuting elements in compact Lie groups*, Mem. Amer. Math. Soc. **157** (2002), No. 157, arXiv:math.GR/9907007.

[107] A. Salam and J. Strathdee, *Supersymmetry and fermion number conservation*, Nucl. Phys. **B87** (1975) 85;

P. Fayet, *Supergauge invariant extension of the Higgs mechanism and a model for the electron and its neutrino*, Nucl. Phys. **B90** (1975) 104.

[108] I. Affleck, M. Dine and N. Seiberg, *Dynamical supersymmetry breaking in supersymmetric QCD*, Nucl. Phys. **B241** (1984) 493.

[109] V.A. Novikov, M.A. Shifman, V.I. Zakharov and A.I. Vainshtein, *Supersymmetric instanton calculus (gauge theories with matter)*, Nucl. Phys. **B260** (1985) 157.

[110] D. Amati, K. Konishi, Y. Meurice, G.C. Rossi and G. Veneziano, *Nonperturbative aspects in supersymmetric gauge theories*, Phys. Repts. **162** (1988) 169.

[111] M. Shifman and A. Vainshtein, *Instantons versus supersymmetry: fifteen years later*, in: [M. Shifman, ITEP lectures on particle physics and field theory, p.485], World Scientific, 1999, `arXiv:hep-th/9902018`.

[112] K. Konishi, *Anomalous supersymmetry transformation of some composite operators in SQCD*, Phys. Lett. **B135** (1984) 439.

[113] T.E. Clark, O. Piguet and K. Sibold, *Absence of radiative corrections to the axial current anomaly in supersymmetric QED*, Nucl. Phys. **B159** (1979) 1.

[114] E.B. Bogomolny, *Stability of classical solutions*, Sov. J. Nucl. Phys. **24** (1976) 441;
M.K. Prasad and C.M. Sommerfeld, *An exact classical solution for the 't Hooft monopole and Julia-Zee dyon*, Phys. Rev. Lett. **35** (1975) 760.

[115] E. Witten and D. Olive, *Supersymmetry algebras that include topological charges*, Phys. Lett. **B78** (1978) 97.

[116] S. Cecotti and C. Vafa, *On classification of $\mathcal{N} = 2$ supersymmetric theories*, Commun. Math. Phys. **158** (1993) 569.

[117] B. Chibisov and M. Shifman, *BPS-saturated walls in supersymmetric theories*, Phys. Rev. D **56** (1997) 7990, `arXiv:hep-th/9706141`.

[118] G. Dvali and M. Shifman, *Domain walls in strongly coupled theories*, Phys. Lett. **B396** (1997) 64, Erratum: Phys. Lett. **407** (1997) 452, `hep-th/9612128`.

[119] A.V. Smilga, *BPS domain walls in supersymmetric QCD: higher unitary groups*, Phys. Rev. **D58** (1998) 065005, `arXiv:hep-th/9711032`.

[120] T. Taylor, G. Veneziano and S. Yankielowicz, *Supersymmetric QCD and its massless limit: an effective Lagrangian analysis*, Nucl. Phys. **B218** (1983) 493.

[121] A. Smilga and A. Veselov, *Complex BPS domain walls and phase transition in mass in supersymmetric QCD*, Phys. Rev. Lett. **79** (1997) 4529, `arXiv:hep-th/9706217`;
Domain walls zoo in supersymmetric QCD, Nucl. Phys. **B515** (1998) 163, `arXiv:hep-th/9810123`.

[122] B. de Carlos and J.M. Moreno, *Domain walls in supersymmetric QCD: From weak to strong coupling*, Phys. Rev. Lett. **83** (1999) 2120, `arxiv:hep-th/9905165 [hep-th]`;
D. Binosi and T. van Veldhuis, *Domain walls in supersymmetric QCD: The taming of the zoo*, Phys. Rev. **D63** (2001) 085016, `arxiv:hep-th/0011113 [hep-th]`;

A. Smilga, *Tenacious domain walls in supersymmetric QCD*, Phys. Rev. **D64** (2001) 125008, arxiv:hep-th/0104195 [hep-th].

[123] I. Affleck, M. Dine and N. Seiberg, *Dynamical supersymmetry breaking in four dimensions and its phenomenological implications*, Nucl. Phys. **B256** (1985) 557.

[124] S.F. Cordes and M. Dine, *Chiral symmetry breaking in supersymmetric O(N) gauge theories*, Nucl. Phys. **B273** (1986) 581;

[125] M.A. Shifman and A.I. Vainshtein, *On gluino condensation in supersymmetric gauge theories with SU(N) and O(N) groups*, Nucl. Phys. **B296** (1988) 445.

[126] A.Yu. Morozov, M.A. Ol'shanetsky and M.A. Shifman, *Gluino condensate in supersymmetric gluodynamics (II)*, Nucl. Phys. **B304** (1988) 291.

[127] A.V. Smilga, *6+1 vacua in supersymmetric QCD with G_2 gauge group*, Phys. Rev. **D58** (1998) 105014, arXiv:hep-th/9801078.

[128] I. Pesando, *Exact results for the supersymmetric G_2 gauge theories*, Mod. Phys. Lett. **A10** (1995) 1871, arXiv:hep-th/9506139;
S.B. Giddings and J.B. Pierre, *Some exact results in supersymmetric theories based on exceptional groups*, Phys. Rev. **D52** (1995) 6065.

[129] K. Intriligator and N. Seiberg, *Phases of $\mathcal{N} = 1$ supersymmetric gauge theories in four dimensions*, Nucl. Phys. **B431** (1994) 551, arXiv:hep-th/9408155.

[130] C. Vafa and E. Witten, *A strong coupling test of S duality*, Nucl. Phys. **B431** (1994) 3, arXiv:hep-th/9408074.

[131] D. Adams, *Hitchhiker's Guide to the Galaxy*, Pan Books, 1979.

[132] A.V. Smilga, *Witten index in 4D supersymmetric gauge theories*, Int. J. Mod. Phys. A **38** (2023) 31, 2330015, arXiv:2308.12941 [hep-th].

[133] A.V. Smilga, *Structure of vacuum in chiral supersymmetric quantum electrodynamics*, Sov. Phys. JETP **64** (1986) 8.

[134] A. Nakamura and F. Palumbo, *Ordering ambiguities in supersymmetric gauge theories*, Phys. Lett. B **147** (1984) 96.

[135] M. de Crombrugghe and V. Rittenberg, *Supersymmetric quantum mechanics*, Ann. Phys. **151** (1983) 99.

[136] S. Pancharatnam, *Generalized theory of interference and its applications. Part I. Coherent pencils*, Proc. Ind. Acad. Sci. **A44** (1956) 247;
M.V. Berry (1984), *Quantal phase factors accompanying adiabatic changes*, Proc. Royal Soc. **A392** (1984) 45.

[137] I. Tamm, *Die verallgemeinerten Kugelfunktionen und die Wellenfunktionen eines Elektrons im Felde eines Magnetpoles*, Z. Phys. **71** (1931) 141.

[138] P.P. Banderet, *Zum Theorie singularer Magnetpole*, Helv. Phys. Acta **19** (1946) 503;
W.V.R. Malkus, *The interaction of the Dirac magnetic monopole with matter*, Phys. Rev. **83** (1951) 899;
Y. Kazama, *Dynamics of electron-monopole system*, Int. J. Theor. Phys. **17** (1978) 249.

[139] K.M. Case, *Singular potentials*, Phys. Rev. **80** (1950) 797;
K. Meetz, *Singular potentials in non-relativistic quantum mechanics*, Nuov. Cim. **34** (1964) 690;

A.M. Perelomov and V.S. Popov, *Collapse onto scattering center in quantum mechanics*, Teor. Mat, Fiz. **4** (1970) 48.

[140] A.V. Smilga, *W boson scattering on monopoles*, Sov. J. Nucl. Phys., **47** (1988) 692.

[141] G. 't Hooft, *Magnetic monopoles in unified gauge theories*, Nucl. Phys. B **79** (1974) 276;

A.M. Polyakov, *Particle spectrum in quantum field theory*, JETP Letters **20** (1974) 194.

[142] B. Julia and A. Zee, *Poles with both magnetic and electric charges in non-abelian gauge theory*, Phys. Rev. D **11** (1975) 2227.

[143] B.Yu. Blok and A.V. Smilga, *Effective zero-mode Hamiltonian in supersymmetric chiral non-Abelian gauge theories*, Nucl. Phys. **B287** (1987) 589.

[144] M. Dine and J.D. Mason, *Supersymmetry and its dynamical breaking*, Rept. Progr. Phys. **74** (2011) 056201, `arXiv:1012.2836 [hep-th]`.

[145] T. ter Veldhuis, *Unexpected symmetries in classical moduli spaces*, Phys. Rev. **D58** (1998) 015010, `arXiv:hep-th/9811132`.

[146] S. Krivonos, O. Lechtenfeld and A. Sutulin, *N-extended supersymmetric Calogero models*, Phys. Lett. B **784** (2018) 137, `arXiv: 1804.10825 [hep-th]`.

[147] I. Hoppe, *Quantum theory of a relativistic surface*, in [Florence 1986, Proceedings, Constraints Theory and Relativistic Dynamics, eds. G. Longhi and L. Lusanna (World Scientific, 1987), p. 267].

[148] B. DeWitt, I. Hoppe and H. Nicolai, *On the quantum mechanics of super-membranes*, Nucl. Phys. **B305** (1988) 545.

[149] A.V. Smilga, *Super-Yang-Mills quantum mechanics and supermembrane spectrum*, in: [Trieste 1989, Proceedings, Supermembranes and physics in 2+1 dimensions, eds. M.J. Duff, C.N. Pope and E. Sezgin (World Scientific, 1990), p. 182], reprint in `arXiv:1406.5987 [hep-th]`.

[150] B. de Witt, M. Lüscher and H. Nicolai, *The supermembrane is unstable*, Nucl. Phys. **B320** (1989) 135.

[151] T. Banks, W. Fischler, S.H. Shenker, L. Susskind, *M theory as a matrix model: a conjecture*, Phys. Rev. **D55** (1997) 5112, `arXiv:hep-th/9610043`.

[152] G.F. Filippov, V.I. Ovcharenko and Yu.F. Smirnov, *Microscopic Theory of Collective Excitations of Atomic Nuclei*, Naukova Dumka, Kiev, 1981 [in Russian].

[153] H.M. Asatryan and G.K. Savvidy, *Configuration manifold of Yang-Mills classical mechanics*, Phys. Lett. **A99** (1983) 290;

Yu. A. Simonov, *QCD Hamiltonian in the polar representation*, Yad. Fiz. **42** (1985) 1311.

[154] M.B. Green and M. Gutperle, *D Particle bound states and the D instanton measure* JHEP 9801 (1998) 005.

[155] J. Hoppe and J. Plefka, *The asymptotic groundstate of SU(3) matrix theory*, `arXiv:hep-th/0002107`.

[156] W. Krauth, H. Nicolai and M. Staudacher, *Monte Carlo approach to M-theory*, Phys. Lett. **B431** (1998) 31, `arXiv:hep-th/9803117`;

W. Krauth and M. Staudacher, *Yang-Mills integrals for orthogonal, symplectic and exceptional groups*, Nucl. Phys. **B584** (2000) 641, hep-th/0004076 [hep-th].

[157] M. Porrati and A. Rozenberg, *Bound states at threshold in supersymmetric quantum mechanics*, Nucl. Phys. **B515** (1998) 184, arXiv:hep-th/9708119.

[158] E.B. Dynkin, *Semisimple subalgebras of semisimple Lie algebras*, Mat. Sbornik **30** (1952) 349;
B. Kostant, *The principal three-dimensional subgroup and the Betti numbers of a complex simple Lie group*, Amer. J. Math. **81** (1959) 973;
N. Jacobson, *Lie Algebras*, Interscience Publishers, 1962;
P. Bala and R.W. Carter, *Classes of unipotent elements in simple algebraic groups*, Math. Proc. Camb. Phil. Soc. **79** (1976) 401;
A. Alexeevsky, *Component groups of centralizers of unipotent elements in simple algebraic groups*, Proc. Tbilisi Math. Inst. **62** (1979) 5.

[159] A. Hanany, B. Kol and A. Rajaraman, *Orientifold points in M theory*, JHEP **10** (1999) 027, hep-th/9909028.

[160] J.M. Maldacena, *The large N limit of superconformal field theories and supergravity*, Adv. Theor. Math. Phys. **2** (1998) 231, arXiv:hep-th/9711200;
S.S. Gubser, I.R. Klebanov and A.M. Polyakov, *Gauge theory correlators from noncritical string theory*, Phys. Lett. **B428** (1998) 105, arXiv:hep-th/0202156.

[161] For a review, see G.W. Semenoff and K. Zarembo, *Wilson loop in SYM theory: From weak to strong coupling*, Nucl. Phys. Proc. Suppl. **108** (2002) 106, arXiv:hep-th/0202156.

[162] N. Itzhaki, J.M. Maldacena, J. Sonnenschein and S. Yankielowicz, Phys. Rev. **D58** (1998) 046004, *Supergravity and the large N limit of theories with sixteen supercharges*, arXiv:hep-th/9802042;
D. Kabat and G. Lifschytz, *Approximations for strongly coupled supersymmetric quantum mechanics*, Nucl. Phys. **B571** (2000) 419, arXiv:hep-th/9910001.

[163] K.N. Anagnostopoulos, M. Hanada, J. Nishimura, S. Takeuchi, *Monte Carlo studies of supersymmetric matrix quantum mechanics with sixteen supercharges at finite temperature*, Phys. Rev. Lett. **100** (2008) 021601, arXiv: 0707.4454 [hep-th];
M. Hanada, Y. Hyakutake, G. Ishikie and J. Nishimura, *Numerical tests of the gauge/gravity duality conjecture for D0-branes at finite temperature and finite N*, Phys. Rev. **D94** (2016) 086010, arXiv:1603.00538 [hep-th], and references therein.

[164] N. Kawahara, J. Nishimura and S. Takeuchi, *Phase structure of matrix quantum mechanics at finite temperature*, JHEP **10** (2007) 097, arXiv:0706.3517 [hep-th].

[165] M. Hanada, S. Matsuura, J. Nishimurad and D. Robles-Llanaa, *Nonperturbative studies of supersymmetric matrix quantum mechanics with 4 and 8 supercharges at finite temperature*, JHEP **02** (2011) 060, 1012.2913 [hep-th].

[166] A.V. Smilga, *Comments on thermodynamics of supersymmetric matrix models*, Nucl. Phys. **B818** (2009) 101, arXiv:0812.4753 [hep-th].

[167] S. Paban, S. Sethi and M. Stern, *Constraints from extended supersymmetry in quantum mechanics*, Nucl.Phys. **B534** (1998) 137, arXiv:hep-th/9805018.

[168] E.T. Akhmedov and A.V. Smilga, *On the relation between effective supersymmetric actions in different dimensions*, Phys. At. Nucl. **66** (2003) 2238, arXiv: hep-th/0202027.

[169] K. Becker and M. Becker, *A two loop test of M(atrix) theory*, Nucl. Phys. **B506** (1997) 48, arXiv:hep-th/9705091.

[170] Y. Okawa and T. Yoneya, *Multibody interactions of D particles in supergravity and matrix theory*, Nucl. Phys. **B538** (1999) 67, arXiv:hep-th/9806108.

[171] R Gregory and R. Laflamme, *Black strings and p-branes are unstable*, Phys. Rev. Lett. **70** (1993) 2837, arXiv:hep-th/9305052.

[172] Y. Hyakutake, *Boosted quantum black hole and black string in M-theory and quantum correction to Gregory-Laflamme instability*, JHEP **09** (2015) 067, arXiv:1503.05083 [gr-qc].

[173] M. Hanada, Y. Hyakutake, J. Nishimura and S. Takeuchi, *Higher derivative corrections to black hole thermodynamics from supersymmetric matrix quantum mechanics*, Phys. Rev. Lett. **102** (2009) 191602, arXiv:0811.3102 [hep-th].

[174] E. Witten, *Quantum field fheory and the Jones polynomial*, Comm. Math. Phys. **121** (1989) 351.

[175] E. Witten, *Supersymmetric index of three-dimensional gauge theory*, in: [Many Faces of the Superworld: Yuri Golfand memorial volume], World Scientific, 2000, p.156], arXiv:hep-th/9903005.

[176] S. Elitzur, G. Moore, A. Schwimmer and N. Seiberg, *Remarks on the canonical quantization of the Chern-Simons-Witten theory*, Nucl. Phys. **B326** (1989) 108.

[177] J.M.F. Labastida and A.V. Ramallo, *Operator formalism for Chern-Simons theories*, Phys. Lett. **B227** (1989) 92.

[178] A.V. Smilga, *Once more on the Witten index of 3d supersymmetric YM-CS theory*, JHEP **05** (2012) 103, arXiv:1202.6566 [hep-th].

[179] A.V. Smilga, *Vacuum structure in 3d supersymmetric gauge theories*, Phys. Usp. **57** (2014) 155, arXiv: 1312.1804 [hep-th].

[180] A.V. Smilga, *Witten index in supersymmetric 3d theories revisited*, JHEP **01** (2010) 086, arXiv:0910.0803 [hep-th].

[181] S. Deser, R. Jackiw and S. Templeton, *Topologically massive gauge theories*, Ann. Phys. (NY) **140** (1982) 372.

[182] J. Gomis, Z. Komargodski and N. Seiberg, *Phases of adjoint QCD_3 and dualities*, SciPost Phys. **5** (2018) 1, arXiv:1710.03258 [hep-th].

[183] R.D. Pisarski and S. Rao, *Topologically massive chromodynamics in the perturbative regime*, Phys. Rev. **D32** (1985) 2081.

[184] P.A.M. Dirac, *Quantised singularities in the electromagnetic field*, Proc. Roy. Soc. **A133** (1931) 141.

[185] K. Gawedzki and A. Kupiainen, *Coset construction from functional integrals*, Nucl. Phys. **B320** (1989) 625.

[186] A.V. Smilga, *Multidimensional Dirac strings and the Witten index of SYMCS theories with groups of higher rank*, JHEP **09** (2014) 008, arXiv:1404.2779 [hep-th].

[187] O. Aharony, A. Hanany, K.A. Intriligator, N. Seiberg and M.J. Strassler, *Aspects of* $\mathcal{N} = 2$ *supersymmetric gauge theories in three dimensions*, Nucl. Phys. **B499** (1997) 67, `arXiv:hep-th/9703110`.

[188] E.A. Ivanov, *Chern-Simons matter systems with manifest* $\mathcal{N} = 2$ *supersymmetry*, Phys. Lett. **B268** (1991) 203.

[189] A.V. Smilga, *Witten index in* $\mathcal{N} = 1$ *and* $\mathcal{N} = 2$ *SYMCS theories with matter*, Nucl. Phys. **B883** (2014) 149, `arXiv:1308.5951 [hep-th]`.

[190] O. Bergman, A. Hanany, A. Karch and B. Kol, *Branes and supersymmetry breaking in 3D gauge theories*, JHEP **10** (1999) 036, `arXiv:hep-th/9908075`;
K. Ohta, *Supersymmetric index and s-rule for Type IIB branes*, JHEP **10** (1999) 006, `arXiv:hep-th/9908120`.

[191] K.A. Intriligator and N. Seiberg, *Aspects of* $\mathcal{N} = 2$ *Chern-Simons-matter theories*, JHEP **07** (2013) 079, `arXiv:1305.1633 [hep-th]`.

[192] F. Benini and A. Zaffaroni, *A topologically twisted index for three-dimensional supersymmetric theories*, JHEP **07** (2015) 127, `arXiv:1504.03698 [hep-th]`.

[193] C. Closset and H. Kim, *Comments on twisted indices in 3d supersymmetric gauge theories*, JHEP **08** (2016) 059, `arXiv:1605.06531 [hep-th]`.

[194] D. Sen, *Supersymmetry in the space-time* $\mathbb{R} \times S^3$, Nucl. Phys. B **284** (1987) 201.

[195] V. Spiridonov and G. Vartanov, *Elliptic hypergeometry of supersymmetric dualities*, Commun. Math. Phys. **304** (2011) 797, `arXiv:0910.5944 [hep-th]`.

[196] C. Römelsberger, *Counting chiral primaries in* $\mathcal{N} = 1$ $d = 4$ *superconformal field theories*, Nucl. Phys. B **747** (2006) 329, `arXiv:hep-th/0510060`;
J. Kinney, J. Maldacena, S. Minwalla and S. Raju, *An index for 4-dimensional super conformal theories*, Commun. Math. Phys. **275** (2007) 209, `arXiv:hep-th/0510251`.

[197] L. Rastelli and S. Razamat, *The supersymmetric index in four dimensions*, J. Phys. A **50** (2017) 443013, `arXiv:1608.02965 [hep-th]`.

[198] A.V. Smilga, *Witten index for weak supersymmetric quantum systems: invariance under deformations*, Int. J. Mod. Phys. A **37** (2022) 18, 2250118, `arXiv:2112.13397 [hep-th]`.

[199] S. Bellucci and A. Nersessian, *(Super)oscillator on* $CP(N)$ *and constant magnetic field*, Phys. Rev. D **67** (2003) 065013, `arXiv:hep-th/0211070`;
Supersymmetric Kähler oscillator in a constant magnetic field, `arXiv:hep-th/0401232`.

[200] V.A. Rubakov and V.P. Spiridonov, *Parasupersymmetric quantum mechanics*, Mod. Phys. Lett. A **3** (1988) 1337.

[201] E. Ivanov and S. Sidorov, *Deformed supersymmetric mechanics*, Class. Quant. Grav. **31** (2014) 075013, arXiv:1307.7690 [hep-th].

[202] A.A. Andrianov, M.V. Ioffe and V.P. Spiridonov, *Higher-derivative super-symmetry and the Witten index*, Phys. Lett. **A174** (1993) 273, arXiv:hep-th/9303005.

[203] A. Turbiner, *Quasiexactly solvable problems and SL(2) group*, Commun. Math. Phys. **118** (1988) 467;
A. G. Ushveridze, *Quasi-Exactly Solvable Models in Quantum Mechanics*, (IOP Publishing, Bristol, 1994).

[204] A.V. Smilga, *Weak supersymmetic su(N|1) quantum systems*, Int. J. Mod. Phys. A, **37** (2022) 17, 2250119, `arXiv:2202.11357 [hep-th]`.

[205] C. Römelsberger, *Calculating the superconformal index and Seiberg duality*, `arXiv:0707.3702 [hep-th]`.

[206] S. Cecotti, P. Fendley, K. Intriligator and C. Vafa, *A new supersymmetric index*, Nucl. Phys. B **286** (1992) 405, `arXiv:hep-th/9204102`).

[207] R. Jackiw and C. Rebbi, *Solitons with fermion number 1/2*, Phys. Rev. D **13** (1976) 3398;
A.J. Niemi and G.W. Semenoff, *Fermion number fractionization in quantum field theory*, Phys. Repts. **135** (1986) 99.

[208] N. Bourbaki, *Lie Groups and Lie Algebras*, Springer, 2004, Chap. 8, §2, Proposition 7.

[209] See e.g. R. Slansky, *Group theory for unified model building*, Phys. Repts. **79** (1981) 1.

Index

Printed in the United States
by Baker & Taylor Publisher Services